Current Topics in Behavioral Neurosciences

Volume 75

Current Topics in Behavioral Neurosciences provides critical and comprehensive discussions of the most significant areas of behavioral neuroscience research, written by leading international authorities. Each volume in the series represents the most informative and contemporary account of its subject available, making it an unrivalled reference source. Each volume will be made available in both print and electronic form.

With the development of new methodologies for brain imaging, genetic and genomic analyses, molecular engineering of mutant animals, novel routes for drug delivery, and sophisticated cross-species behavioral assessments, it is now possible to study behavior relevant to psychiatric and neurological diseases and disorders on the physiological level. The Behavioral Neurosciences series focuses on translational medicine and cutting-edge technologies. Preclinical and clinical trials for the development of new diagnostics and therapeutics as well as prevention efforts are covered whenever possible. Special attention is also drawn on epigenetical aspects, especially in psychiatric disorders.

CTBN series is indexed in PubMed and Scopus.

Jocelien Olivier • Bart A. Ellenbroek
Editors

Effects of Drug Exposure on Brain Development

Springer

Editors
Jocelien Olivier
Groningen Institute for Evolutionary Life Sciences
University of Groningen
Groningen, The Netherlands

Bart A. Ellenbroek
School of Psychological Sciences
Victoria University of Wellington
Wellington, New Zealand

ISSN 1866-3370 ISSN 1866-3389 (electronic)
Current Topics in Behavioral Neurosciences
ISBN 978-3-032-08680-8 ISBN 978-3-032-08681-5 (eBook)
https://doi.org/10.1007/978-3-032-08681-5

This Springer imprint is published by the registered company Springer Nature Switzerland AG
The registered company address is: Gewerbestrasse 11, 6330 Cham, Switzerland

Preface

The developing brain is exquisitely sensitive to its environment, and among the most profound environmental influences are drugs—both licit and illicit—that may be encountered during critical periods of life. In recent decades, the intersection of neuroscience, pharmacology, and developmental psychology has illuminated the far-reaching consequences of drug exposure before birth, during childhood, and even prior to conception. Yet, as scientific understanding grows, so too does the complexity of the questions we face: How do different classes of drugs—ranging from alcohol and nicotine to antidepressants, antipsychotics, opioids, and cannabinoids—shape the architecture and function of the brain? What are the mechanisms by which these substances exert their effects, and how do genetic, epigenetic, and environmental factors interact to determine outcomes? Most importantly, how can this knowledge inform clinical practice, public health policy, and the support of individuals and families affected by drug exposure?

This volume of Current Topics in Behavioural Neuroscience, *Effects of Prenatal and Postnatal Drug Exposure on Brain Development*, brings together leading experts and the latest research to provide a comprehensive, nuanced exploration of these questions. The rationale for this volume is rooted in both urgency and hope: urgency, because the use of drug among pregnant individuals and young people is unavoidable, hundreds of thousands babies, children, and adolescents are yearly exposed to it, and hope, because advances in research are revealing not only the risks but also the possibilities for intervention, resilience, and recovery.

The volume begins with an in-depth look at *prenatal antidepressant exposure* and its impact on the developing brain. Antenatal mood disturbances are common, and serotonin reuptake inhibitors (SRIs) are frequently prescribed during pregnancy. The opening chapter synthesizes neuroimaging and clinical studies, revealing that both maternal depression and SRI exposure are associated with alterations in brain structure and function, from the newborn period through adolescence. These changes are not uniform; rather, they are shaped by a constellation of factors including the severity of maternal illness, genetic variability, and the timing and duration of exposure. The chapter highlights the methodological challenges in disentangling the effects of medication from those of the underlying mood disorder, yet it also points to emerging evidence that prenatal SRI exposure can have distinct, sometimes subtle, effects on grey and white matter development, functional connectivity, and neurobehavioural outcomes. The narrative underscores the importance of balancing the mental health needs of the mother with the developmental needs of the child, and calls for nuanced, individualized approaches to treatment and research.

Building on this foundation, the volume turns to the *impact of selective serotonin reuptake inhibitors (SSRIs) on brain development*, with a particular focus on the gut-brain axis and the role of the microbiome. Drawing from both clinical and preclinical studies, the authors explore how maternal SSRI use during pregnancy can alter the maternal gut microbiome, disrupt serotonin signalling, and influence offspring behaviour and brain gene expression. Animal models reveal that perinatal SSRI exposure can lead to changes in myelination and social behaviour during juvenile age and in adulthood, often in a sex-specific manner. The chapter weaves together molecular, cellular, and behavioural findings, illustrating how early-life drug exposure can set the stage for long-term neurodevelopmental trajectories. The narrative also highlights the complexity of interpreting these effects, given the interplay between maternal illness, drug action, and environmental factors.

The discussion then shifts to *valproate (VPA)*, a widely used antiepileptic and mood-stabilizing drug with well-documented teratogenic effects. The chapter provides a comprehensive review of both clinical and animal studies, demonstrating that prenatal and early postnatal exposure to VPA increases the risk of neurodevelopmental disorders, particularly autism spectrum disorder (ASD). Through detailed analysis of single and multiple administration studies in rodents, the authors show how the timing, dosage, and duration of VPA exposure critically shape behavioural and neuroanatomical outcomes. The chapter also delves into the pharmacodynamics of VPA, including its effects on GABAergic neurotransmission, ion channel function, and histone deacetylase inhibition, which together contribute to its complex impact on neural development. Importantly, the narrative does not shy away from the limitations of animal models, emphasizing the need for careful translation of findings to human populations and for ongoing refinement of research methodologies.

The theme of *antipsychotic drug exposure* is explored next, with a focus on both prenatal and early postnatal administration. The chapter traces the evolution of antipsychotic use from first-generation drugs like haloperidol to second-generation agents such as risperidone and olanzapine, highlighting the dramatic increase in paediatric prescriptions over recent decades. Through a synthesis of preclinical studies, the authors reveal that early-life antipsychotic exposure can lead to persistent changes in dopamine receptor density, neurotransmitter systems, and behavioural outcomes, including alterations in locomotor activity, cognitive function, and social behaviour. The narrative underscores the importance of developmental timing, with prenatal exposure often leading to hypodopaminergic states and postnatal exposure resulting in hyperdopaminergic profiles. The chapter also addresses the challenges of dosing, treatment duration, and the interpretation of animal data, ultimately advocating for cautious, evidence-based use of antipsychotics in young populations.

The next contribution focuses on *opioids*, a class of drugs at the centre of a global public health crisis. The chapter provides a sweeping review of the effects of gestational opioid exposure, from illicit substances like morphine to medications for opioid use disorder (MOUD) such as methadone and buprenorphine. Drawing on both human and animal studies, the authors document the acute and long-term consequences of prenatal opioid exposure, including neonatal opioid withdrawal syndrome (NOWS), neurodevelopmental delays, cognitive and behavioural impairments, and alterations in brain structure and function. The narrative is careful to distinguish between the effects of illicit opioid use and those of MOUD, noting that while MOUD can mitigate some risks, it is not without its own consequences. The chapter also explores the impact of opioid exposure on maternal brain function and caregiving behaviours, highlighting the intricate interplay between pharmacology, neurobiology, and the social environment.

Moving beyond the traditional focus on maternal exposures, the next chapter discusses the *intergenerational and transgenerational effects of paternal alcohol consumption*. The authors synthesize evidence from human and animal studies showing that preconceptual paternal alcohol use can influence offspring development through epigenetic mechanisms, including DNA methylation, histone modification, and non-coding RNA regulation. The narrative traces the effects of paternal alcohol exposure on birth outcomes, cognitive and behavioural phenotypes, and even gene expression patterns in subsequent generations. This chapter challenges readers to consider the broader context of drug exposure, recognizing that the legacy of substance use can extend far beyond the maternal-fetal interactions, the individual and across generations.

Lastly, the book addresses the *impact of cannabis (THC) and nicotine on attention and neurodevelopment*. Through a comprehensive review of clinical and preclinical studies, the authors demonstrate that prenatal exposure to cannabis is associated with attention deficits, increased risk for ADHD, and alterations in brain structure and function. The narrative highlights the complexity of cannabis effects, which can vary by dose, timing, and genetic vulnerability, and which may be compounded by co-exposure to other substances. Nicotine, too, is shown to have dose-dependent effects on attention, with prenatal exposure linked to cognitive and behavioural problems, including ADHD and conduct disorder. The chapter underscores the importance of considering both maternal and paternal exposures, as well as the potential for sex-specific and long-lasting effects.

Throughout the book, several unifying themes emerge. First, the *timing, dose, and duration* of drug exposure are critical determinants of outcome, with sensitive periods of brain development conferring heightened vulnerability. Second, the *mechanisms of drug action* are diverse and multifaceted, encompassing neurotransmitter systems, neurotrophic factors, epigenetic modifications, and interactions with the microbiome. Third, results can be influenced by sex, highlighting the need for taking along sex as a biological variable. Fourth, the *translation of animal research to human populations* requires careful consideration of species differences, experimental design, and the broader social and environmental context. Finally, the book highlights the *potential for intervention and resilience*, noting that not all individuals exposed to drugs will experience adverse outcomes, and that supportive environments, early interventions, and targeted therapies can mitigate risks.

In bringing together these diverse strands of research, *Effects of Prenatal and Postnatal Drug Exposure on Brain Development* aims to provide a resource for scientists, clinicians, policymakers, and families alike. The chapters that follow offer both cautionary tales and reasons for optimism, revealing the profound ways in which drugs can shape the developing brain, but also the remarkable capacity for adaptation and recovery. As our understanding deepens, so too does our responsibility to translate knowledge into action—supporting those affected by drug exposure, informing public health strategies, and advancing the science of brain development for generations to come.

Groningen, The Netherlands Jocelien Olivier
Wellington, New Zealand Bart A. Ellenbroek

Contents

Prenatal Antidepressant Exposure and the Developing Brain: A Review of Neuroimaging Findings

Kayleigh S. J. Campbell and Tim F. Oberlander

Contents

Abstract Antenatal mood disturbances are experienced by as many as 20% of pregnant mothers and are commonly treated with serotonin reuptake inhibitor (SRI) antidepressants. Both maternal depression and SRIs during pregnancy are associated with low birth weight and infant neurobehavioral disturbances, as well as longer-term impacts on child neurodevelopment, behavior, and mental health. As maternal depression and its pharmacotherapy are inherently interrelated prenatal exposures, distinguishing how these early life factors uniquely impact child development remains methodologically challenging. Over the past several years, however, advanced neuroimaging has been successfully used to identify neural correlates of prenatal depression and SRI antidepressant exposure on the developing brain, extending from the early newborn period through adolescence. In this review, we examine the use of magnetic resonance imaging and electroencephalography to study child brain structure or function, with a specific focus on prenatal antidepressants as the primary exposure in relation to either typical development or exposure to

K. S. J. Campbell and T. F. Oberlander (✉)
BC Children's Hospital Research Institute, Vancouver, BC, Canada

Department of Pediatrics, University of British Columbia, Vancouver, BC, Canada
e-mail: toberlander@bcchr.ca

Curr Topics Behav Neurosci (2025) 75: 1–22
https://doi.org/10.1007/7854_2025_591
Published Online: 14 May 2025

maternal depressed mood alone. We include both cross-sectional and longitudinal neuroimaging studies, as well as those that link early brain findings with cognitive or behavioral outcome in childhood. We also discuss factors that may shape neurodevelopmental risk (e.g., maternal mental illness severity, sex differences, genetic variability) and present suggestions for future research that will advance our understanding of child brain development in the context of maternal mood disturbances during pregnancy.

Keywords Brain–behavior relationships · Brain development · Child · Electroencephalography (EEG) · Genetic variability · Infant · Maternal depression · Magnetic resonance imaging (MRI) · Neonate · Neuroimaging · Prenatal antidepressant exposure · Selective serotonin reuptake inhibitors (SSRIs) · Sex differences

1 Introduction

Globally, the prevalence of depression during pregnancy ranges from 15% to 65% (Dadi et al. 2020). Risk for antenatal depression is shaped by complex interactions between genetics and several environmental factors, such as history of mental disorders, adverse life events, poor social support, socioeconomic status, smoking, and other maternal comorbidities (Lancaster et al. 2010). Antenatal depression also carries an increased risk for postnatal depression (Hutchens and Kearney 2020), highlighting the inherent chronicity of maternal mental illness and potential for long-term detrimental effects on health and quality of life, with consequences for the developing child, family, and broader community.

Not only does untreated or poorly managed antenatal depression present significant risk to maternal and fetal health, but exposed offspring also have a lasting susceptibility to adverse neurodevelopmental, behavioral, and long-term mental health consequences, which may even confer transgenerational risk via epigenetic programming (Babenko et al. 2015). Antenatal depression is associated with higher rates of preterm delivery, impaired intrauterine growth, and low birth weight (Ghimire et al. 2021), as well as sleep problems and difficult temperaments in infancy and toddlerhood (Kim et al. 2020). Across childhood, prenatal depression exposure has also been associated with poorer cognitive development (Tarabulsy et al. 2014) and greater externalizing and internalizing problems (Rogers et al. 2020), which may underlie the nearly twofold increase in depression risk across adolescence and adulthood (Tirumalaraju et al. 2020). Therefore, effective management of antenatal depression is an urgent public health issue; yet, treatment decisions continue to be a challenge for mothers and clinicians, largely given the uncertainty on risks and benefits associated with pharmacotherapy.

Serotonin reuptake inhibitor (SRI) antidepressants are commonly prescribed during pregnancy for the management of maternal depressed mood with the intention that they, by extension, buffer the developing child from the adverse effects of prenatal depression exposure. However, prenatal SRI exposure is associated with a similar set of perinatal (Kautzky et al. 2022) and child developmental outcomes, including early motor delays (Grove et al. 2018), internalizing behaviors (Brandlistuen et al. 2015; Park et al. 2021), as well as increased risk for depressive and anxiety disorders into young adulthood (Bliddal et al. 2023; Rommel et al. 2021). However, relationships between prenatal SRI exposure and child development are multifaceted and inconsistent, and it is unclear whether observed associations are driven by the underlying maternal mental illness, rather than the medication itself (Suarez et al. 2022). This leaves critical, clinically urgent questions about the nature of developmental risk associated with these two distinct – yet inherently interrelated – prenatal exposures, both of which may influence early brain development.

Advanced neuroimaging technologies, such as magnetic resonance imaging (MRI) and electroencephalography (EEG), have opened a window into in vivo human brain development, and their increasing application in research has yielded critical insight into key neurobiological mechanisms underlying a wide range of neurodevelopmental and behavioral outcomes that emerge in childhood. In recent years, these methodologies have been successfully used to characterize early brain outcomes associated with both prenatal depression and SRI exposure across childhood. Emerging neuroimaging studies have also begun to elucidate key neural correlates associated with subsequent behavioral outcome in children prenatally exposed to maternal depression and/or SRIs, which may be essential for understanding how factors, such as genetic variability and postnatal environment, shape individual risk or resilience in this developmental context. Thus, neuroimaging offers tremendous opportunity for distinguishing and understanding the specific and potentially nuanced impacts of prenatal depression and SRI exposures on the developing brain.

In this chapter, we review the use of advanced neuroimaging to study the impacts of prenatal SRI antidepressant exposure on child brain development, extending from the newborn period to adolescence. We also discuss findings linking early structural or functional brain changes with subsequent neurodevelopment and behavioral outcome, as well as factors that may shape observed associations, such as sex differences and child genetics. As SRI use during pregnancy occurs in the context of maternal mood disturbances, we particularly highlight neuroimaging studies that have attempted to distinguish prenatal SRI exposure from maternal depression alone. Critically, neuroimaging findings from included studies should be interpreted as complementary to the rapidly evolving literature linking maternal depression, anxiety, and stress during pregnancy with child brain outcome (for review see (Cattarinussi et al. 2021; Demers et al. 2021)). While such studies are beyond the scope of this chapter, we highlight notable consistencies with the broader child neuroimaging literature on prenatal depression exposure.

2 SRI Pharmacotherapy During Pregnancy

Non-pharmacologic treatments, such as psychotherapy, mindfulness practices, or certain health-lifestyle behaviors, are often recommended for mild-moderate maternal mood disturbances; yet, these options may not be efficacious or widely accessible. Pharmacotherapy is often an unavoidable consideration, particularly among those with greater illness severity (Kennedy et al. 2016). Selective serotonin reuptake inhibitors (SSRIs) and serotonin-norepinephrine reuptake inhibitors (SNRIs), together referred to herein as SRI antidepressants, are widely prescribed psychotropic medications during pregnancy. A recent meta-analysis reported that antenatal SRI use had an international pooled prevalence of 3.0–4.7% across a 26-year period (1989–2015) among developed nations; however, rates varied by geographic region possibly due to differences in prescribing practices, healthcare organization, and treatment-seeking behaviors: European countries and Australia-Asia had prevalence rates of ≤2%, while North American rates were 2–9% (Molenaar et al. 2020).

SRI antidepressants are primarily assumed to enhance central serotonergic neurotransmission, forming the basis of their therapeutic benefit (Willner et al. 2013). SRIs are competitive inhibitors of the serotonin transporter (SERT) and act by blocking the reuptake of intrasynaptic serotonin (5-hydroxytryptamine; 5-HT), thereby increasing the duration and magnitude of serotonergic activity on postsynaptic 5-HT receptors. While SRIs are generally well tolerated during pregnancy, they readily cross the placenta and enter the intrauterine environment. Indeed, high fetal SRI concentrations are detected in umbilical cord blood at delivery, with drug-specific estimates ranging from 50% to 80% of that in maternal circulation (Schoretsanitis et al. 2021). However, there is high inter-individual variability in both fetal and maternal SRI levels, and the extent of fetal drug exposure likely depends on several pregnancy-related and pharmacological factors, such as placental function and hepatic metabolism.

Depressed mothers often face high levels of decisional conflict on whether to stop, continue, or start SRI pharmacotherapy during pregnancy (Battle et al. 2013). Weighing risks of adverse outcome against relapse is complicated by limited evidence on the risks and benefits associated with SRI medication itself, relative to risks associated with the underlying depressive disorder. Disentangling these distinct prenatal exposures is a significant methodological challenge, since individuals who start or continue antidepressants during pregnancy often have an increased propensity for antenatal depression, including greater initial mental illness severity and/or lifetime history of depression (Lancaster et al. 2010; Petersen et al. 2011). This *confounding by indication* makes drawing conclusions on developmental risk attributable to the medication alone extremely challenging. As randomizing SRI exposure to account for depressive symptoms is not ethically feasible, some strategies may help minimize confounding bias, such as restricted study enrolment, stringent statistical methods (e.g., high-dimensional propensity scores, discordant sibling analysis), or use of specific comparator groups; however, residual confounding cannot be eliminated in observational studies.

3 Early Brain Development, Prenatal Depression, and SRI Exposure

Development before birth occurs at an extraordinary pace. From a single cell at conception, a massive yet precisely orchestrated interplay of genetic and environmental factors produce all biological systems needed for life *ex utero* (Silbereis et al. 2016). During this time, the brain undergoes its most dramatic development across the lifespan, forming from a tubular structure into a highly complex and efficiently organized organ by term birth (≥37-weeks), continuing to develop such that the basic architecture for mature functioning is largely present by the second postnatal year (Gilmore et al. 2018). The prenatal period is, therefore, a critical time for environmental influences: neurobiological variations originating from in utero exposures have the potential to alter the course of brain development with long-term health implications (Barker 1995). Understanding how the developing brain responds to maternal depressed mood and its pharmacotherapy may uncover important prenatal origins for variation in neurodevelopment, behavior, and subsequent vulnerability to psychopathology.

How maternal psychology influences early brain development has been a research focus for decades. While there are no direct neural connections between the mother and fetus, plausible biological mechanisms involve placental transfer of maternal stress-related molecules, such as glucocorticoids, monoamines, and proinflammatory cytokines (Monk et al. 2019). Maternal depression may also impact placental functioning itself (Bronson and Bale 2016), resulting in disrupted nutrient and oxygen exchange and transplacental barrier integrity that further contributes to in utero exposure to stress-related factors. These changes may have a number of consequences in the fetus, such as neuroendocrine dysregulation, altered neurotransmitter or inflammatory signaling, and/or epigenetic changes that may collectively impact early brain development (Monk et al. 2019).

Critically, this intrauterine environment is the background on which prenatal antidepressant exposure occurs. In particular, 5-HT – the target neurotransmitter of SRIs – has widespread neurotrophic roles in the differentiation, growth, migration, and myelination of developing brain circuitry (Deneris and Gaspar 2018), long before it regulates mood, cognition, social behavior, sleep, and several other physiologic functions (Pourhamzeh et al. 2022). Indeed, preclinical evidence suggests that early developmental SRI exposure alters central 5-HT signaling, with resulting excess synaptic 5-HT levels inducing widespread changes in gene expression and brain circuitry abnormalities in the offspring (Glover and Clinton 2016; Homberg et al. 2010; Simpson et al. 2011). Human findings also provide indirect evidence of altered in utero 5-HT signaling, as SRIs appear to inhibit placental SERT (Horackova et al. 2021), and cord blood studies in SRI-exposed newborns report altered levels of 5-HT and its primary metabolite (5-HIAA), as well as biomarkers of CNS integrity such as reelin and S100B (Bellissima et al. 2020; Brummelte et al. 2013; Laine et al. 2003). These changes were subsequently linked with newborn neurobehavioral disturbances, including poorer neuromotor function, increased irritability, and sleep disruptions. Thus, altered 5-HT signaling reflects a putative mechanism through which prenatal SRI exposure may impact the developing brain.

4 Neuroimaging as a Window into the Developing Brain

Advanced neuroimaging tools, including MRI and EEG, are indispensable for non-invasively studying the inner structure and function of the human brain. Neuroscientific research using MRI and EEG has exploded over the last two decades, as these tools provide detailed and accurate tissue characterization and neurophysiologic specificity, with a range of complementary techniques sensitive to different aspects of brain development, injury, and pathology.

MRI is a powerful modality for visualizing the brain, with high spatial resolution and contrast versatility allowing for multidimensional investigation of in vivo brain structure and function. Newborns can be safely scanned at high field strengths (1.5–3.0 Tesla), as MRI uses non-ionizing radiation and maintains low levels of heat deposition (Malik et al. 2015). MRI is widely utilized for studying brain development and is currently the most sensitive neuroimaging tool for detecting and quantifying abnormalities. The most common approaches include: (1) T_1- and T_2-weighted imaging, which provide complementary anatomical visualization and are typically used to examine brain morphometry, volumetrics, or abnormalities. These two modalities differ in their sensitivity to various tissue characteristics, with T_1 imaging being sensitive to fat content and T_2 to water content; (2) diffusion MRI, which is used to characterize tissue microstructure and delineate anatomical connectivity; and (3) functional MRI (fMRI), which measures cerebral blood flow changes that occur with brain activity to study the synchronization of neural processing across distinct brain regions.

In contrast, EEG uses electrodes placed on the scalp to provide direct measures of cortical electrophysiology, reflecting spontaneous and synchronized neuronal activity generated by postsynaptic impulses of cortical neurons (Kirschstein and Köhling 2009). While EEG has lower spatial resolution compared to MRI, this modality offers precise, millisecond tracking of cortical dynamics. EEG records multiple distinct waveforms (i.e., delta, theta, alpha, beta, gamma), each with specific frequency bands and power spectra that associate with various cortical processes occurring at rest, or in response to behavioral tasks or stimulus presentation.

5 Prenatal SRI Exposure and Child Brain Development

In this section, we review human MRI and EEG studies that investigate child brain structure or function in association with prenatal SRI antidepressants as the primary exposure. From the included studies, we also review child brain findings associated with maternal depressed mood alone, particularly in comparison with prenatal SRI exposure, where possible. However, similar neuroimaging findings may occur with prenatal exposure to other aspects of maternal mental health, such as anxiety, chronic stress, traumatic experiences, daily hassles, pregnancy-specific or disaster-related stress, possibly due to the comorbidity of these psychopathologies and/or construct overlap between assessment tools. While this literature is beyond the scope of this chapter, we highlight notable consistencies throughout and refer the reader to recent reviews covering this broader topic (Cattarinussi et al. 2021; Demers et al. 2021).

5.1 Gray Matter Morphology

Prenatal SRI exposure may impact the anatomical development of the brain. Before birth, the brain undergoes dramatic morphological changes, with explosive synaptogenesis, dendritic expansion, and glial cell infiltration supporting the formation of specialized cortical layers and subcortical gray matter (GM) structures, as well as an overall increase in brain surface area, volume, and density (Ouyang et al. 2019). These processes continue postnatally, such that the brain reaches approx. 80% of its adult size by the second postnatal year (Gilmore et al. 2018). Preclinical work suggests that altered 5-HT levels in the developing rodent brain, through either genetic manipulation or pharmacologic blockade of SERT via SRI exposure, can disrupt these critical morphological processes. For example, reduced dendritic complexity, cortical thickness, and GM density have been reported in multiple regions, including the somatosensory cortex (notably, layer IV), infralimbic cortex, and hippocampus (Glover and Clinton 2016; Homberg et al. 2010), which may underlie long-term impairments in sensory perception and anxiety- and depression-like behaviors consistently seen in SRI-exposed rodent offspring (Ramsteijn et al. 2020). In humans, variability in GM volume and cortical thickness has also been linked with emotional reactivity in infants (Suh et al. 2019) and risk for major depressive disorder in adults (Zhang et al. 2023), highlighting the importance of understanding the potential impact of prenatal SRI exposure on developing brain morphology.

Neuroimaging studies examining structural morphometry within the first 2-years typically use T_2-weighted MRI, as this modality provides superior tissue contrast due to differences in infant brain properties (i.e., higher water content, lower myelination), compared to older children and adults (Dubois et al. 2021). To date, three studies have used T_2 MRI to investigate regional GM volumes in infants with prenatal SRI exposure (Jha et al. 2016; Lugo-Candelas et al. 2018; Sethna et al. 2021), but report mixed findings. At approximately 1-month of age, larger volumes in the right amygdala, insula, superior frontal gyrus, and left occipital gyrus were observed in SRI-exposed infants ($n = 16$) compared to both healthy controls and infants of depressed mothers with high prenatal depressive symptom levels, even when controlling for maternal comorbid psychiatric disorders and infant genotype for the *SLC6A4* promotor region (encoding SERT) (Lugo-Candelas et al. 2018). Whereas another study found no differences in GM volume at 1-month, aside from a smaller right thalamus in SRI-exposed infants ($n = 27$) compared to matched controls of non-depressed mothers (Jha et al. 2016). Interestingly, both studies failed to identify volume differences in infants with prenatal depression exposure alone, even though mothers experienced clinically significant depressive symptoms in these samples (Jha et al. 2016; Lugo-Candelas et al. 2018), and prior work links higher prenatal symptoms, as well as genetic risk for depression, with altered limbic volumes in the newborn period (Acosta et al. 2020; Groenewold et al. 2022; Qiu et al. 2017). At 3–6 months of age, a third T_2 MRI study observed larger subcortical and smaller midbrain volumes in infants of mothers with a diagnosed depressive disorder, regardless of prenatal antidepressant status (Sethna et al. 2021). However, these findings were independent of maternal symptom severity, so it is unclear whether infant GM changes related to depression exposure or familial transmission of a neural phenotype associated with vulnerability to depression, since longitudinal follow-up was not conducted.

More recently, two studies report longer-term associations between prenatal SRI exposure and brain morphometry in middle-late childhood using T_1 MRI (Koc et al. 2023; Moreau et al. 2023). Drawn from participants of the Adolescent Brain Cognitive Development (ABCD) study, one of the largest multisite longitudinal prospective studies of brain development and child health through to adulthood, Moreau et al. conducted a comprehensive investigation of subcortical volume, cortical thickness, and surface area in children with prenatal SRI exposure at 9–10 years of age ($n = 280$). While there were no morphologic differences in regions commonly associated with depression (i.e., limbic/paralimbic structures), small but significantly greater differences in left superior parietal surface area and lateral occipital cortical thickness were observed in children with prenatal SRI exposure compared to controls, even when accounting for maternal depressive symptoms, as well as child depression, SRI use, and polygenic risk score for depression (Moreau et al. 2023). However, this study was based on retrospective report of maternal SRI use and did not include a measure of depressed mood specific to the prenatal period. In contrast, the Generation R study identified smaller total cerebral GM volume predominantly across corticolimbic regions in children with prenatal SRI exposure from 7-to-15 years of age ($n = 41$), but not in children of mothers with prenatal depression, postnatal depression, or preconception SRI use, compared to non-exposed controls (Koc et al. 2023). The authors also observed smaller total cerebral white matter (WM), amygdala, and fusiform gyrus volumes in children with prenatal SRI exposure, but only in middle-to-late childhood. Instead, these regions underwent a period of catch-up growth to no longer differ from non-exposed controls by 15-years. However, child psychological factors (mood, medication use) were not accounted for and brain-behavioral correlates were not studied, so the functional significance of these longitudinal brain findings remains unclear.

In sum, studies examining GM morphology following prenatal SRI exposure have yielded mixed findings, both in infancy and in later childhood. While some evidence points to regionally specific alterations in limbic and cortical structures that may vary with child age, results are inconsistent across studies and limited by confounding, retrospective reporting, or lack of behavioral follow-up. Additional longitudinal neuroimaging research is needed to clarify developmental significance of regional GM variation in SRI-exposed offspring.

5.2 White Matter Microstructure and Connectivity

Developing WM may be particularly sensitive to adverse in utero environmental exposures. Early developmental processes center on the formation, migration, and organization of neural pathways that form a vast network of axons facilitating interregional connectivity in the brain. Billions of new neurons are generated by the third-trimester and this period, until approximately 2-years of age, sees the greatest pace of synaptogenesis across the lifespan (Huttenlocher and Dabholkar 1997). Most major WM pathways are present by birth, with myelination and competitive pruning progressively refining the efficiency and specialization of brain networks to support increasingly complex functional capacities and behaviors in the developing child (Cao et al. 2017; Ouyang et al. 2019). Critically, serotonin modulates a number of these processes in developing WM (Deneris and Gaspar 2018), and preclinical models of developmental SRI exposure provide evidence of abnormal brain circuitry, particularly among thalamocortical, corticolimbic, and interhemispheric fibers (Homberg et al. 2010; Simpson et al. 2011).

WM microstructure and anatomical connectivity can be examined with diffusion MRI, which may provide important clues toward understanding of how prenatal SRI exposure impacts the developing brain. This modality is sensitive to molecular diffusion properties that reveal key information on the geometric and compositional characteristics of in vivo brain tissue (Le Bihan and Johansen-Berg 2012), such as fractional anisotropy (FA) and mean, axial, and radial diffusivities (MD, AD, RD). These indices are commonly used to describe WM integrity (i.e., cohesion, organization, myelination), with lower FA and higher diffusivity associated with immature or atypical development of WM tracts (Ouyang et al. 2019).

Five studies have used diffusion MRI to investigate associations between prenatal SRI exposure and WM microstructure and connectivity (Campbell et al. 2021; Jha et al. 2016; Koc et al. 2024; Lugo-Candelas et al. 2018; Podrebarac et al. 2017) but, again, findings are inconsistent. In a cohort of very preterm infants (born <34 weeks' gestation), prenatal SRI exposure ($n = 14$) was associated with higher FA and lower MD, RD, and AD in superior WM tracts (high centrum semiovale), as well as lower N-acetylaspartate and higher lactate levels in the calcarine region, together suggesting altered neuronal integrity and impaired oxidative metabolism in these regions, compared to non-exposed preterm infants (Podrebarac et al. 2017). However, maternal mood was not accounted for in this study, and the extent to which these findings were driven by prematurity could not be determined. In term-born infants scanned at approx. 1-month, Lugo-Candelas et al. report increased right amygdala–insula connectivity in SRI-exposed infants ($n = 16$) compared to controls, as well as infants of unmedicated but clinically depressed mothers. The authors found no other differences in structural connectivity, FA, or MD at the global brain network-level (Lugo-Candelas et al. 2018). In contrast, Jha et al. observed widespread WM microstructural disruptions in SRI-exposed infants ($n = 27$) compared to matched controls at 1-month, particularly lower MD and RD in corticofugal and corticothalamic projection tracts, as well as in limbic association pathways including the cingulum, fornix, and uncinate fasciculus (Jha et al. 2016). However, this study was limited by lack of direct comparison between prenatal SRI- and unmedicated depression-exposed infants, making it unclear whether observed microstructural alterations were driven by maternal factors contributing to increased propensity for depression during pregnancy (and hence need for pharmacotherapy), rather than SRI exposure. Indeed, studies investigating prenatal depression exposure alone also report poorer WM integrity across similar regions in the infant brain (Donnici et al. 2023; Lautarescu et al. 2022), and possibly in a sex-dependent manner (Dean et al. 2018; Hashempour et al. 2023).

Prenatal SRI exposure may also have sex-dependent effects on the developing brain (Kott and Brummelte 2019; Simpson et al. 2011). To explore this unknown, Campbell et al. undertook a data-driven investigation to determine whether prenatal SRI exposure had sex-specific associations with early brain microstructure. Using the corpus callosum as an a priori region of interest, latent patterns in callosal FA were found to have an 82% accuracy at discriminating neonates based on prenatal SRI exposure status and sex (Campbell et al. 2021). Specifically, SRI-exposed males had lower FA and higher RD in the anterior genu and isthmus-splenium compared to females. Whereas, in both sexes, SRI-exposed infants ($n = 24$) had lower FA and higher MD in a region extending from the genu-to-rostral body compared to non-exposed male and female infants, which was independent of prenatal maternal mood symptoms. While limited to newborn interhemispheric fibers, these findings highlight sex as a moderator of prenatal SRI-related effects on early WM development and may have key implications for sex-specific cognitive or behavioral outcomes that warrant investigation. Interestingly, disrupted microstructure in anterior callosal fibers was also observed in SRI-exposed children from 7-to-15 years of age (Koc et al. 2024), potentially pointing to a conserved neural phenotype of prenatal SRI exposure across child development. In this study, lower FA was observed in the forceps minor of the corpus callosum in both prenatal SRI ($n = 37$) and depression-only exposed children compared to non-exposed controls at 7-years. This was followed by a catch-up period of increasing FA between 7- and 15-years in prenatal depression-exposed, but not SRI-exposed children. However, neither prenatal SRI nor depression-related findings differed by child sex, in contrast to the findings in newborns, even though neuroimaging spanned over puberty (Ingalhalikar et al. 2014).

Together, diffusion MRI studies to date suggest that prenatal SRI exposure is associated with altered WM microstructure and connectivity, with evidence pointing to widespread but regionally variable effects in infancy, as well as sex-dependent vulnerabilities. Notably, findings from both newborn and school-aged/early adolescent samples implicate frontal interhemispheric WM as a potential site of prenatal SRI-related impacts across early brain development, though further study is needed to clarify consistency of findings and their functional significance.

5.3 Functional Connectivity

Beyond differences in brain structure, emerging neuroimaging evidence also suggests that prenatal SRI exposure may impact developing brain function, reflecting possible neurobiological mechanisms underlying observed behavioral outcomes in SRI-exposed children. Functional networks emerge as early as mid-gestation (Turk et al. 2019) and are continually and dynamically shaped by the organization, consolidation, and maturation of neural circuitry to support increasingly efficient global communication architecture and specialized higher-order functional processes that unfold across development (Cao et al. 2017). fMRI enables the investigation of functional connectivity by capturing spatiotemporal changes in blood oxygenation levels associated with neural activity across anatomically disparate brain regions, providing insight into the development of functional networks and their relation to cognitive processes, emotion regulation, and behavior (Lee et al. 2013).

Four studies, reporting on two cohorts, have used fMRI to assess resting-state functional connectivity in sleeping, non-sedated infants with prenatal SRI exposure (Rotem-Kohavi et al. 2018, 2019; Salzwedel et al. 2016, 2020). In the first postnatal week, Rotem-Kohavi et al. observed hyperconnectivity in temporal and parietal regions of the putative auditory resting-state network in SRI-exposed neonates ($n = 20$) compared to non-exposed controls and neonates of mothers with third-trimester depressive symptoms (Rotem-Kohavi et al. 2018). Subsequently, using a graph theoretical approach, higher hub values were observed in the left anterior cingulate, insula, caudate, and amygdala in unmedicated depression-exposed neonates compared to controls (Rotem-Kohavi et al. 2019), suggesting greater importance of these regions for integrating information within and across functional networks. Interestingly, these findings may have consistencies with higher corticolimbic functional connectivity previously reported in infants with prenatal depression exposure (Marr et al. 2023; Qiu et al. 2015). They also found that hub values predicted 10% of the variance in infant temperament at 6-months of age, particularly with mid-cingulate and temporal regions associating with smiling and orienting behaviors in SRI- and depression-exposed infants (Rotem-Kohavi et al. 2019), suggesting that altered newborn functional network organization may lie on a developmental pathway for specific social-emotional behaviors in the first postnatal year.

In another set of studies, Salzwedel et al. examined resting-state functional connectivity 2–6 weeks after birth in infants with in utero polydrug exposure (Salzwedel et al. 2016, 2020). First, when investigating the thalamus as a region of interest, combined prenatal cocaine and SRI exposure ($n = 14$) was uniquely associated with increased thalamocortical functional connectivity compared to healthy controls and cocaine-only exposed infants. These findings predicted poorer cognitive scores at 3-months and were independent of maternal depressive symptoms (Salzwedel et al. 2016). This potential drug-drug interaction may further substantiate an impact of altered 5-HT signaling on the developing brain, since cocaine also acts on the serotonergic system and has high binding affinity for SERT (Ravna et al. 2003). Second, when conducting a whole-brain approach and broadening prenatal polydrug exposure to include SRIs, alcohol, nicotine, opioids, cocaine, and marijuana, prenatal SRIs were found to exhibit the largest single-drug effect on infant functional connectivity compared to any other drug and maternal depressed mood symptoms. Over 10% of functional connections associated with prenatal SRI exposure, with (primarily) hyperconnectivity in orbital/medial/lateral frontal and primary visual networks compared to healthy controls (Salzwedel et al. 2020). Further, they also found that functional connectivity in SRI-exposed infants ($n = 15$) had the strongest correspondence with lower cognitive, language, and motor scores at 3-months of age, again, compared to all other prenatal drug exposures as well as maternal depressed mood.

Current fMRI studies suggest that prenatal SRI exposure may influence functional connectivity in early infancy, with evidence pointing to hyperconnectivity in sensory and frontal networks. While findings vary by analytic approach and sample characteristics, associations between prenatal SRIs and network-level differences may predict infant behavioral or developmental outcomes, supporting a potential pathway from altered early brain function to later neurodevelopmental variability.

5.4 Electrocortical Activity

Synchronized EEG patterns emerge in the third-trimester, with predominantly high-amplitude low-frequency oscillatory activity (e.g., delta and theta waves) transitioning to more organized and complex patterning over the course of childhood in parallel with the functional maturation of cortical networks (Vasung et al. 2019). Importantly, abnormalities in specific EEG patterns have cognitive and behavioral correlates across childhood, which may serve as markers of neurodevelopmental disorders and emerging mental health problems (Amidfar and Kim 2021). For example, alpha asymmetry associates with emotional processing and regulation in infants and children, and prior EEG studies consistently link prenatal depression exposure with greater right frontal alpha activation across the first postnatal year (Field et al. 2010; Lusby et al. 2016; Soe et al. 2016), pointing to a neural pathway for affective and behavioral problems in children of depressed mothers.

Four EEG studies have investigated prenatal SRI exposure and its association with cortical patterning across multiple brain regions (Grieve et al. 2019; Hermansen et al. 2017; Tokariev et al. 2022; Videman et al. 2016), particularly in its relation to infant sleep state and child cognitive function. Using advanced computational EEG analyses, two studies on the same cohort report widespread disruptions in cortical activity in SRI-exposed newborns during the first week and at 2.5-weeks after birth ($n = 22$) (Tokariev et al. 2022; Videman et al. 2016). Compared to newborns of non-SRI-treated mothers with lower depressive symptoms, SRI-exposed newborns had reduced frontal alpha activity at 1-week, as well as lower cross-frequency integration, interhemispheric synchrony, and global intermittency at 2.5-weeks (Videman et al. 2016). Further analyses of sleep network dynamics revealed increased frontal-occipital theta and decreased midline alpha connectivity during the transition from active-to-quiet sleep, as well as reduced phase-amplitude coupling in left temporal, right midfrontal, perisylvian, and central sulcal areas in SRI-exposed newborns during quiet sleep at 2.5-weeks. These findings were then associated with cognitive scores at 2-years of age (Tokariev et al. 2022). However, maternal mood was not accounted for in these studies, even though mothers of SRI-exposed newborns had greater prenatal symptom levels and lifetime history of depression, and post hoc analyses revealed small, yet significant, associations with maternal mood (Videman et al. 2016). Nonetheless, these newborn EEG findings demonstrate consistencies with another study by Grieve et al., who also report disrupted cortical patterning during quiet sleep state, with increased delta brush burst activity across left temporal, central, right parietal, and bilateral occipital areas in SRI-exposed infants at 6-weeks of age ($n = 10$) compared to control and depression-only exposed infants (Grieve et al. 2019). Together, these findings suggest that prenatal SRIs may disrupt emerging sleep architecture, but it is unknown whether similar disturbances persist beyond early infancy.

However, evidence of a lasting developmental impact of prenatal SRI exposure was observed by Hermansen et al., who examined EEG responses (event-related potentials, ERPs) in relation to executive functions (flanker task) in children at 5–6 years of age. While there were no differences in child interference suppression during either congruent or incongruent flanker trial conditions, only control group children displayed a clear association between processing speed (reaction time) and N2 ERP latencies in parietal brain regions. Children with either prenatal SRI ($n = 21$) or depression-only exposure did not demonstrate this expected brain-behavior relationship. Since ERP latency generally associates with relative cognitive demand of a task, the authors suggest these findings may indicate reduced neural efficiency or greater resource allocation required for optimal task performance in exposed children (Hermansen et al. 2017). However, it is unclear whether these outcomes were specific to SRI or depression exposure in the prenatal period or related to ongoing maternal mood disturbances postnatally.

Although limited, existing EEG findings suggest alterations in cortical function and sleep-related brain dynamics in early infancy that may extend to school-age neural efficiency, though further work is needed to disentangle the effects of prenatal SRIs from maternal mood.

6 Summary and Future Directions

Neuroimaging offers a suite of powerful tools to study the developing brain, and its application for investigating the impacts of prenatal SRI exposure over recent years has revealed several associations with brain morphology, microstructure, connectivity, and cortical activity across childhood. To date, most studies have been conducted in infants, with reports of widespread disruptions in anatomical and functional connectivity (Jha et al. 2016; Salzwedel et al. 2020), as well as cortical activity during quiet sleep state (Grieve et al. 2019; Tokariev et al. 2022), following prenatal SRI exposure. Studies also report several regional differences in brain volume, microstructure, and/or function across corticolimbic (Jha et al. 2016; Lugo-Candelas et al. 2018), thalamocortical (Jha et al. 2016; Salzwedel et al. 2016), interhemispheric callosal (Campbell et al. 2021; Videman et al. 2016), frontal and occipital regions (Grieve et al. 2019; Lugo-Candelas et al. 2018; Rotem-Kohavi et al. 2019; Salzwedel et al. 2020; Tokariev et al. 2022) in SRI-exposed infants, with differences in anterior callosal microstructure (Koc et al. 2024), amygdala, superior parietal and occipital volumes (Koc et al. 2023; Moreau et al. 2023) possibly persisting into late childhood. Moreover, effects of prenatal SRI exposure appear to vary by sex in the newborn brain (Campbell et al. 2021; Salzwedel et al. 2020), but not later in childhood (Koc et al. 2023, 2024). Overall, findings across studies are mixed, likely due to methodological heterogeneity, differences in child age at MRI, presence of confounding sample characteristics (e.g., polydrug exposure, prematurity), and small sample sizes in most reports.

Interestingly, findings from several studies suggest that effects of prenatal SRI exposure on the developing brain may be beyond those of maternal depression alone. For instance, SRI-related effects remained significant even after adjusting for maternal mood symptoms during pregnancy (Campbell et al. 2021; Moreau et al. 2023; Salzwedel et al. 2020; Videman et al. 2016), and studies using comparator groups report greater differences in child brain outcome with prenatal SRI exposure compared to children of non-SRI-treated (unmedicated) depressed mothers, ranging from mild-moderate antenatal symptoms (Hermansen et al. 2017; Rotem-Kohavi et al. 2018, 2019) to clinically significant symptoms and/or diagnosed major depressive disorder (Koc et al. 2023; Lugo-Candelas et al. 2018). Whereas others found that SRIs either had no effect (Sethna et al. 2021) or may even attenuate the effects of prenatal depression exposure in regions involved in emotion regulation (Koc et al. 2024; Rotem-Kohavi et al. 2019). However, not all studies controlled for maternal depressive symptoms during pregnancy (Moreau et al. 2023; Podrebarac et al. 2017; Tokariev et al. 2022), some failed to directly compare prenatal SRI- with depression-only exposed child brain outcomes (Jha et al. 2016; Koc et al. 2024), and none have explored potential SRI-mood interactions, thereby limiting conclusions that can be drawn on the impact of prenatal SRI exposure alone.

Confounding by indication continues to challenge our understanding of prenatal SRI exposure and its impact on the developing brain, particularly given that maternal depression is a chronic illness that often spans both pre- and postnatal periods. Fetal MRI may offer a promising novel approach for distinguishing how maternal SRI treatment might differ from depression, during the period of ongoing intrauterine exposure and prior to the increasing influence of the postnatal environment. Indeed, prior work links maternal depression with altered fetal brain volume, cortical gyrification, and metabolite levels (Wu et al. 2020); however, fetal MRI has not yet been conducted to investigate prenatal SRI exposure. The use of fetal MRI in longitudinal study designs may assist in identifying prenatal factors that promote or reduce vulnerability to these exposures.

Beyond cross-sectional differences in brain structure or function, understanding the impact of prenatal SRI exposure ultimately requires linking neuroimaging findings with subsequent child developmental outcome. While brain-behavior relationships have not been widely explored in this context, preliminary evidence suggests that early brain changes observed in SRI-exposed infants may underlie temperamental variability at 6-months of age (Rotem-Kohavi et al. 2019) or cognitive development at 3- and 24-months (Salzwedel et al. 2016, 2020; Tokariev et al. 2022). However, aside from one study that examined ERPs in relation to perceptual processing in middle childhood (Hermansen et al. 2017), it is unknown whether these relationships translate to meaningful differences in cognitive function, behavior, or mental health outcome in SRI-exposed offspring beyond early life.

Recent findings from two large cohorts, the ABCD (USA) (Moreau et al. 2023) and Generation R (The Netherlands) studies (Koc et al. 2023, 2024), highlight how prenatal SRI exposure may have long-term impacts on brain structure into middle-late childhood and adolescence. But with a lack of early life MRI data, it is unknown when child brain changes emerged and whether they reflect prenatal SRI exposure, or some combination of genetic and environmental factors that unfold across postnatal development. Longitudinal multimodal neuroimaging studies are urgently needed to identify long-term brain trajectories associated with prenatal SRI exposure, including their dynamic interaction with factors such as child genetics, mood, behavior, and everyday environment, ideally in a natural setting.

Toward understanding the interplay between gene × environment, an emerging area of research involves the use of polygenic risk scores, which measure individual-level genetic liability for a given trait. Recently, its application to neuroimaging has aimed to disentangle offspring polygenic inheritance associated with depression risk from prenatal depression as an environmental exposure, which were found to interact in their effect on infant limbic volumes (Acosta et al. 2020). However, in the context of prenatal SRI exposure, neither infant genetic variability for *SLC6A4* (Lugo-Candelas et al. 2018) nor child polygenic risk for depression (Moreau et al. 2023) associated with differences in brain morphology in SRI-exposed infants at 1-month or children at 9–10 years of age. While this preliminary evidence may further support a prenatal drug-related effect, additional studies considering how genetic susceptibility shapes child brain outcome in the context of prenatal SRI and depression exposure are needed.

Recent work in the field of "neuroimaging-omics" describes novel approaches for integrated multi-scale and multi-modality analyses of brain connectome with high-throughput biological (e.g., epigenome, transcriptome, metabolome, microbiome) datasets (Chan et al. 2022). Multi-omics allows for a systems-level developmental perspective, rather than studying the brain in isolation, and holds immense promise for identifying potential biomarkers and unraveling the intricate neurobiological mechanisms underlying the impact of prenatal depression exposure and its pharmacotherapy. Continued innovation in deep learning computational capabilities will pave the way for more complex and nuanced developmental models incorporating multi-omics data and their relationship with individualized phenotypes and behavioral characteristics (Ressler and Williams 2021). To achieve this, multi-site collaboration is ultimately needed for the establishment of large, diverse, and well-characterized biological, neuroimaging, and behavioral datasets spanning all stages of child development.

7 Conclusions

This chapter reviews neuroimaging studies on the effects of prenatal SRI antidepressant exposure on the developing brain from the newborn period to mid-adolescence. While evidence is presently mixed as there are relatively few studies on this topic, widespread structural and functional brain changes have been associated with prenatal SRI exposure during infancy, with some consistency among regions or pathways involved in cognitive and emotional processing. Recent evidence also suggests potential long-term effects of prenatal SRI exposure on child brain development into adolescence, but longitudinal brain trajectories from early life are lacking, and it remains unclear whether brain outcome at any stage relates to behavioral disturbances in SRI-exposed children. However, it is important to recognize that prenatal SRI exposure occurs in the context of maternal mood disturbances, and our understanding of these two distinct, yet inherently interconnected prenatal exposures remains challenged by the very nature that depression is a chronic illness embedded within an uncertain clinical context and everyday setting. While it remains essential that maternal mental well-being is optimally supported using all nonpharmacological and/or pharmacological options available, it ultimately may not be possible to fully disentangle medication effects from their underlying indication. Future research might then shift focus away from searching for "main effects" of prenatal SRI exposure toward studying the complex interactions between multiple developmental factors that shape both risk and resilience in children of depressed mothers, SRI-treated or not. Such an outlook may contribute timely research findings that could guide novel evidence-based interventions that benefit mothers and optimize child development.

References

Acosta H, Kantojärvi K, Hashempour N, Pelto J, Scheinin NM, Lehtola SJ et al (2020) Partial support for an interaction between a polygenic risk score for major depressive disorder and prenatal maternal depressive symptoms on infant right amygdalar volumes. Cereb Cortex 30(12):6121–6134. https://doi.org/10.1093/cercor/bhaa158

Amidfar M, Kim Y-K (2021) EEG correlates of cognitive functions and neuropsychiatric disorders: a review of oscillatory activity and neural synchrony abnormalities. Curr Psychiatry Res Rev 16(4):228–243. https://doi.org/10.2174/2666082216999201209130117

Babenko O, Kovalchuk I, Metz GAS (2015) Stress-induced perinatal and transgenerational epigenetic programming of brain development and mental health. Neurosci Biobehav Rev 48:70–91. https://doi.org/10.1016/j.neubiorev.2014.11.013

Barker DJP (1995) The fetal and infant origins of disease. Eur J Clin Investig 25(7):457–463. https://doi.org/10.1111/j.1365-2362.1995.tb01730.x

Battle CL, Salisbury AL, Schofield CA, Ortiz-Hernandez S (2013) Perinatal antidepressant use: understanding women's preferences and concerns. J Psychiatr Pract 19(6):443–453. https://doi.org/10.1097/01.pra.0000438183.74359.46

Bellissima V, Visser G, Ververs T, Pluchinotta F, Varrica A, Baryshnikova E et al (2020) Antenatal maternal antidepressants drugs treatment affects S100B levels in maternal-fetal biological fluids in a dose dependent manner. Clin Chim Acta 501:20–26. https://doi.org/10.1016/j.cca.2019.11.027

Bliddal M, Wesselhoeft R, Strandberg-Larsen K, Ernst MT, Weissman MM, Gingrich JA et al (2023) Prenatal antidepressant exposure and emotional disorders until age 22: a danish register study. Child Adolesc Psychiatry Ment Health 17(1):73. https://doi.org/10.1186/s13034-023-00624-9

Brandlistuen RE, Ystrom E, Eberhard-Gran M, Nulman I, Koren G, Nordeng H (2015) Behavioural effects of fetal antidepressant exposure in a Norwegian cohort of discordant siblings. Int J Epidemiol 44(4):1397–1407. https://doi.org/10.1093/ije/dyv030

Bronson SL, Bale TL (2016) The placenta as a mediator of stress effects on neurodevelopmental reprogramming. Neuropsychopharmacology 41:207–218. https://doi.org/10.1038/npp.2015.231

Brummelte S, Galea LAM, Devlin AM, Oberlander TF (2013) Antidepressant use during pregnancy and serotonin transporter genotype (SLC6A4) affect newborn serum reelin levels. Dev Psychobiol 55(5):518–529. https://doi.org/10.1002/dev.21056

Campbell KSJ, Williams LJ, Bjornson BH, Weik E, Brain U, Grunau RE et al (2021) Prenatal antidepressant exposure and sex differences in neonatal corpus callosum microstructure. Dev Psychobiol 63(6):e22125. https://doi.org/10.1002/dev.22125

Cao M, Huang H, He Y (2017) Developmental connectomics from infancy through early childhood. Trends Neurosci 40(8):494–506. https://doi.org/10.1016/j.tins.2017.06.003

Cattarinussi G, Aarabi MH, Sanjari Moghaddam H, Homayoun M, Ashrafi M, Soltanian-Zadeh H, Sambataro F (2021) Effect of parental depressive symptoms on offspring's brain structure and function: a systematic review of neuroimaging studies. Neurosci Biobehav Rev 131 (May):451–465. https://doi.org/10.1016/j.neubiorev.2021.09.046

Chan YH, Wang C, Soh WK, Rajapakse JC (2022) Combining neuroimaging and omics datasets for disease classification using graph neural networks. Front Neurosci 16(May). https://doi.org/10.3389/fnins.2022.866666

Dadi AF, Miller ER, Bisetegn TA, Mwanri L (2020) Global burden of antenatal depression and its association with adverse birth outcomes: an umbrella review. BMC Public Health 20(1):173. https://doi.org/10.1186/s12889-020-8293-9

Dean DC, Planalp EM, Wooten W, Kecskemeti SR, Adluru N, Schmidt CK et al (2018) Association of prenatal maternal depression and anxiety symptoms with infant white matter microstructure. JAMA Pediatr 172(10):973–981. https://doi.org/10.1001/jamapediatrics.2018.2132

Demers CH, Aran Ö, Glynn LM, Davis EP (2021) Prenatal programming of neurodevelopment: structural and functional changes. In: Prenatal stress and child development. Springer International Publishing, Cham, pp 193–242. https://doi.org/10.1007/978-3-030-60159-1_9

Deneris E, Gaspar P (2018) Serotonin neuron development: shaping molecular and structural identities. Wiley Interdiscip Rev Dev Biol 7(1):11–16. https://doi.org/10.1002/wdev.301

Donnici C, Tomfohr-Madsen L, Long X, Manning KY, Giesbrecht G, Lebel C (2023) Prenatal depressive symptoms are associated with altered structural brain networks in infants and moderated by infant sleep. J Affect Disord 339(June):118–126. https://doi.org/10.1016/j.jad.2023.06.054

Dubois J, Alison M, Counsell SJ, Hertz-Pannier L, Hüppi PS, Benders MJNL (2021) MRI of the neonatal brain: A reivew of methodological challenges and neuroscientific advances. Journal of Magnetic Resonance Imaging 53(5):1318–1343. https://doi.org/10.1002/jmri.27192

Field T, Diego M, Hernandez-Reif M, Figueiredo B, Deeds O, Ascencio A et al (2010) Comorbid depression and anxiety effects on pregnancy and neonatal outcome. Infant Behav Dev 33(1): 23–29. https://doi.org/10.1016/j.infbeh.2009.10.004

Ghimire U, Papabathini SS, Kawuki J, Obore N, Musa TH (2021) Depression during pregnancy and the risk of low birth weight, preterm birth and intrauterine growth restriction- an updated meta-analysis. Early Hum Dev 152(September 2020):105243. https://doi.org/10.1016/j.earlhumdev.2020.105243

Gilmore JH, Knickmeyer RC, Gao W (2018) Imaging structural and functional brain development in early childhood. Nat Rev Neurosci 19(3):123–137. https://doi.org/10.1038/nrn.2018.1

Glover ME, Clinton SM (2016) Of rodents and humans: a comparative review of the neurobehavioral effects of early life SSRI exposure in preclinical and clinical research. Int J Dev Neurosci 51(1):50–72. https://doi.org/10.1016/j.ijdevneu.2016.04.008

Grieve PG, Fifer WP, Cousy NP, Monk CE, Stark RI, Gingrich JA, Myers MM (2019) Neonatal infant EEG bursts are altered by prenatal maternal depression and serotonin selective reuptake inhibitor use. Clin Neurophysiol 130(11):2019–2025. https://doi.org/10.1016/j.clinph.2019.08.021

Groenewold NA, Wedderburn CJ, Pellowski JA, Fouché J-P, Michalak L, Roos A et al (2022) Subcortical brain volumes in young infants exposed to antenatal maternal depression: findings from a South African birth cohort. NeuroImage Clin 36(September):103206. https://doi.org/10.1016/j.nicl.2022.103206

Grove K, Lewis AJ, Galbally M (2018) Prenatal antidepressant exposure and child motor development: a meta-analysis. Pediatrics 142(1). https://doi.org/10.1542/peds.2018-0356

Hashempour N, Tuulari JJ, Merisaari H, Acosta H, Lewis JD, Pelto J et al (2023) Prenatal maternal depressive symptoms are associated with neonatal left amygdala microstructure in a sex-dependent way. Eur J Neurosci 57(10):1671–1688. https://doi.org/10.1111/ejn.15989

Hermansen TK, Yrttiaho S, Røysamb E, Melinder A (2017) Perceptual interference processing in preschool children, with and without prenatal exposure to selective serotonin reuptake inhibitors. Psychopharmacology 234(3):339–351. https://doi.org/10.1007/s00213-016-4467-7

Homberg JR, Schubert D, Gaspar P (2010) New perspectives on the neurodevelopmental effects of SSRIs. Trends Pharmacol Sci 31(2):60–65. https://doi.org/10.1016/j.tips.2009.11.003

Horackova H, Karahoda R, Cerveny L, Vachalova V, Ebner R, Abad C, Staud F (2021) Effect of selected antidepressants on placental homeostasis of serotonin: maternal and fetal perspectives. Pharmaceutics 13(8). https://doi.org/10.3390/pharmaceutics13081306

Hutchens BF, Kearney J (2020) Risk factors for postpartum depression: an umbrella review. J Midwifery Womens Health 65(1):96–108. https://doi.org/10.1111/jmwh.13067

Huttenlocher PR, Dabholkar AS (1997) Regional differences in synaptogenesis in human cerebral cortex. J Comp Neurol 387(2):167–178. https://doi.org/10.1002/(SICI)1096-9861(19971020)387:2<167::AID-CNE1>3.0.CO;2-Z

Ingalhalikar M, Smith A, Parker D, Satterthwaite TD, Elliott MA, Ruparel K, Hakonarson H, Gur RE, Gur RC, Verma R (2014) Sex differences in the structural connectome of the human brain. Proceedings of the National Academy of Sciences 111(2):823–828. https://doi.org/10.1073/pnas.1316909110

Jha SC, Meltzer-brody S, Steiner RJ, Cornea E, Woolson S, Ahn M et al (2016) Antenatal depression, treatment with selective serotonin reuptake inhibitors, and neonatal brain structure: a propensity-matched cohort study. Psychiatry Res Neuroimaging 253:43–53. https://doi.org/10.1016/j.pscychresns.2016.05.004

Kautzky A, Slamanig R, Unger A, Höflich A (2022) Neonatal outcome and adaption after in utero exposure to antidepressants: a systematic review and meta-analysis. Acta Psychiatr Scand 145(1):6–28. https://doi.org/10.1111/acps.13367

Kennedy SH, Lam RW, McIntyre RS, Tourjman SV, Bhat V, Blier P et al (2016) Canadian network for mood and anxiety treatments (CANMAT) 2016 clinical guidelines for the management of adults with major depressive disorder: section 3. Pharmacological treatments. Can J Psychiatr 61(9):540–560. https://doi.org/10.1177/0706743716659417

Kim Y, Bird A, Peterson E, Underwood L, Morton SMB, Grant CC (2020) Maternal antenatal depression and early childhood sleep: potential pathways through infant temperament. J Pediatr Psychol 45(2):203–217. https://doi.org/10.1093/jpepsy/jsaa001

Kirschstein T, Köhling R (2009) What is the source of the EEG? Clin EEG Neurosci 40(3): 146–149. https://doi.org/10.1177/155005940904000305

Koc D, Tiemeier H, Stricker BH, Muetzel RL, Hillegers M, El Marroun H (2023) Prenatal antidepressant exposure and offspring brain morphologic trajectory. JAMA Psychiatry 80(12):1208. https://doi.org/10.1001/jamapsychiatry.2023.3161

Koc D, El Marroun H, Stricker BH, Muetzel RL, Tiemeier H (2024) Intrauterine exposure to antidepressants or maternal depressive symptoms and offspring brain white matter trajectories from late childhood to adolescence. Biol Psychiatry Cognit Neurosci Neuroimaging 9(2): 217–226. https://doi.org/10.1016/j.bpsc.2023.10.009

Kott J, Brummelte S (2019) Trick or treat? Evaluating contributing factors and sex-differences for developmental effects of maternal depression and its treatment. Horm Behav 111(December 2018):31–45. https://doi.org/10.1016/j.yhbeh.2019.01.004

Laine K, Heikkinen T, Ekblad U, Kero P (2003) Effects of exposure to selective serotonin reuptake inhibitors during pregnancy on serotonergic symptoms in newborns and cord blood monoamine and prolactin concentrations. Arch Gen Psychiatry 60(7):720–726. https://doi.org/10.1001/archpsyc.60.7.720

Lancaster CA, Gold KJ, Flynn HA, Yoo H, Marcus SM, Davis MM (2010) Risk factors for depressive symptoms during pregnancy: a systematic review. Am J Obstet Gynecol 202(1): 5–14. https://doi.org/10.1016/j.ajog.2009.09.007

Lautarescu A, Bonthrone AF, Pietsch M, Batalle D, Cordero-Grande L, Tournier JD et al (2022) Maternal depressive symptoms, neonatal white matter, and toddler social-emotional development. Transl Psychiatry 12(1):1–12. https://doi.org/10.1038/s41398-022-02073-y

Le Bihan D, Johansen-Berg H (2012) Diffusion MRI at 25: exploring brain tissue structure and function. NeuroImage 61(2):324–341. https://doi.org/10.1016/j.neuroimage.2011.11.006

Lee MH, Smyser CD, Shimony JS (2013) Resting-state fMRI: a review of methods and clinical applications. Am J Neuroradiol 34(10):1866–1872. https://doi.org/10.3174/ajnr.A3263

Lugo-Candelas C, Cha J, Hong S, Bastidas V, Weissman M, Fifer WP et al (2018) Associations between brain structure and connectivity in infants and exposure to selective serotonin reuptake inhibitors during pregnancy. JAMA Pediatr 172(6):525–533. https://doi.org/10.1001/jamapediatrics.2017.5227

Lusby CM, Goodman SH, Yeung EW, Bell MA, Stowe ZN (2016) Infant EEG and temperament negative affectivity: coherence of vulnerabilities to mothers' perinatal depression. Dev Psychopathol 28(4pt1):895–911. https://doi.org/10.1017/S0954579416000614

Malik SJ, Beqiri A, Price AN, Teixeira JN, Hand JW, Hajnal JV (2015) Specific absorption rate in neonates undergoing magnetic resonance procedures at 1.5 T and 3 T. NMR Biomed 28(3): 344–352. https://doi.org/10.1002/nbm.3256

Marr MC, Graham AM, Feczko E, Nolvi S, Thomas E, Sturgeon D et al (2023) Maternal perinatal stress trajectories and negative affect and amygdala development in offspring. Am J Psychiatry 180(10):766–777. https://doi.org/10.1176/appi.ajp.21111176

Molenaar NM, Bais B, Lambregtse-van den Berg MP, Mulder CL, Howell EA, Fox NS et al (2020) The international prevalence of antidepressant use before, during, and after pregnancy: a systematic review and meta-analysis of timing, type of prescriptions and geographical variability. J Affect Disord 264(December 2019):82–89. https://doi.org/10.1016/j.jad.2019.12.014

Monk C, Lugo-candelas C, Trumpff C (2019) Prenatal developmental origins of future psychopathology: mechanisms and pathways. Annu Rev Clin Psychol 15:1–28

Moreau AL, Voss M, Hansen I, Paul SE, Barch DM, Rogers CE, Bogdan R (2023) Prenatal selective serotonin reuptake inhibitor exposure, depression, and brain morphology in middle childhood: results from the ABCD study. Biol Psychiatry Glob Open Sci 3(2):243–254. https://doi.org/10.1016/j.bpsgos.2022.02.005

Ouyang M, Dubois J, Yu Q, Mukherjee P, Huang H (2019) Delineation of early brain development from fetuses to infants with diffusion MRI and beyond. NeuroImage 185(October 2017): 836–850. https://doi.org/10.1016/j.neuroimage.2018.04.017

Park M, Hanley GE, Guhn M, Oberlander TF (2021) Prenatal antidepressant exposure and child development at kindergarten age: a population-based study. Pediatr Res 89(6):1515–1522. https://doi.org/10.1038/s41390-020-01269-6

Petersen I, Gilbert RE, Evans SJW, Man SL, Nazareth I (2011) Pregnancy as a major determinant for discontinuation of antidepressants: an analysis of data from The Health Improvement Network. J Clin Psychiatry 72(7):979–985. https://doi.org/10.4088/JCP.10m06090blu

Podrebarac SK, Duerden EG, Chau V, Grunau RE, Synnes A, Oberlander TF, Miller SP (2017) Antenatal exposure to antidepressants is associated with altered brain development in very preterm-born neonates. Neuroscience 342:252–262. https://doi.org/10.1016/j.neuroscience.2016.11.025

Pourhamzeh M, Moravej FG, Arabi M, Shahriari E, Mehrabi S, Ward R et al (2022) The roles of serotonin in neuropsychiatric disorders. Cell Mol Neurobiol 42(6):1671–1692. https://doi.org/10.1007/s10571-021-01064-9

Qiu A, Anh TT, Li Y, Chen H, Rifkin-Graboi A, Broekman BFP et al (2015) Prenatal maternal depression alters amygdala functional connectivity in 6-month-old infants. Transl Psychiatry 5(2):e508–e508. https://doi.org/10.1038/tp.2015.3

Qiu A, Shen M, Buss C, Chong YS, Kwek K, Saw SM et al (2017) Effects of antenatal maternal depressive symptoms and socio-economic status on neonatal brain development are modulated by genetic risk. Cereb Cortex 27(5):3080–3092. https://doi.org/10.1093/cercor/bhx065

Ramsteijn AS, Van de Wijer L, Rando J et al (2020) Perinatal selective serotonin reuptake inhibitor exposure and behavioral outcomes: a systematic review and meta-analyses of animal studies. Neurosci Biobehav Rev 114:53–69. https://doi.org/10.1016/j.neubiorev.2020.04.010

Ravna AW, Sylte I, Dahl SG (2003) Molecular mechanism of citalopram and cocaine interactions with neurotransmitter transporters. J Pharmacol Exp Ther 307(1):34–41. https://doi.org/10.1124/jpet.103.054593

Ressler KJ, Williams LM (2021) Big data in psychiatry: multiomics, neuroimaging, computational modeling, and digital phenotyping. Neuropsychopharmacology 46(1):1–2. https://doi.org/10.1038/s41386-020-00862-x

Rogers A, Obst S, Teague SJ, Rossen L, Spry EA, MacDonald JA et al (2020) Association between maternal perinatal depression and anxiety and child and adolescent development: a meta-analysis. JAMA Pediatr 3125:1082. https://doi.org/10.1001/jamapediatrics.2020.2910

Rommel A-S, Momen NC, Molenaar NM, Liu X, Munk-Olsen T, Bergink V (2021) Long-term prenatal effects of antidepressant use on the risk of affective disorders in the offspring: a register-based cohort study. Neuropsychopharmacology 46(8):1518–1525. https://doi.org/10.1038/s41386-021-01005-6

Rotem-Kohavi N, Williams LJ, Virji-Babul N, Bjornson BH, Brain U, Werker JF et al (2018) Alterations in resting-state networks following in utero selective serotonin reuptake inhibitor exposure in the neonatal brain. Biol Psychiatry Cognit Neurosci Neuroimaging 4(1):39–49. https://doi.org/10.1016/j.bpsc.2018.08.004

Rotem-Kohavi N, Williams LJ, Muller AM, Abdi H, Virji-Babul N, Bjornson BH et al (2019) Hub distribution of the brain functional networks of newborns prenatally exposed to maternal depression and SSRI antidepressants. Depress Anxiety 36(8):753–765. https://doi.org/10.1002/da.22906

Salzwedel A, Grewen K, Goldman B, Gao W (2016) Thalamocortical functional connectivity and behavioral disruptions in neonates with prenatal cocaine exposure. Neurotoxicol Teratol 56:16–25. https://doi.org/10.1016/j.ntt.2016.05.009

Salzwedel A, Chen G, Chen Y, Grewen K, Gao W (2020) Functional dissection of prenatal drug effects on baby brain and behavioral development. Hum Brain Mapp 41(17):4789–4803. https://doi.org/10.1002/hbm.25158

Schoretsanitis G, Westin AA, Stingl JC, Deligiannidis KM, Paulzen M, Spigset O (2021) Antidepressant transfer into amniotic fluid, umbilical cord blood & breast milk: a systematic review & combined analysis. Prog Neuro-Psychopharmacol Biol Psychiatry 107(December 2020): 110228. https://doi.org/10.1016/j.pnpbp.2020.110228

Sethna V, Siew J, Gudbrandsen M, Pote I, Wang S, Daly E et al (2021) Maternal depression during pregnancy alters infant subcortical and midbrain volumes. J Affect Disord 291(April):163–170. https://doi.org/10.1016/j.jad.2021.05.008

Silbereis JC, Pochareddy S, Zhu Y, Li M, Sestan N (2016) The cellular and molecular landscapes of the developing human central nervous system. Neuron 89(2):248. https://doi.org/10.1016/j.neuron.2015.12.008

Simpson KL, Weaver KJ, de Villers-Sidani E, Lu JY-F, Cai Z, Pang Y et al (2011) Perinatal antidepressant exposure alters cortical network function in rodents. Proc Natl Acad Sci USA 108(45):18465–18470. https://doi.org/10.1073/pnas.1109353108

Soe NN, Wen DJ, Poh JS, Li Y, Broekman BFP, Chen H et al (2016) Pre- and post-natal maternal depressive symptoms in relation with infant frontal function, connectivity, and behaviors. PLoS One 11(4):e0152991. https://doi.org/10.1371/journal.pone.0152991

Suarez EA, Bateman BT, Hernández-Díaz S, Straub L, Wisner KL, Gray KJ et al (2022) Association of antidepressant use during pregnancy with risk of neurodevelopmental disorders in children. JAMA Intern Med 02120(11):1149–1160. https://doi.org/10.1001/jamainternmed.2022.4268

Suh JS, Schneider MA, Minuzzi L, MacQueen GM, Strother SC, Kennedy SH, Frey BN (2019) Cortical thickness in major depressive disorder: a systematic review and meta-analysis. Prog Neuro-Psychopharmacol Biol Psychiatry 88(July 2018):287–302. https://doi.org/10.1016/j.pnpbp.2018.08.008

Tarabulsy GM, Pearson J, Vaillancourt-Morel M-P, Bussières E-L, Madigan S, Lemelin J-P et al (2014) Meta-analytic findings of the relation between maternal prenatal stress and anxiety and child cognitive outcome. J Dev Behav Pediatr 35(1):38–43. https://doi.org/10.1097/DBP.0000000000000003

Tirumalaraju V, Suchting R, Evans J, Goetzl L, Refuerzo J, Neumann A et al (2020) Risk of depression in the adolescent and adult offspring of mothers with perinatal depression: a systematic review and meta-analysis. JAMA Netw Open 3(6):e208783. https://doi.org/10.1001/jamanetworkopen.2020.8783

Tokariev A, Oberlander VC, Videman M, Vanhatalo S (2022) Cortical cross-frequency coupling is affected by in utero exposure to antidepressant medication. Front Neurosci 16(March):1–12. https://doi.org/10.3389/fnins.2022.803708

Turk E, van den Heuvel MI, Benders MJ, de Heus R, Franx A, Manning JH et al (2019) Functional connectome of the fetal brain. J Neurosci 39(49):9716–9724. https://doi.org/10.1523/JNEUROSCI.2891-18.2019

Vasung L, Abaci Turk E, Ferradal SL, Sutin J, Stout JN, Ahtam B et al (2019) Exploring early human brain development with structural and physiological neuroimaging. NeuroImage 187: 226–254. https://doi.org/10.1016/j.neuroimage.2018.07.041

Videman M, Tokariev A, Saikkonen H, Stjerna S, Heiskala H, Mantere O, Vanhatalo S (2016) Newborn brain function is affected by fetal exposure to maternal serotonin reuptake inhibitors. Cereb Cortex 27(6):bhw153. https://doi.org/10.1093/cercor/bhw153

Willner P, Scheel-Krüger J, Belzung C (2013) The neurobiology of depression and antidepressant action. Neurosci Biobehav Rev 37(10):2331–2371. https://doi.org/10.1016/j.neubiorev.2012.12.007

Wu Y, Lu Y-C, Jacobs M et al (2020) Association of prenatal maternal psychological distress with fetal brain growth, metabolism, and cortical maturation. JAMA Netw Open 3:e1919940. https://doi.org/10.1001/jamanetworkopen.2019.19940

Zhang Y, Banihashemi L, Samolyk A, Taylor M, English G, Schmithorst VJ et al (2023) Early infant prefrontal gray matter volume is associated with concurrent and future infant emotionality. Transl Psychiatry 13(1):125. https://doi.org/10.1038/s41398-023-02427-0

Maternal SSRI Use During Pregnancy: Links Between Gut Microbiome, Offspring Behaviour, and Brain Transcriptomics

Mayerli A. Prado Rivera, Joëlle D. Jagersma, and Jocelien D. A. Olivier

Contents

Abstract Selective serotonin reuptake inhibitors (SSRIs) are commonly prescribed to treat maternal depression during pregnancy, yet their potential impact on foetal brain development remains a concern. Although maternal illness is a known confounder, growing evidence from both clinical and preclinical studies suggests that perinatal SSRI exposure may independently influence neurodevelopmental outcomes. This review draws on rodent studies to explore how developmental SSRI exposure affects the gut microbiome, maternal behaviour, myelination, and offspring social behaviour. Elevated serotonin levels caused by SSRIs can alter both brain development and the maternal gut microbiota, with possible long-term effects on offspring. Behaviourally, SSRI-exposed offspring often show reduced social play, altered social interactions, and sex-specific effects on aggression and sexual behaviour, in which males appear more sensitive to these effects than females. Maternal

M. A. Prado Rivera and J. D. A. Olivier (✉)
Neurobiology, Groningen Institute for Evolutionary Life Sciences, University of Groningen, Groningen, The Netherlands
e-mail: j.d.a.olivier@rug.nl

J. D. Jagersma
Neurobiology, Groningen Institute for Evolutionary Life Sciences, University of Groningen, Groningen, The Netherlands

Department of Otorhinolaryngology-Head and Neck Surgery, University of Groningen, University Medical Center Groningen, Groningen, The Netherlands

Curr Topics Behav Neurosci (2025) 75: 23–41
https://doi.org/10.1007/7854_2025_602
Published Online: 29 August 2025

care is only modestly affected. Overall, developmental SSRI exposure in the absence of maternal illness can disrupt brain development and social behaviour in offspring, potentially through gut-brain axis mechanisms and altered myelination.

Keywords Brain development · Maternal gut microbiome · Myelination · Pregnancy · Selective Serotonin Reuptake Inhibitors · Social behavior

1 Introduction

Depression is a mental disorder affecting 264 million people worldwide (WHO 2017). Pregnant women are not spared: 20% of pregnant women experiences depressive symptoms. One to 3% of pregnant women require antidepressants (Molenaar et al. 2020). Selective serotonin reuptake inhibitors (SSRIs) are the most commonly prescribed antidepressants during pregnancy. However, perinatal SSRI use by mothers can have long-lasting adverse effects on the development of their children and may result in higher anxiety levels (Hanley et al. 2015) and lower scores on motor-, social-emotional-, and adaptive behavioural tests (Hanley et al. 2013). Moreover, maternal SSRI exposure during pregnancy is associated with increased risk of developing mental and behavioural disorders such as autism spectrum disorder, attention-deficit/hyperactivity disorder, and mental retardation in the offspring (Rotem-Kohavi and Oberlander 2017; Halvorsen et al. 2019); however, when adjusting for maternal illness (depression or anxiety), these links mostly disappear (Andrade 2017; Kaplan et al. 2017). This suggests that maternal illness is a key factor influencing risks of developmental disorders associated with antidepressant exposure during pregnancy.

SSRIs prevent serotonin (5-HT) reuptake, thereby increasing extracellular levels of 5-HT (Fuller and Wong 1987). Approximately 90–95% of 5-HT is produced in the gastrointestinal tract and plays a critical role in regulating gut motility, secretion, and microbial composition (Gershon and Tack 2007). By increasing 5-HT levels in the gut, a disruption of the homeostasis can take place which has been shown to alter microbial diversity and abundance in both animal and in vitro studies (Yano et al. 2015; Cussotto et al. 2019). These SSRI-induced changes in the gut microbiota may have downstream effects on neurodevelopment, particularly during sensitive periods such as pregnancy (Clarke et al. 2013). Serotonin plays an important role in many processes, one of them being a key neurotrophic factor in human brain development (Yan et al. 1997; Lee 2009; Miceli et al. 2013). As such, SSRI-induced elevations in 5-HT levels may interfere with normal neurodevelopmental processes.

Since all SSRIs cross the placenta, concerns have arisen about their possible impact during pregnancy. Although the exact mechanisms remain unclear, studies have linked prenatal SSRI exposure to various adverse perinatal outcomes, including reduced foetal head growth (El Marroun et al. 2012), low birth weight (Chambers et al. 1996), neural tube defects (Alwan et al. 2007; Malm et al. 2011), cardiac malformations, particularly atrial and ventricular septal defects (Bérard et al. 2015, 2017b), craniosynostosis (Bérard et al. 2015), persistent pulmonary hypertension (Kieler et al. 2012; Bérard et al. 2017a), and lower APGAR scores at birth (Jensen et al. 2013). In addition, human newborns exposed to SSRIs during pregnancy have lower levels of reelin, which is a marker for neurodevelopment (Brummelte et al. 2013). Also, alterations in white and grey matter structure (Podrebarac et al. 2017), volume (Koc et al. 2023), and connectivity (Videman et al. 2017; Lugo-Candelas et al. 2018; Rotem-Kohavi et al. 2019) occur in the brain of SSRI-exposed offspring, demonstrating that altering 5-HT signalling during development may result in lasting brain changes, and affect behaviour later in life. However, the mechanisms by which neurodevelopment in the offspring is influenced remain elusive.

Findings from animal studies have reinforced the hypothesis that early SSRI exposure may disrupt normal brain development, including hippocampal neurogenesis and epigenetic programming (Vorhees et al. 1994; Maciag et al. 2006; Glover et al. 2015; Glover and Clinton 2016; Sprowles et al. 2016). Outcomes of animal studies resemble and complement correlational studies in humans. Animal experiments have advantages over human research, including a high degree of control over drug dosing and exposure period. Animal models are also indispensable in elucidating long-term effects of SSRIs on offspring (Zucker 2017), while avoiding the confounding influence of underlying mental disorders, and enable to investigate pathways and mechanisms responsible for these effects. Results are rapidly acquired, as rodents develop much faster than humans while brain development follows a remarkably similar route in both species (Semple et al. 2013). Moreover, placental transfer of SSRIs is similar in humans, mice (Noorlander et al. 2008), and rats (Olivier et al. 2011). Also in rodents, the importance of gut microbiota–brain communication for brain development, behaviour, and function has become increasingly evident in recent decades. This includes its role in regulating neural processes such as neurogenesis and myelination (Cryan et al. 2019). Beyond its role in 5-HT synthesis, the maternal microbiome is increasingly recognized as a key regulator of foetal brain development (Hoban et al. 2016). In this chapter, we aim to give the reader an overview of the alterations in the maternal gut microbiome due to SSRI treatment and describe subsequent effects on offspring social behaviour. Finally, the potential regulating neural processes altered by SSRIs, mainly myelination, will be described. The main focus will be on animal research exposed to SSRIs without underlying maternal illness (or stress).

2 Role of the Gut-Microbiome in Brain Development

In mice lacking microbiota (germ-free), it was shown that blood-brain barrier permeability, brain volume, neural circuitry, myelination, and microglia were altered, stressing that disruptions of the microbial colonization during development can have far-reaching influences on health later in life (Cryan et al. 2019). Germ-free mice have increased myelin-related gene expression in the prefrontal cortex (PFC), resulting in hypermyelination of the PFC (Hoban et al. 2016). Myelin-associated glycoprotein (MAG) and myelin basic protein (MBP) are both proteins involved in myelination and were both found to be upregulated in the PFC of germ-free mice. The perinatal fluoxetine exposure studies from our lab, and others, demonstrated sex-specific brain effects (see section: Effects SSRI use during pregnancy on offspring social behaviour). These data parallel with several germ-free studies that showed more robust brain alterations in males compared to females (Clarke et al. 2013; Hoban et al. 2016). Given its antimicrobial properties (Rgen et al. 2011; Munoz-Bellido et al. 1996), fluoxetine may inhibit certain bacteria, leading to an increased abundance of other bacteria. Known routes of gut–brain communications include neuroactive microbial metabolites (Nicholson et al. 2012). The increase in the abundance of bacteria that produce metabolites involved with myelination will coincide with increased myelination. However, this step has not been proven yet.

Convincing evidence for the direct role of the maternal microbiome in modulating the neurodevelopment in the offspring was provided by Vuong et al. (2020). They showed that antibiotic treatment of the mother resulted in reduced axogenesis-related gene expression and impaired outgrowth of thalamic axons. In addition, they showed that specific metabolites, produced in the gut, are specifically driving this neurodevelopmental effect in the offspring, and more importantly can reverse the deficits in axon outgrowth and promote foetal thalamocortical axogenesis. However, exactly how maternal SSRI treatment during pregnancy influences early neurodevelopment to impart lasting alterations in offspring behaviour remains poorly understood.

With respect to SSRIs influencing the gut microbiome, it has been shown that fluoxetine modulates bacterial colonization in the (non-pregnant) gut (Fung et al. 2019). When pregnant mice were treated with fluoxetine from embryonic day 7.5 till embryonic day 14.5, no overt effects were found in the maternal microbiome. However, select rare taxa were altered including short-chain fatty acid-producing bacteria (Vuong et al. 2021). This is opposite to the findings from our lab, where we treated control and depressive-like female rats from gestational day (G)1 until postnatal day (PND)21 (6 weeks) with fluoxetine. We found a clear effect of fluoxetine in the relative abundance of several taxa and specifically in depressive-like animals (Ramsteijn et al. 2020). With a targeted metabolomics approach, we found that specific amino acids, such as serine and aspartic acid, were reduced most strongly in depressive-like females treated with fluoxetine. Altogether, these results show that the maternal gut microbiome can influence the foetal brain development during pregnancy. The next step would be to investigate which maternal metabolites contribute to alterations in foetal brain development as a result of maternal SSRI treatment during pregnancy.

3 Effects of SSRIs on Maternal Behaviour

Maternal care has been associated with increased serotonergic activity (Pawluski et al. 2019). The increase of 5-HT levels by SSRIs can therefore potentially influence maternal behaviour. Because maternal behaviour is known to influence offspring social behaviour (Starr-Phillips and Beery 2014), it is important to address how maternal behaviour is influenced by SSRI treatment. Dams treated with 10 mg/kg fluoxetine throughout pregnancy (from G1 until birth) delivered smaller litter sizes compared to saline-treated dams (Svirsky et al. 2016). In addition, on the first day, nest quality was lower in the fluoxetine treatment group compared to the controls, but this was restored after day 1 (Svirsky et al. 2016). Sprague-Dawley dams treated with 5 mg/kg fluoxetine from PND1–PND28 spent more time engaged in arched-back nursing compared with vehicle-treated dams, although the overall time devoted to nursing or licking their pups did not differ between groups (Pawluski et al. 2012). Fluoxetine treatment from G8 to PND21 did not affect the duration of nursing or licking behaviours directed toward the pups (Gemmel et al. 2018). When we treated dams with 10 mg/kg fluoxetine from G1 to PND21, the live birth index was decreased compared to those treated with vehicle (Houwing et al. 2019b). We also observed a higher percentage in the frequency of fluoxetine-treated dams on the nest compared to controls (Prado Rivera et al. 2026), perhaps to compensate for the lower birth weight we found in the fluoxetine-treated pups. Similarly, fluoxetine administration during gestation was associated with increased frequencies of maternal behaviours such as touching, sniffing, and hovering over the pups on PND1 (Johns et al. 2005). Altogether it seems that changes in maternal care due to fluoxetine treatment during pregnancy are mostly minor, and if negative, short-lasting. Mainly more active nursing have been reported as a result of SSRI treatment during the perinatal period.

4 Effects of SSRI Use During Pregnancy on Offspring Social Behaviour

Serotonin is a central neuromodulator for social behaviour, and because maternal SSRI use during pregnancy, in mothers with underlying maternal illness, was linked to autism spectrum disorders in offspring (Rotem-Kohavi and Oberlander 2017; Andrade 2017; Kaplan et al. 2017; Halvorsen et al. 2019), many rodent studies have examined the effects of perinatal SSRI exposure on offspring social behaviour. In one study tested for social interaction, female mice born to dams treated with fluoxetine (10 mg/kg) throughout pregnancy (from G1 until birth) showed a significant increase in exploration time of a 4-week-old stimulus mouse compared to a novel object, an effect not observed in male offspring (Svirsky et al. 2016). However, when the same mice were tested in a recognition memory task involving a familiar and an unfamiliar conspecific, no significant effects of prenatal fluoxetine exposure were detected in recognition memory (Svirsky et al. 2016). When rats were treated with 20 mg/kg fluoxetine from PND0 to PND4, they displayed increased occurrence of sniff and contact behaviours, as well as total social interaction time compared to control rats (Ko et al. 2014). In a semi-natural environment with 4 adult male +4 adult females living together, perinatally fluoxetine-treated (G1-PND21) females were socially interacting more passively compared to controls, an effect not found in males (Houwing et al. 2019a). These findings suggest that perinatal SSRI exposure can have positive and sex-specific effects on social interaction in rodent offspring.

When Sprague-Dawley rats were treated with 5 mg/kg fluoxetine from G1 to PND21, they showed increased socially aggressive play (Gemmel et al. 2017). The same lab found that fluoxetine in Sprague-Dawley increased social investigation in adult females and increased social play (pouncing, nape attacks) in adult males (Gemmel et al. 2019). In our lab, however, we found that fluoxetine treatment from G1 to PND21 (in Wistar rats) led to reduced juvenile play behaviour in both male and female offspring. However, in adulthood, only male offspring exposed to fluoxetine continued to exhibit reduced social interaction compared to controls, an effect not observed in females (Houwing et al. 2019b). Postnatal exposure to citalopram (PND8–21) also led to reduced social play behaviour, which was significant in male, but not in female rats (Simpson et al. 2011). When Sprague-Dawley rats were treated with 12 mg/kg/day of fluoxetine from G11 until birth, male offspring of fluoxetine-exposed dams showed reduced social interaction, as indicated by decreased distance moved in the chamber containing a social conspecific during a three-chamber test. In contrast, female offspring did not exhibit this effect (Chen et al. 2021). Rodriguez-Porcel and colleagues came to comparable results. After injecting Long Evans rats with citalopram or fluoxetine from PND8–PND21, they found that the response to novel conspecific animal and preference for investigating a novel animal versus a novel object was significantly decreased compared to controls in both males and females (Rodriguez-Porcel et al. 2011). Similarly, boxing and pinning, part of juvenile play behaviour, were significantly reduced in males with both citalopram and fluoxetine, but not (or less) in females (Rodriguez-Porcel et al. 2011). Overall, these findings indicate that perinatal SSRI exposure can alter play behaviour in a sex- and strain-dependent manner, with male offspring generally showing more consistent and pronounced reductions in social and play behaviours compared to females.

Regarding aggression levels, prenatal fluoxetine treatment (from G1 to birth) has been shown to significantly increase both the number and duration of aggressive attacks exhibited by adult male mice during a social memory test (Svirsky et al. 2016). In C57Bl/6 mice, offspring of dams exposed to 25 mg/kg fluoxetine from gestational day 4–14 exhibited increased aggressive behaviour in males (Kiryanova et al. 2016). In contrast, other studies have reported reduced aggression following perinatal SSRI exposure; for instance, mice exposed to fluoxetine from G0 to PND21 (resident-intruder test; Lisboa et al. 2007), and rats exposed postnatally to citalopram from PND1 to PND19 (aggressive behavior in box exposed to electric shock; Manhaes De Castro et al. 2001), both demonstrated increased latency to initiate an attack against an intruder. Similarly, in our laboratory, male rats treated with fluoxetine from G1 to PND 21 also showed increased attack latency and exhibited fewer offensive behaviours in a resident-intruder test (Houwing et al. 2020). In a semi-natural environment with 8 animals living together (4 male +4 females), perinatally fluoxetine-treated (G1-PND21) females were for a significantly shorter duration involved in conflicts in the burrow compared to control animals (Houwing et al. 2019a). Taken together, these studies suggest that perinatal SSRI exposure can lead to both increases and decreases in aggressive behaviour, with outcomes varying by sex, species, timing of exposure, and testing context, though a general trend toward reduced aggression is observed in many cases.

With respect to sexual behaviours, postnatal administration of clomipramine, a tricyclic antidepressant, from PND8 to PND21 was found to disrupt male sexual behaviour, with a reduced proportion of males achieving ejaculation (Mirmiran et al. 1981). Among those that did ejaculate, the time to first ejaculation was significantly prolonged. Although the total numbers of mounts and intromissions remained comparable between treated and control groups, the higher mount-to-intromission ratio in clomipramine-exposed animals suggested reduced sexual efficiency. In line with this, we found that prenatal exposure to fluoxetine (from G11 until birth) had no detectable effect on sexual behaviour in adulthood in sexually trained animals (Olivier et al. 2011). Also when Swiss dams were treated daily, by gavage, with 7.5 mg/kg of fluoxetine from G0-PND21, their offspring showed no difference in anogenital distance and did not alter the number of mounts, intromissions, and ejaculations (Gouvêa et al. 2008). However, Maciag et al. (2006) reported that postnatal exposure to citalopram (PND8–PND21) led to notable impairments in sexual behaviour, including diminished mounting activity, fewer intromissions, and decreased rates of ejaculation (Maciag et al. 2006). When Long-Evans rats received 10 mg/kg citalopram or 5 mg/kg fluoxetine from PND8-PND21, this also significantly reduced the number of mounts, intromissions, and ejaculations. Also a dose-dependent effect of citalopram injected from PND8 to PND21 caused a decrease in the number of mounts and a trend in reduced intromissions and ejaculation behaviours of male Long-Evans rats (Harris et al. 2012). In addition, developmental fluoxetine treatment from PND1 to PND21, as shown by Rayen et al. (2013), resulted in reduced anogenital distance in juvenile males, fewer intromissions, and longer latencies to both first intromission and ejaculation in sexually naive male offspring (Rayen et al. 2013). When the effects of perinatal fluoxetine from G1 to PND21 was tested on sexual behaviour in a semi-natural environment (with primed females), we found that fluoxetine females spent less time showing paracopulatory behaviour and were pursued by other rats for a shorter duration compared with control females. As a consequence, fluoxetine females also showed fewer lordosis responses than control females (Houwing et al. 2019a). However, when the total female sexual behaviour was scored during the whole oestrus period (from first lordosis to the last), no effects were found on adult female sexual behaviour. Fluoxetine and control females displayed similar paracopulatory behaviour and lordosis (Hegstad et al. 2020). When female Sprague-Dawley rat dams were chronically treated with 5 mg/kg of fluoxetine from PND1 to PND21, their female offspring showed significantly more proceptive behaviours (hopping and darting) than their controls and had a higher lordosis quotient (Rayen et al. 2014). Fluoxetine-exposed males mounted and intromitted more often than control-treated males in the semi-natural environment (Houwing et al. 2019a). This was also found when sexual behaviour was scored for the entire sexually active period of males (Baann et al. 2025). It seems that the timing of SSRI exposure, the execution of sexual test, species, and strains may influence the outcome of sexual performance after perinatal SSRI exposure, with most negative effects on sexual behaviour found with postnatal exposure, and neutral or positive effects found when rodents were also prenatally exposed.

In summary, perinatal SSRI exposure can lead to lasting and sex-dependent alterations in social, play, aggressive, and sexual behaviours in rodents. While effects vary across studies depending on strain, sex, timing, and testing context, a general pattern emerges: males are more consistently affected than females, with reduced social and play behaviours, altered aggression (often reduced), and impaired sexual behaviours particularly following postnatal exposure. Prenatal or combined prenatal-postnatal exposure tends to yield more variable outcomes, including neutral or occasionally facilitative effects, especially in sexual and social domains. Overall, the main outcome is that early-life SSRI exposure can disrupt the developmental trajectory of key social and reproductive behaviours, with males showing greater vulnerability.

5 Effects on SSRI Use During Pregnancy on Offspring Brain Gene Expression

During early development, the serotonin transporter (SERT) is broadly expressed in both serotonergic and non-serotonergic neurons throughout the brain. In contrast, after birth, its expression becomes largely restricted to serotonergic neurons, shown both in humans and in rodents (Verney et al. 2002; Narboux-Nême et al. 2008). As a result, SSRIs have markedly different effects during early brain development than they do in adulthood. Rodent studies provide a valuable opportunity to investigate molecular changes occurring during critical periods of neurodevelopment that may underlie long-term effects of early-life interventions (Ehrlich et al. 2015; Boulle et al. 2016a, b; Meyer et al. 2018). Behavioural abnormalities resulting from neonatal antidepressant exposure are associated with widespread alterations in the 5-HT system. These include decreased firing activity of serotonergic neurons in the dorsal raphe nucleus (Kinney et al. 1997), lower expression levels of tryptophan hydroxylase (TPH2) and the SERT in the raphe region (Maciag et al. 2006), reduced serotonergic fibre density in the hippocampus (Weaver et al. 2010), and disrupted monoamine concentrations across various brain areas (Hilakivi et al. 1987; Feenstra et al. 1996; Vijayakumar and Meti 1999; Yannielli et al. 1999). In addition, mice exposed to SSRIs from PND2 to PND11 show reduced serotonergic fibre innervation in the medial prefrontal cortex (mPFC) (Rebello et al. 2014).

In males, transcriptome-wide analyses with micro-arrays showed in Sprague-Dawley rats that escitalopram exposure (from G7 to weaning) led to no differences in gene expression in the amygdala, hippocampus, and hypothalamus (Bourke et al. 2013). However, when Sprague-Dawley rats were treated with fluoxetine from PND2 to PND21, life-long changes in hippocampal *Hdac4* expression were found, which may contribute to a downregulation of specific genes (*mTOR* and *Gnai1*) following fluoxetine exposure (Sarkar et al. 2014). Another microarray study demonstrated that paroxetine administration (10 mg/kg) from 1 week prior to conception through weaning significantly altered gene expression in the offspring's brain. Here, brain tissues were collected at PNDs 7, 14, 21, and 75 for analysis. In the amygdala of paroxetine-exposed low-responder rats, gene expression changes were minimal at PND7, peaked with 187 differentially expressed genes at PND21, and declined to 80 genes by PND75, with most of the changes representing downregulation at the later time points. Genes included '*haemoglobin complex*' and '*intracellular organelle*'. In the hippocampus, 722 genes were differentially expressed at PND7, decreasing to 80 by PND21, but increasing again to 360 by PND75. While early postnatal stages (PND7 and PND14) were marked by widespread downregulation, the majority of gene expression changes observed at PND75 were upregulations. Genes were involved in '*structural constituent of ribosome*', '*oxidoreductase activity, acting on a haeme group of donors, oxygen as acceptor*', '*synaptic vesicle*', '*negative regulation of kinase activity*', and '*neurogenesis*' (Glover et al. 2015). These data show that the most pronounced paroxetine-induced changes occurred in the hippocampus during the first two postnatal weeks, whereas in the amygdala, these changes became more prominent around PND21 (Glover et al. 2015). This suggests that the transcriptomic response to SSRIs is both region-specific and developmentally timed.

Simpson et al. (2011) hypothesized that since 5-HT plays a critical role in early cortical barrel development and sensory processing, and given that many pathologies observed in autism spectrum disorder (ASD) involve altered corpus callosum morphology and disrupted corticocortical connectivity (Courchesne and Pierce 2005; Welsh et al. 2005), perinatal citalopram exposure may lead to structural changes in the corpus callosum. Indeed, they showed that perinatal exposure to the SSRI citalopram resulted in impaired social behaviour and disruptions in both raphe circuitry and cortical network activity in rats. Notably, these effects appear to be more pronounced in males, suggesting a sex-dependent vulnerability (Simpson et al. 2011). Their findings further suggested that early citalopram exposure interfered with oligodendrocyte maturation in the corpus callosum, as indicated by abnormal myelin development and reduced interhemispheric connectivity.

Interestingly, in our lab we exposed rats to fluoxetine (10 mg/kg) from G1 to PND21 and showed with RNAseq that 21-day-old offspring rats have altered patterns in genes related to '*oligodendrocyte development, differentiation, and myelination*' (Ramsteijn et al. 2022). In males exposed to fluoxetine, genes related to myelination were upregulated in the prefrontal cortex but downregulated in the basolateral amygdala. In contrast, fluoxetine-exposed females showed no significant changes in myelin-related gene expression in the prefrontal cortex, while the basolateral amygdala exhibited a downregulation pattern similar to that observed in males. Differential gene expression analysis revealed that perinatal fluoxetine exposure altered the expression of key myelin-related genes, including myelin-associated glycoprotein (MAG), myelin basic protein (MBP), claudin-11 (CLDN11), and 2′,3′-cyclic-nucleotide 3′-phosphodiesterase (CNP). Notably, expression levels of MAG and MBP were found to correlate with their DNA methylation status, suggesting that epigenetic mechanisms may play a role in mediating fluoxetine-induced disruptions in myelination during development.

Oligodendrocyte-driven axonal myelination is essential for proper brain development and function, it supports the formation of neural circuits, facilitates rapid signal transmission, and provides critical metabolic support to neurons (Stadelmann et al. 2019). In humans, myelination begins during mid-to-late gestation (Lebel and Deoni 2018), whereas in rodents, it starts shortly after birth (Downes and Mullins 2014). By PND21, the limbic system is among the last brain regions to complete myelination (Downes and Mullins 2014). Efficient myelination is a highly coordinated process that relies on the interaction of multiple key proteins, including MAG, MBP, CLDN11, and CNP, to form and maintain the multi-layered lipid sheath surrounding axons (Stadelmann et al. 2019). MAG anchors the myelin sheath to the axon and MBP promotes the compaction of membrane layers, ensuring the structural integrity of the sheath. CLDN11 forms tight junctions in the outer myelin layers, contributing to its barrier function. In contrast, CNP opposes MBP by creating cytoplasmic channels that allow for molecular diffusion within the sheath. These findings suggest that developmental SSRI exposure may disrupt myelination processes during a particularly sensitive window in brain development, potentially leading to long-term consequences for neural connectivity and function.

The finding that perinatal fluoxetine exposure influences myelin-related gene expression is supported by a study of Kroeze et al. (2015). They administered fluoxetine to female rats from PND1 to PND21 and reported downregulation of myelin-associated genes such as ciliary neurotrophic factor and transferrin in the adult hippocampus. In a related study, the same group investigated developmental gene expression in the medial prefrontal cortex of male SERT-knockout rats, a model that mimics SSRI-induced SERT inhibition (Kroeze 2017). They found that the expression of CLDN11, MAG, and MBP was decreased at PND8, increased at PND14, and again decreased at PND21 compared to wild-type controls. Interestingly, these dynamics contrast with findings of increased myelin-related gene expression in the prefrontal cortex following fluoxetine exposure at PND21 at the same brain region (Ramsteijn et al. 2022). However, the study by Simpson et al. (2011) showed that perinatal citalopram treatment led to both hypo- and hypermethylation in the corpus callosum in adulthood. Side-by-side hypo- and hypermyelination can take place depending on cellular activity, molecular signals, timing, and environmental influences during development (Mitew et al. 2018, Mount and Monje 2017), and these local imbalances may disrupt circuit synchrony. Taken together, these findings suggest that the impact of developmental SSRI exposure on myelination is highly region-, age-, and sex-dependent.

6 Link between Gut-Microbiome, Social Behaviour, and Myelination

Early in brain development, radial glial cells generate diverse neuronal and glial cell types (Benamer et al. 2020; Zonouzi et al. 2019). Forebrain radial glia cells give rise to oligodendrocyte progenitor cells (OPCs), which are immature cells that produce myelinating oligodendrocytes (Kessaris et al. 2006; Bergles and Richardson 2015). In the rodent forebrain, OPCs are first generated at embryonic day 12 (Kessaris et al. 2006) and in humans around 7.5 weeks of gestation (Buchet et al. 2011). OPCs become uniformly distributed after birth and persist in the adult central nervous system. OPCs differentiate by first forming pre-myelinating oligodendrocytes, which are first detected around PND9 in mouse, PND11 in rat, or 17–27 weeks in humans (Trapp et al. 1997; Fard et al. 2017). Pre-myelinating oligodendrocytes either differentiate into myelinating oligodendrocytes or die. During myelination, oligodendrocytes flatten, wrap around the axons, and elongate to form a myelin internode (Snaidero et al. 2014). Extrinsic growth factors (such as FGF2 and BDNF) regulate myelin thickness. Once wrapping is complete, MBP promotes actin depolymerization to facilitate myelin sheath compaction (Snaidero et al. 2017).

Serotonin alterations during development have been shown to affect myelination processes. As discussed above, early life exposure to citalopram induced both hypo- and hypermyelination in the corpus collosum (Simpson et al. 2011). In addition, myelin genes were upregulated in the prefrontal cortex and downregulated in the basolateral amygdala after perinatal fluoxetine exposure (Ramsteijn et al. 2022). Moreover, males were more severely affected than females. Related to this, social behavioural outcomes show that male rats are more vulnerable to the effects of perinatal SSRIs than females (e.g. Houwing et al. 2019b), which may be due to different timing of 5-HT peaks in males (at PND3) and females (gradual elevation at PND5) (Connell et al. 2004). It remains to be established whether males are more compromised by perinatal SSRI treatment than their female counterparts due to this timing exposure.

In germ-free mice, alterations in offspring social behaviour have been observed alongside hypermyelination (Hoban et al. 2016). This raises the possibility that similar mechanisms may be involved in perinatal SSRI-exposed animals. Microbial molecules are transferred from mother to offspring during both gestation and lactation, supporting the development of the brain and highlighting their critical role in early-life development (Braniste et al. 2014; Thion et al. 2018). In addition, oligodendroglia provide metabolic support to neurons via myelin sheaths (Fünfschilling et al. 2012; Lee et al. 2012). Metabolite availability therefore greatly influences axon growth and health. Whether the observed alterations in myelin-related gene expression and social behaviour are mediated by the maternal microbiome remains unclear. Future research should investigate whether fluoxetine-induced changes in the maternal gut microbiome and its metabolic output contribute to disrupted myelination in the offspring. At the same time, alternative mechanisms influencing offspring behaviour should also be explored.

7 Conclusion

Perinatal exposure to SSRIs has profound and lasting effects on offspring brain development, social behaviour, and molecular signalling pathways. While SSRIs are crucial for managing maternal depression, which itself carries risks for foetal development, their influence on the developing serotonergic system introduces additional complexity. This chapter highlights growing evidence from animal studies indicating that SSRI exposure during critical neurodevelopmental windows, particularly gestation and early postnatal periods, alters brain structure and function in offspring, often in a sex-specific manner.

One of the most consistently affected systems is myelination. Fluoxetine exposure has been linked to region-specific changes in myelin-related gene expression, disrupted oligodendrocyte maturation, and altered interhemispheric connectivity, especially in the prefrontal cortex and corpus callosum. These effects may be driven by SSRI-induced disruptions to the maternal gut microbiome, which plays a pivotal role in shaping neurodevelopment via microbial metabolites. Findings suggest that fluoxetine can alter microbiota composition and reduce neuroactive metabolite levels, potentially impairing processes such as axonogenesis and myelination.

Importantly, animal models reveal that neurobiological changes often manifest behaviourally. Offspring exposed to SSRIs during development display reduced social play, impaired social investigation, and altered aggression levels, effects that mirror features observed in people with neurodevelopmental disorders. These findings emphasize the importance of exposure timing, dosage, and maternal condition, as well as the need to differentiate drug effects from the underlying illness.

In conclusion, while SSRIs remain vital for maternal mental health, their developmental impact on offspring cannot be overlooked. Future research must further dissect the molecular and microbial mechanisms underlying these effects, ideally identifying intervention windows or protective strategies that mitigate risks without compromising maternal health.

References

Alwan S, Reefhuis J, Rasmussen SA, Olney RS, Friedman JM (2007) Use of selective serotonin-reuptake inhibitors in pregnancy and the risk of birth defects. N Engl J Med 356(26):2684–2692. https://doi.org/10.1056/NEJMOA066584

Andrade C (2017) Antidepressant exposure during pregnancy and risk of autism in the offspring, 1: meta-review of meta-analyses. J Clin Psychiatry 78(8):e1047–e1051. https://doi.org/10.4088/JCP.17F11903

Baann AK, Houwing DJ, Olivier JDA, Heijkoop R, Snoeren EMS (2025) Perinatal fluoxetine exposure increases male rat sexual behavior. Neuropharmacology 279, 110635. https://doi.org/10.1016/j.neuropharm.2025.110635

Benamer N, Vidal M, Balia M, Cecilia Angulo M (2020) Myelination of parvalbumin interneurons shapes the function of cortical sensory inhibitory circuits. Nat Commun 11:5151. https://doi.org/10.1038/s41467-020-18984-7

Bérard A, Zhao JP, Sheehy O (2015) Sertraline use during pregnancy and the risk of major malformations. Am J Obstet Gynecol 212(6):795.e1–795.e12. https://doi.org/10.1016/J.AJOG.2015.01.034
Bérard A, Sheehy O, Zhao JP, Vinet É, Bernatsky S, Abrahamowicz M (2017a) SSRI and SNRI use during pregnancy and the risk of persistent pulmonary hypertension of the newborn. Br J Clin Pharmacol 83(5):1126–1133. https://doi.org/10.1111/BCP.13194
Bérard A, Zhao JP, Sheehy O (2017b) Antidepressant use during pregnancy and the risk of major congenital malformations in a cohort of depressed pregnant women: an updated analysis of the Quebec Pregnancy Cohort. BMJ Open 7(1). https://doi.org/10.1136/BMJOPEN-2016-013372
Bergles DE, Richardson WD (2015) Oligodendrocyte development and plasticity. Cold Spring Harb Perspect Biol 8(2). https://doi.org/10.1101/CSHPERSPECT.A020453
Boulle F, Pawluski JL, Homberg JR, Machiels B, Kroeze Y, Kumar N, Steinbusch HWM, Kenis G, van den Hove DLA (2016a) Developmental fluoxetine exposure increases behavioral despair and alters epigenetic regulation of the hippocampal BDNF gene in adult female offspring. Horm Behav 80:47–57. https://doi.org/10.1016/j.yhbeh.2016.01.017
Boulle F, Pawluski JL, Homberg JR, Machiels B, Kroeze Y, Kumar N, Steinbusch HWM, Kenis G, Van den Hove DLA (2016b) Prenatal stress and early-life exposure to fluoxetine have enduring effects on anxiety and hippocampal BDNF gene expression in adult male offspring. Dev Psychobiol 58(4):427–438. https://doi.org/10.1002/dev.21385
Bourke CH, Stowe ZN, Neigh GN, Olson DE, Owens MJ (2013) Prenatal exposure to escitalopram and/or stress in rats produces limited effects on endocrine, behavioral, or gene expression measures in adult male rats. Neurotoxicol Teratol 39:100–109. https://doi.org/10.1016/j.ntt.2013.07.008
Braniste V, Al-Asmakh M, Kowal C, Anuar F, Abbaspour A, Tóth M, Korecka A, Bakocevic N, Ng LG, Kundu P (2014) The gut microbiota influences blood-brain barrier permeability in mice. Sci Transl Med 6(263):263ra158
Brummelte S, Galea LAM, Devlin AM, Oberlander TF (2013) Antidepressant use during pregnancy and serotonin transporter genotype (SLC6A4) Affect newborn serum reelin levels. Dev Psychobiol 55(5):518–529. https://doi.org/10.1002/dev.21056
Buchet D, Garcia C, Deboux C, Nait-Oumesmar B, Baron-Van Evercooren A (2011) Human neural progenitors from different foetal forebrain regions remyelinate the adult mouse spinal cord. Brain 134(Pt 4):1168–1183. https://doi.org/10.1093/BRAIN/AWR030
Chambers CD, Johnson KA, Dick LM, Felix RJ, Jones KL (1996) Birth outcomes in pregnant women taking fluoxetine. N Engl J Med 335(14):1010–1015. https://doi.org/10.1056/NEJM199610033351402
Chen MX, Cheng S, Lei L, Zhang XF, Liu Q, Lin A, Wallis CU, Lukowicz MJ, Sham PC, Li Q, Ao LJ (2021) The effects of maternal SSRI exposure on the serotonin system, prefrontal protein expression and behavioral development in male and female offspring rats. Neurochem Int 146: 105041. https://doi.org/10.1016/J.NEUINT.2021.105041
Clarke G, Grenham S, Scully P, Fitzgerald P, Moloney RD, Shanahan F, Dinan TG, Cryan JF (2013) The microbiome-gut-brain axis during early life regulates the hippocampal serotonergic system in a sex-dependent manner. Mol Psychiatry. https://doi.org/10.1038/mp.2012.77
Connell S, Karikari C, Hohmann CF (2004) Sex-specific development of cortical monoamine levels in mouse. Dev Brain Res 151(1–2):187–191. https://doi.org/10.1016/j.devbrainres.2004.03.008
Courchesne E, Pierce K (2005) Brain overgrowth in autism during a critical time in development: implications for frontal pyramidal neuron and interneuron development and connectivity. Int J Dev Neurosci 23(2–3):153–170. https://doi.org/10.1016/J.IJDEVNEU.2005.01.003
Cryan JF, O'Riordan KJ, Cowan CSM, Sandhu KV, TFS B, Boehme M, Codagnone MG, Cussotto S, Fulling C, Golubeva AV, Guzzetta KE, Jaggar M, Long-Smith CM, Lyte JM, Martin JA, Molinero-Perez A, Moloney G, Morelli E, Morillas E, Cruz-Pereira JS, Peterson VL, Rea K, Ritz NL, Sherwin E, Spichak S, Teichman EM, van de Wouw M, Paula Ventura-Silva A, Wallace-Fitzsimons SE, Hyland N, Clarke G, Dinan TG (2019) The microbiota-gut-brain axis. Physiol Rev 99:1877–2013. https://doi.org/10.1152/physrev.00018.2018

Cussotto S, Strain CR, Fouhy F, Strain RG, Peterson VL, Clarke G, Stanton C, Dinan TG, Cryan JF (2019) Differential effects of psychotropic drugs on microbiome composition and gastrointestinal function. Psychopharmacology 236(5):1671–1685. https://doi.org/10.1007/S00213-018-5006-5

Downes N, Mullins P (2014) The development of myelin in the Brain of the juvenile rat. Toxicol Pathol 42(5):913–922. https://doi.org/10.1177/0192623313503518

Ehrlich DE, Neigh GN, Bourke CH, Nemeth CL, Hazra R, Ryan SJ, Rowson S, Jairam N, Sholar CA, Rainnie DG, Stowe ZN, Owens MJ (2015) Prenatal stress, regardless of concurrent escitalopram treatment, alters behavior and amygdala gene expression of adolescent female rats. Neuropharmacology 97:251–258. https://doi.org/10.1016/j.neuropharm.2015.05.012

El Marroun H, Jaddoe VWV, Hudziak JJ, Roza SJ, Steegers EAP, Hofman A, Verhulst FC, White TJH, Stricker BHC, Tiemeier H (2012) Maternal use of selective serotonin reuptake inhibitors, fetal growth, and risk of adverse birth outcomes. Arch Gen Psychiatry 69(7):706–714. https://doi.org/10.1001/ARCHGENPSYCHIATRY.2011.2333

Fard MK, Van der Meer F, Sánchez P, Cantuti-Castelvetri L, Mandad S, Jäkel S, Fornasiero EF, Schmitt S, Ehrlich M, Starost L, Kuhlmann T, Sergiou C, Schultz V, Wrzos C, Brück W, Urlaub H, Dimou L, Stadelmann C, Simons M (2017) BCAS1 expression defines a population of early myelinating oligodendrocytes in multiple sclerosis lesions. Sci Transl Med 9(419). https://doi.org/10.1126/SCITRANSLMED.AAM7816

Feenstra MGP, Van Galen H, Te Riele PJM, Botterblom MHA, Mirmiran M (1996) Decreased hypothalamic serotonin levels in adult rats treated neonatally with clomipramine. Pharmacol Biochem Behav 55(4):647–652. https://doi.org/10.1016/S0091-3057(96)00276-6

Fuller RW, Wong DT (1987) Serotonin reuptake blockers in vitro and in vivo. J Clin Psychopharmacol 7(6):36S–43S. https://doi.org/10.1097/00004714-198712001-00004

Fünfschilling U, Supplie LM, Mahad D, Boretius S, Saab AS, Edgar J, Brinkmann BG, Kassmann CM, Tzvetanova ID, Möbius W, Diaz F, Meijer D, Suter U, Hamprecht B, Sereda MW, Moraes CT, Frahm J, Goebbels S, Nave K-A (2012) Glycolytic oligodendrocytes maintain myelin and long-term axonal integrity. Nature 485:517–521. https://doi.org/10.1038/nature11007

Fung TC, Vuong HE, Luna CDG, Pronovost GN, Aleksandrova AA, Riley NG, Vavilina A, Mcginn J, Rendon T, Forrest LR, Hsiao EY (2019) Intestinal serotonin and fluoxetine exposure modulate bacterial colonization in the gut. Nat Microbiol 4. https://doi.org/10.1038/s41564-019-0540-4

Gemmel M, Hazlett M, Bögi E, De Lacalle S, Hill L, Kokras N, Hammond G, Dalla C, Charlier T, Pawluski J (2017) Perinatal fluoxetine effects on social play, the HPA system, and hippocampal plasticity in pre-adolescent male and female rats: interactions with pre-gestational maternal stress. Psychoneuroendocrinology 84:159–171. https://doi.org/10.1016/J.PSYNEUEN.2017.07.480

Gemmel M, Harmeyer D, Bögi E, Fillet M, Hill LA, Hammond GL, Charlier TD, Pawluski JL (2018) Perinatal fluoxetine increases hippocampal neurogenesis and reverses the lasting effects of pre-gestational stress on serum corticosterone, but not on maternal behavior, in the rat dam. Behav Brain Res 339:222–231. https://doi.org/10.1016/J.BBR.2017.11.038

Gemmel M, De Lacalle S, Mort SC, Hill LA, Charlier TD, Pawluski JL (2019) Perinatal fluoxetine has enduring sexually differentiated effects on neurobehavioral outcomes related to social behaviors. Neuropharmacology 144:70–81. https://doi.org/10.1016/J.NEUROPHARM.2018.10.009

Gershon MD, Tack J (2007) The serotonin signaling system: from basic understanding to drug development for functional GI disorders. Gastroenterology 132(1):397–414. https://doi.org/10.1053/J.GASTRO.2006.11.002/ASSET/CA67746A-B129-4657-8483-4874E98CAAAD/MAIN.ASSETS/GR4.JPG

Glover ME, Clinton SM (2016) Of rodents and humans: a comparative review of the neurobehavioral effects of early life SSRI exposure in preclinical and clinical research. Int J Dev Neurosci 51:50–72. https://doi.org/10.1016/J.IJDEVNEU.2016.04.008

Glover ME, Pugh PC, Jackson NL, Cohen JL, Fant AD, Akil H, Clinton SM (2015) Early-life exposure to the SSRI paroxetine exacerbates depression-like behavior in anxiety/depression-prone rats. Neuroscience 284:775–797. https://doi.org/10.1016/j.neuroscience.2014.10.044

Gouvêa TS, Morimoto HK, de Faria MJSS, Moreira EG, Gerardin DCC (2008) Maternal exposure to the antidepressant fluoxetine impairs sexual motivation in adult male mice. Pharmacol Biochem Behav 90(3):416–419. https://doi.org/10.1016/J.PBB.2008.03.025

Halvorsen A, Hesel B, Østergaard SD, Danielsen AA (2019) In utero exposure to selective serotonin reuptake inhibitors and development of mental disorders: a systematic review and meta-analysis. Acta Psychiatr Scand 139(6):493–507. https://doi.org/10.1111/acps.13030

Hanley GE, Brain U, Oberlander TF (2013) Infant developmental outcomes following prenatal exposure to antidepressants, and maternal depressed mood and positive affect. Early Hum Dev 89(8):519–524. https://doi.org/10.1016/j.earlhumdev.2012.12.012

Hanley GE, Brain U, Oberlander TF (2015) Prenatal exposure to serotonin reuptake inhibitor antidepressants and childhood behavior. Pediatr Res 78(2):174–180. https://doi.org/10.1038/pr.2015.77

Harris SS, MacIag D, Simpson KL, Lin RCS, Paul IA (2012) Dose-dependent effects of neonatal SSRI exposure on adult behavior in the rat. Brain Res 1429:52–60. https://doi.org/10.1016/J.BRAINRES.2011.10.025

Hegstad J, Huijgens PT, Houwing DJ, Olivier JDA, Heijkoop R, Snoeren EMS (2020) Female rat sexual behavior in a seminatural environment is not affected by perinatal fluoxetine exposure. Psychoneuroendocrinology. https://doi.org/10.1017/CBO9781107415324.004

Hilakivi LA, Hilakivi I, Ahtee L, Haikala H, Attila M (1987) Effect of neonatal nomifensine exposure on adult behavior and brain monoamines in rats. J Neural Transm 70(1–2):99–116. https://doi.org/10.1007/BF01252512

Hoban AE, Stilling RM, Ryan FJ, Shanahan F, Dinan TG, Claesson MJ, Clarke G, Cryan JF (2016) Regulation of prefrontal cortex myelination by the microbiota. Transl Psychiatry 6:e774. https://doi.org/10.1038/tp.2016.42

Houwing DJ, Heijkoop R, Olivier JDA, Snoeren EMS (2019a) Perinatal fluoxetine exposure changes social and stress-coping behavior in adult rats housed in a seminatural environment. Neuropharmacology 151:84–97. https://doi.org/10.1016/j.neuropharm.2019.03.037

Houwing DJ, Staal L, Swart JM, Ramsteijn AS, Wöhr M, de Boer SF, Olivier JDA (2019b) Subjecting dams to early life stress and perinatal fluoxetine treatment differentially alters social behavior in young and adult rat offspring. Front Neurosci 13:1–15. https://doi.org/10.3389/fnins.2019.00229

Houwing DJ, Esquivel-Franco DC, Ramsteijn AS, Schuttel K, Struik EL, Arling C, de Boer SF, Olivier JDA (2020) Perinatal fluoxetine treatment and dams' early life stress history have opposite effects on aggressive behavior while having little impact on sexual behavior of male rat offspring. Psychopharmacology 237(9):2589–2600. https://doi.org/10.1007/s00213-020-05535-7

Jensen HM, Grøn R, Lidegaard Ø, Pedersen LH, Andersen PK, Kessing LV (2013) Maternal depression, antidepressant use in pregnancy and Apgar scores in infants. Br J Psychiatry 202(5):347–351. https://doi.org/10.1192/BJP.BP.112.115931

Johns JM, Joyner PW, McMurray MS, Elliott DL, Hofler VE, Middleton CL, Knupp K, Greenhill KW, Lomas LM, Walker CH (2005) The effects of dopaminergic/serotonergic reuptake inhibition on maternal behavior, maternal aggression, and oxytocin in the rat. Pharmacol Biochem Behav 81(4):769–785. https://doi.org/10.1016/J.PBB.2005.06.001

Kaplan YC, Keskin-Arslan E, Acar S, Sozmen K (2017) Maternal SSRI discontinuation, use, psychiatric disorder and the risk of autism in children: a meta-analysis of cohort studies. Br J Clin Pharmacol 83(12):2798–2806. https://doi.org/10.1111/BCP.13382

Kessaris N, Fogarty M, Iannarelli P, Grist M, Wegner M, Richardson WD (2006) Competing waves of oligodendrocytes in the forebrain and postnatal elimination of an embryonic lineage. Nat Neurosci 9(2):173–179. https://doi.org/10.1038/NN1620

Kieler H, Artama M, Engeland A, Ericsson O, Furu K, Gissler M, Nielsen RB, Nrøgaard M, Stephansson O, Valdimarsdottir U, Zoega H, Haglund B (2012) Selective serotonin reuptake inhibitors during pregnancy and risk of persistent pulmonary hypertension in the newborn: population based cohort study from the five Nordic countries. BMJ 344(7842). https://doi.org/10.1136/BMJ.D8012

Kinney GG, Vogel GW, Feng P (1997) Decreased dorsal raphe nucleus neuronal activity in adult chloral hydrate anesthetized rats following neonatal clomipramine treatment: implications for endogenous depression. Brain Res 756(1–2):68–75. https://doi.org/10.1016/S0006-8993(97)00119-4

Kiryanova V, Meunier SJ, Vecchiarelli HA, Hill MN, Dyck RH (2016) Effects of maternal stress and perinatal fluoxetine exposure on behavioral outcomes of adult male offspring. Neuroscience 320:281–296. https://doi.org/10.1016/J.NEUROSCIENCE.2016.01.064

Ko MC, Lee LJH, Li Y, Lee LJ (2014) Long-term consequences of neonatal fluoxetine exposure in adult rats. Dev Neurobiol 74(10):1038–1051. https://doi.org/10.1002/DNEU.22185

Koc D, Tiemeier H, Stricker BH, Muetzel RL, Hillegers M, El MH (2023) Prenatal antidepressant exposure and offspring Brain morphologic trajectory. JAMA Psychiatry. https://doi.org/10.1001/JAMAPSYCHIATRY.2023.3161

Kroeze Y (2017) Effects of serotonin signaling on behavior and gene transcription. Radboud University Nijmegen

Kroeze Y, Peeters D, Boulle F, van den Hove DLA, van Bokhoven H, Zhou H, Homberg JR (2015) Long-term consequences of chronic fluoxetine exposure on the expression of myelination-related genes in the rat hippocampus. Transl Psychiatry 5(9):e642–e642. https://doi.org/10.1038/tp.2015.145

Lebel C, Deoni S (2018) The development of brain white matter microstructure. NeuroImage 182 (June 2017):207–218. https://doi.org/10.1016/j.neuroimage.2017.12.097

Lee L-J (2009) Neonatal fluoxetine exposure affects the neuronal structure in the somatosensory cortex and somatosensory-related behaviors in adolescent rats. Neurotox Res 15:212–223. https://doi.org/10.1007/s12640-009-9022-4

Lee Y, Morrison BM, Li Y, Lengacher S, Farah MH, Hoffman PN, Liu Y, Tsingalia A, Jin L, Zhang P-W, Pellerin L, Magistretti PJ, Rothstein JD (2012) Oligodendroglia metabolically support axons and contribute to neurodegeneration. Nature 487:443–448. https://doi.org/10.1038/nature11314

Lisboa SFS, Oliveira PE, Costa LC, Venâncio EJ, Moreira EG (2007) Behavioral evaluation of male and female mice pups exposed to fluoxetine during pregnancy and lactation. Pharmacology 80(1):49–56. https://doi.org/10.1159/000103097

Lugo-Candelas C, Cha J, Hong S, Bastidas V, Weissman M, Fifer WP, Myers M, Talati A, Bansal R, Peterson BS, Monk C, Gingrich JA, Posner J (2018) Associations between brain structure and connectivity in infants and exposure to selective serotonin reuptake inhibitors during pregnancy. JAMA Pediatr 172(6):525–533. https://doi.org/10.1001/jamapediatrics.2017.5227

Maciag D, Simpson KL, Coppinger D, Lu Y, Wang Y, Lin RCS, Paul IA (2006) Neonatal antidepressant exposure has lasting effects on behavior and serotonin circuitry. Neuropsychopharmacology 31(1):47–57. https://doi.org/10.1038/sj.npp.1300823

Malm H, Artama M, Gissler M, Ritvanen A (2011) Selective serotonin reuptake inhibitors and risk for major congenital anomalies. Obstet Gynecol 118(1):111–120. https://doi.org/10.1097/AOG.0B013E318220EDCC

Manhaes De Castro R, Barreto Medeiros JM, Mendes Da Silva C, Ferreira LMP, Guedes RCA, Cabrai Filho JE, Costa JA (2001) Reduction of intraspecific aggression in adult rats by neonatal treatment with a selective serotonin reuptake inhibitor. Brazilian J Med Biol Res = Rev Bras Pesqui Medicas e Biol 34(1):121–124. https://doi.org/10.1590/S0100-879X2001000100015

Meyer LR, Dexter B, Lo C, Kenkel E, Hirai T, Roghair RD, Haskell SE (2018) Perinatal SSRI exposure permanently alters cerebral serotonin receptor mRNA in mice but does not impact adult behaviors. J Matern Neonatal Med 31(11):1393–1401. https://doi.org/10.1080/14767058.2017.1317342

Miceli S, Negwer M, Van Eijs F, Kalkhoven C, Van Lierop I, Homberg J, Schubert D, Feldmeyer D, Tropea D (2013) High serotonin levels during brain development alter the structural input-output connectivity of neural networks in the rat somatosensory layer IV. Front Cell Neurosci 7(88):1–15. https://doi.org/10.3389/fncel.2013.00088

Mirmiran M, Van De Poll NE, Corner MA, Van Oyen HG, Bour HL (1981) Suppression of active sleep by chronic treatment with chlorimipramine during early postnatal development: effects upon adult sleep and behavior in the rat. Brain Res 204(1):129–146. https://doi.org/10.1016/0006-8993(81)90657-0

Mitew S, Gobius I, Fenlon LR, McDougall SJ, Hawkes D, Lulu Xing Y, Bujalke H, Gundlach AL, Richards LJ, Kilpatrick TJ, Merson TD, Emery B (2018) Pharmacogenetic stimulation of neuronal activity increases myelination in an axon-specific manner. Nat Commun 9:306. https://doi.org/10.1038/s41467-017-02719-2

Molenaar NM, Bais B, Lambregtse-van den Berg MP, Mulder CL, Howell EA, Fox NS, Rommel AS, Bergink V, Kamperman AM (2020) The international prevalence of antidepressant use before, during, and after pregnancy: a systematic review and meta-analysis of timing, type of prescriptions and geographical variability. J Affect Disord 264:82–89. https://doi.org/10.1016/J.JAD.2019.12.014

Mount CW, Monje M (2017) Wrapped to adapt: experience-dependent myelination. Neuron 95: 743–756. https://doi.org/10.1016/j.neuron.2017.07.009

Munoz-Bellido JL, Mufioz-Criado S, Garcfa-Rodrfguez JA (1996) In-vitro activity of psychiatric drugs against Corynebacterium urealyticum {Corynebacterium group D2)

Narboux-Nême N, Pavone LM, Avallone L, Zhuang X, Gaspar P (2008) Serotonin transporter transgenic (SERTcre) mouse line reveals developmental targets of serotonin specific reuptake inhibitors (SSRIs). Neuropharmacology 55(6):994–1005. https://doi.org/10.1016/J.NEUROPHARM.2008.08.020

Nicholson JK, Holmes E, Kinross J, Burcelin R, Gibson G, Jia W, Pettersson S (2012) Host-gut microbiota metabolic interactions. Science 336(6086):1262–1267. https://doi.org/10.1126/SCIENCE.1223813

Noorlander CW, Ververs FFT, Nikkels PGJ, van Echteld CJA, Visser GHA, Smidt MP (2008) Modulation of serotonin transporter function during fetal development causes dilated heart cardiomyopathy and lifelong behavioral abnormalities. PLoS One 3(7):e2782. https://doi.org/10.1371/journal.pone.0002782

Olivier JDA, Vallès A, Van Heesch F, Afrasiab-Middelman A, Roelofs JJPM, Jonkers M, Peeters EJ, Korte-Bouws GAH, Dederen JP, Kiliaan AJ, Martens GJ, Schubert D, Homberg JR (2011) Fluoxetine administration to pregnant rats increases anxiety-related behavior in the offspring. Psychopharmacology 217(3):419–432. https://doi.org/10.1007/s00213-011-2299-z

Pawluski JL, Charlier TD, Fillet M, Houbart V, Crispin HT, Steinbusch HW, van den Hove DL (2012) Chronic fluoxetine treatment and maternal adversity differentially alter neurobehavioral outcomes in the rat dam. Behav Brain Res 228(1):159–168. https://doi.org/10.1016/J.BBR.2011.11.043

Pawluski JL, Li M, Lonstein JS (2019) Serotonin and motherhood: from molecules to mood. Front Neuroendocrinol 53:100742. https://doi.org/10.1016/J.YFRNE.2019.03.001

Podrebarac SK, Duerden EG, Chau V, Grunau RE, Synnes A, Oberlander TF, Miller SP (2017) Antenatal exposure to antidepressants is associated with altered brain development in very preterm-born neonates. Neuroscience 342:252–262. https://doi.org/10.1016/j.neuroscience.2016.11.025

Prado-Rivera MA, Ebbing JJ, Yurova MY, Olivier JDA (2026) Perinatal fluoxetine does not impair cognition in dams and offspring. Behav Brain Res 496:115812

Ramsteijn AS, Jašarević E, Houwing DJ, Bale TL, Olivier JDA (2020) Antidepressant treatment with fluoxetine during pregnancy and lactation modulates the gut microbiome and metabolome in a rat model relevant to depression. Gut Microbes 11(4):735–753. https://doi.org/10.1080/19490976.2019.1705728

Ramsteijn AS, Verkaik-Schakel RN, Houwing DJ, Plösch T, Olivier JDA (2022) Perinatal exposure to fluoxetine and maternal adversity affect myelin-related gene expression and epigenetic regulation in the corticolimbic circuit of juvenile rats. Neuropsychopharmacology 47(9): 1620–1632. https://doi.org/10.1038/S41386-022-01270-Z

Rayen I, Steinbusch HWM, Charlier TD, Pawluski JL (2013) Developmental fluoxetine exposure and prenatal stress alter sexual differentiation of the brain and reproductive behavior in male rat offspring. Psychoneuroendocrinology 38(9):1618–1629. https://doi.org/10.1016/J.PSYNEUEN.2013.01.007

Rayen I, Steinbusch HWM, Charlier TD, Pawluski JL (2014) Developmental fluoxetine exposure facilitates sexual behavior in female offspring. Psychopharmacology 231(1):123–133. https://doi.org/10.1007/S00213-013-3215-5

Rebello TJ, Yu Q, Goodfellow NM, Caffrey Cagliostro MK, Teissier A, Morelli E, Demireva EY, Chemiakine A, Rosoklija GB, Dwork AJ, Lambe EK, Gingrich JA, Ansorge MS (2014) Postnatal day 2 to 11 constitutes a 5-HT-sensitive period impacting adult mPFC function. J Neurosci 34(37):12379–12393. https://doi.org/10.1523/JNEUROSCI.1020-13.2014

Rgen J, Bohnert A, Szymaniak-Vits M, Schuster S, Kern WV (2011) Efflux inhibition by selective serotonin reuptake inhibitors in Escherichia coli. J Antimicrob Chemother. https://doi.org/10.1093/jac/dkr258

Rodriguez-Porcel F, Green D, Khatri N, Harris SS, May WL, Lin RCS, Paul IA (2011) Neonatal exposure of rats to antidepressants affects behavioral reactions to novelty and social interactions in a manner analogous to autistic spectrum disorders. Anat Rec (Hoboken) 294(10):1726–1735. https://doi.org/10.1002/AR.21402

Rotem-Kohavi N, Oberlander TF (2017) Variations in neurodevelopmental outcomes in children with prenatal SSRI antidepressant exposure. Birth Defects Res 109(12):909–923. https://doi.org/10.1002/BDR2.1076

Rotem-Kohavi N, Williams LJ, Virji-Babul N, Bjornson BH, Brain U, Werker JF, Grunau RE, Miller SP, Oberlander TF (2019) Alterations in resting-state networks following in utero selective serotonin reuptake inhibitor exposure in the neonatal Brain. Biol Psychiatry Cogn Neurosci Neuroimaging 4(1):39–49. https://doi.org/10.1016/j.bpsc.2018.08.004

Sarkar A, Chachra P, Kennedy P, Pena CJ, Desouza LA, Nestler EJ, Vaidya VA (2014) Hippocampal HDAC4 contributes to postnatal fluoxetine-evoked depression-like behavior. Neuropsychopharmacology 39(9):2221–2232. https://doi.org/10.1038/npp.2014.73

Semple BD, Blomgren K, Gimlin K, Ferriero DM, Noble-Haeusslein LJ, Neurobiol Author manuscript P (2013) Brain development in rodents and humans: identifying benchmarks of maturation and vulnerability to injury across species NIH Public Access Author Manuscript. Prog Neurobiol 0:1–16. https://doi.org/10.1016/j.pneurobio.2013.04.001

Simpson KL, Weaver KJ, de Villers-Sidani E, Lu JY-F, Cai Z, Pang Y, Rodriguez-Porcel F, Paul IA, Merzenich M, Lin RCS (2011) Perinatal antidepressant exposure alters cortical network function in rodents. Proc Natl Acad Sci 108(45):18465–18470

Snaidero N, Möbius W, Czopka T, Hekking LHP, Mathisen C, Verkleij D, Goebbels S, Edgar J, Merkler D, Lyons DA, Nave KA, Simons M (2014) Myelin membrane wrapping of CNS axons by PI(3,4,5)P3-dependent polarized growth at the inner tongue. Cell 156(1–2):277–290. https://doi.org/10.1016/J.CELL.2013.11.044

Snaidero N, Velte C, Myllykoski M, Raasakka A, Ignatev A, Werner HB, Erwig MS, Möbius W, Kursula P, Nave KA, Simons M (2017) Antagonistic functions of MBP and CNP establish cytosolic channels in CNS myelin. Cell Rep 18(2):314. https://doi.org/10.1016/J.CELREP.2016.12.053

Sprowles JLN, Hufgard JR, Gutierrez A, Bailey RA, Jablonski SA, Williams MT, Vorhees CV (2016) Perinatal exposure to the selective serotonin reuptake inhibitor citalopram alters spatial learning and memory, anxiety, depression, and startle in Sprague-Dawley rats. Int J Dev Neurosci 54:39–52. https://doi.org/10.1016/J.IJDEVNEU.2016.08.007

Stadelmann C, Timmler S, Barrantes-Freer A, Simons M (2019) Myelin in the central nervous system: structure, function, and pathology. Physiol Rev 99(3):1381–1431. https://doi.org/10.1152/physrev.00031.2018

Starr-Phillips EJ, Beery AK (2014) Natural variation in maternal care shapes adult social behavior in rats. Dev Psychobiol 56(5):1017–1026. https://doi.org/10.1002/DEV.21182

Svirsky N, Levy S, Avitsur R (2016) Prenatal exposure to selective serotonin reuptake inhibitors (SSRI) increases aggression and modulates maternal behavior in offspring mice. Dev Psychobiol 58(1):71–82. https://doi.org/10.1002/DEV.21356

Thion MS, Low D, Silvin A, Chen J, Grisel P, Schulte-Schrepping J, Blecher R, Ulas T, Squarzoni P, Hoeffel G, Coulpier F, Siopi E, David FS, Scholz C, Shihui F, Lum J, Amoyo AA, Larbi A, Poidinger M, Buttgereit A, Lledo PM, Greter M, Chan JKY, Amit I, Beyer M, Schultze JL, Schlitzer A, Pettersson S, Ginhoux F, Garel S (2018) Microbiome influences prenatal and adult microglia in a sex-specific manner. Cell 172(3):500–516.e16. https://doi.org/10.1016/J.CELL.2017.11.042

Trapp BD, Nishiyama A, Cheng D, Macklin W (1997) Differentiation and death of premyelinating oligodendrocytes in developing rodent brain. J Cell Biol 137(2):459–468. https://doi.org/10.1083/JCB.137.2.459

Verney C, Lebrand C, Gaspar P (2002) Changing distribution of monoaminergic markers in the developing human cerebral cortex with special emphasis on the serotonin transporter. Anat Rec 267(2):87–93. https://doi.org/10.1002/AR.10089

Videman M, Tokariev A, Saikkonen H, Stjerna S, Heiskala H, Mantere O, Vanhatalo S (2017) Newborn Brain function is affected by fetal exposure to maternal serotonin reuptake inhibitors. Cereb Cortex 27(6):3208–3216. https://doi.org/10.1093/cercor/bhw153

Vijayakumar M, Meti BL (1999) Alterations in the levels of monoamines in discrete brain regions of clomipramine-induced animal model of endogenous depression. Neurochem Res 24(3): 345–349. https://doi.org/10.1023/A:1020992314534

Vorhees CV, Acuff-smith KD, Schilling MA, Fisher JE, Moran MS, Buelke-sam J (1994) A developmental neurotoxicity evaluation of the effects of prenatal exposure to fluoxetine in rats. Fundam Appl Toxicol 23(2):194–205. https://doi.org/10.1006/FAAT.1994.1098

Vuong HE, Pronovost GN, Williams DW, Coley EJL, Siegler EL, Qiu A, Kazantsev M, Wilson CJ, Rendon T, Hsiao EY (2020) The maternal microbiome modulates fetal neurodevelopment in mice. Nature 586:281–286. https://doi.org/10.1038/s41586-020-2745-3

Vuong H, Coley E, Kazantsev M, Cooke M, Rendon T, Paramo J, Hsiao E (2021) Interactions between maternal fluoxetine exposure, the maternal gut microbiome and fetal neurodevelopment in mice. Behav Brain Res 410:113353. https://doi.org/10.1016/J.BBR.2021.113353

Weaver KJ, Paul IA, Lin RCS, Simpson KL (2010) Neonatal exposure to citalopram selectively alters the expression of the serotonin transporter in the hippocampus: dose-dependent effects. Anat Rec (Hoboken) 293(11):1920–1932. https://doi.org/10.1002/AR.21245

Welsh JP, Ahn ES, Placantonakis DG (2005) Is autism due to brain desynchronization? Int J Dev Neurosci 23(2–3):253–263. https://doi.org/10.1016/J.IJDEVNEU.2004.09.002

World Health Organization (2017) Depression and other common mental disorders. World Health Organization

Yan W, Wilson CC, Haring JH (1997) Effects of neonatal serotonin depletion on the development of rat dentate granule cells. Dev Brain Res 98(2):177–184. https://doi.org/10.1016/S0165-3806(96)00176-9

Yannielli PC, Kargieman L, Gregoretti L, Cardinali DP (1999) Effects of neonatal clomipramine treatment on locomotor activity, anxiety-related behavior and serotonin turnover in Syrian hamsters. Neuropsychobiology 39(4):200–206. https://doi.org/10.1159/000026584

Yano JM, Yu K, Donaldson GP, Shastri GG, Ann P, Ma L, Nagler CR, Ismagilov RF, Mazmanian SK, Hsiao EY (2015) Indigenous bacteria from the gut microbiota regulate host serotonin biosynthesis. Cell. https://doi.org/10.1016/j.cell.2015.02.047

Zonouzi M, Berger D, Jokhi V, Kedaigle A, Lichtman J, Arlotta P (2019) Individual oligodendrocytes show bias for inhibitory axons in the neocortex. Cell Rep 27(10):2799–2808.e3. https://doi.org/10.1016/J.CELREP.2019.05.018

Zucker I (2017) Risk mitigation for children exposed to drugs during gestation: a critical role for animal preclinical behavioral testing. Neurosci Biobehav Rev 77:107–121. https://doi.org/10.1016/j.neubiorev.2017.03.005

Pre- and Postnatal Valproate Exposure Affects Brain Development

Michaela A. Pettie, Linda C. M. Wilkin-Krug, and Bart Ellenbroek

Contents

Abstract This review investigates the teratogenic impact of valproic acid (VPA) on brain development, focusing on dose-dependent and timing-related behavioural and neurological outcomes in rats and mice, with an emphasis on using it as a model for autism spectrum disorders (ASD). Single and multiple administration methods (e.g., oral gavage and intraperitoneal injections) across various rat and mouse strains consistently report behavioural alterations (i.e., altered social interaction and locomotor activity) and neuronal changes, particularly in the hippocampus and cerebellum. We underscore the importance of understanding dose-related changes and the critical role of VPA exposure in determining the long-term neurodevelopmental effects. While animal models provide valuable insights into the pre- and postnatal effects of drug exposure, this chapter also addresses the limitation of extrapolating such findings to humans given the face and construct validity of the model. Overall, this review emphasises VPA's utility in modelling ASD-like behaviours and the need for ongoing research to refine these models for better applicability to humans.

B. Ellenbroek (✉)
Behavioural Neurogenetics Laboratory, School of Psychology, Te Herenga Waka – Victoria University of Wellington, Wellington, New Zealand
e-mail: bart.ellenbroek@vuw.ac.nz

Curr Topics Behav Neurosci (2025) 75: 43–68
https://doi.org/10.1007/7854_2025_603
Published Online: 29 August 2025

Keywords Autism spectrum disorder · Locomotor activity · Neurodevelopment · Social behaviour · Teratogenicity · Ultrasonic vocalizations · Valproate

1 Introduction

The intricate processes that govern brain development can be influenced by a variety of environmental factors, including toxins, drugs, nutrition, cognitive stimulation, and emotional experiences. These factors can support neurodevelopment or pose risks via adequate nutrition, stimulation, or adverse life events, respectively. Both humans and animals are particularly sensitive to the impacts of environmental factors during the perinatal period, which spans the time from conception to the end of the first year of life. Drug exposure during this time can have enduring consequences, affecting cognition and behaviour. One such drug that is known to adversely impact brain development is valproate (VPA). As we will discuss in this chapter, VPA taken during pregnancy significantly increases the risk of the child developing autism spectrum disorders (ASD), and when VPA is administered to animals during gestation, their offspring exhibit ASD-like behaviours (i.e., decreased communication and social behaviours). Examining the impact of this drug exposure during the sensitive and critical perinatal development period is crucial for understanding the origins of various neurodevelopmental disorders, like ASD, while also providing insights for other developmental conditions, such as attention deficit hyperactivity disorder (ADHD).

While we will briefly discuss the mechanisms through which environmental factors can lead to developmental disorders, the main focus of this chapter is on the critical impact of pre- and postnatal VPA exposure on brain development, particularly in relation to ASD. Through a review of clinical and preclinical research, the chapter highlights VPA's unique teratogenic and neurodevelopmental effects, serving as a model for understanding mechanisms that underlie ASD. The following discussion underscores the specificity of VPA's effects, the timing of exposure during pregnancy or gestation, and the importance of this model in translational research aimed at elucidating the pathophysiology of ASD.

2 Overview of Valproate

Valproate, also known as valproic acid, VPA, and 2-propylpentanoic acid (valproate, VPA, 2-propylpentanoic acid), is a medication sold under the brand names Belvo, Depakote, Dyzantil, Convulex, and Syonell (Gill et al. 2011). It is prescribed in three main forms: valproic acid, sodium valproate (the sodium salt of valproic acid), and divalproex sodium (a 1:1 ratio of valproic acid and sodium valproate; Delage et al. 2023). VPA was originally synthesised in 1881, but its anticonvulsant properties were not understood until 1963 (López-Muñoz et al. 2018). Clinical trials in the 1970s demonstrated its efficacy in treating manic episodes for those with bipolar disorder and migraines (Richens and Ahmad 1975). More recently, VPA has gained attention for its potential in cancer treatment. VPA has such wide-ranging therapeutic effects because of its complex pharmacodynamic activity, which includes enhanced GABAergic neurotransmission, blocking of sodium and calcium ion channels and histone deacetylase (HDAC) inhibition. Below in the Pharmacodynamics section we will discuss these effects of valproate in more detail. VPA is still widely used as antiepileptic medication in the treatment of epilepsy, and it is becoming increasingly more common in the treatment of neurological (e.g., non-epileptic seizures, migraines, and pain) and psychiatric disorders (e.g., bipolar and anxiety disorders) (Bollino et al. 2015; Johannessen 2000; Lloyd 2013). As mentioned in the introduction, prenatal exposure to VPA has been linked to the development of ASD in humans (Christensen et al. 2013; Genton et al. 2006; Rasalam et al. 2005). Additionally, the pre- and postnatal exposure to VPA has been shown to induce ASD-like behaviours in animals, as demonstrated by reduced communication and social interaction (Pettie 2020; Ranger and Ellenbroek 2016), which will be covered in more detail later in this chapter. VPA exposure serves as a reliable model for investigating the neural mechanisms underlying ASD-related social deficits and behaviours (Tartaglione et al. 2019). In the subsequent sections, we will explore the pre- and postnatal effects of VPA on brain development, emphasising its role as a key tool in ASD research.

3 Teratological Effect of VPA

Before we discuss the mechanisms by which VPA mediates its effects on brain development, it is first important to cover teratogenic compounds. Teratogens are agents that disrupt typical foetal development, leading to various behavioural and cognitive functional deficits in humans and animals (Alwan and Chambers 2015). Teratogens can be chemical (e.g., drugs, alcohol), physical (e.g., trauma to the body), or biological (e.g., viruses, toxoplasma). The outcomes of teratogen exposure depend on the type, timing, and duration of exposure during critical developmental periods (Demirtaş 2020). Studying the effects of teratogens in animals is crucial for identifying key developmental processes that underpin typical behaviour and cognition by revealing which stages and pathways are most vulnerable to disruption. By observing specific deficits caused by teratogens, researchers can infer the feature of normal development. For instance, examining how specific teratogens impact brain regions involved in learning and memory, such as the hippocampus, highlights the importance of the regions for cognitive functions (Demirtaş 2020). In the case of VPA as teratogen, VPA exposure during development has broad impacts on neurodevelopment, and this has been linked to neurodevelopmental disorders (i.e., ASD, ADHD), making it a valuable model for studying the complex processes involved in typical brain development (Gąssowska-Dobrowolska et al. 2020). As such, studying teratogens like VPA highlights key vulnerabilities in brain development and the origins of neurodevelopmental disorders.

The first trimester of pregnancy is crucial for the development of the central nervous system, including the brain and spinal cord. During this period, the neural tube is open, making the central nervous system highly vulnerable to teratogenic effects (Defoort et al. 2006). This stage is marked by rapid cell growth and cellular organisation and the absence of fully developed blood-brain barrier, which normally provides some protection from teratogens (Demirtaş 2020). VPA's teratogenic impact is particularly significant during this open neural tube period, where it can disrupt neural tube closure and extend the window of vulnerability (Akimova et al. 2017). Although VPA exposure during other stages of development can lead to functional behaviour and cognitive changes, its effects are most severe when the neural tube remains open (Rodier et al. 1996). For humans the neural tube closes during the third and fourth weeks of pregnancy, and organogenesis, which extends until the 8 week of pregnancy, presents a period of heightened vulnerability (Demirtaş 2020; Kim et al. 2011). For rats and mice whose gestation is 21–22 days, their vulnerable period of neural tube closure is approximately gestational day 10–12, but the period of vulnerability to teratogens extends for gestational days 8–15 (Ergaz et al. 2016). These timing differences relate to the importance of understanding the developmental windows when examining VPA's teratogenic effects.

3.1 Teratogenic Effects of VPA in Humans

In humans, maternal VPA use during pregnancy significantly increases the risk of children developing ASD (Adab et al. 2004; Blotière et al. 2019; Christensen et al. 2013; Christianson et al. 1994; Dean et al. 2002; Genton et al. 2006; Harden 2013; Moore et al. 2000; Smith and Brown 2014; Vajda et al. 2019; Viinikainen et al. 2006; Williams and Hersh 1997). Specifically, VPA increases the risk of ASD by 4.42% when used during pregnancy (Christensen et al. 2013). Rasalam et al. (2005) identified that children who had been exposed to VPA during gestation were significantly more likely to be diagnosed with ASD compared with children who had been exposed to other anticonvulsants, showing a distinctive pattern of cognitive delays, especially in language and motor skills. Additionally, when VPA is the sole medication taken during pregnancy, the risk of ASD for the children is seven times higher (Bromley et al. 2008). These findings underscore the role of VPA as a teratogenic compound that leads to neural tube defects, cognitive impairment, developmental deficits, and other birth abnormalities (Ergaz et al. 2016; Lloyd 2013; Roullet et al. 2013). The disruption to central nervous system formation and development likely contributes to the neurodevelopmental changes associated with ASD.

3.2 Early Studies on the Teratogenic Effects in Animals

The discovery of VPA's teratogenic effects transformed our understanding of its role in neurodevelopmental disorders, highlighting the drug's potential to induce significant and lasting changes in the brain. Investigations in the 1980s primarily focused on VPA's teratogenic effects, and, in doing so, provided critical insights into the behavioural outcomes in pups prenatally exposed to VPA. These studies laid the groundwork for the development of the well-established ASD animal model.

To our knowledge, Nau et al. (1981b) is the earliest published examination of VPA teratogenicity, which involved administering chronic VPA doses to pregnant rats to assess potential effects on foetal development, including defects or mutations. The work of Nau and colleagues established suitable doses for rodents and methods for chronic VPA administration. Administering doses of 0, 200, 400, 800 mg/kg of VPA via subcutaneous injections from gestational days (GD) 7–15 (mid-pregnancy) in mice, Nau et al. (1981b) observed that higher doses (400–800 mg/kg) resulted in significantly reduced foetal weight, increased incidence of malformation, and increased resorptions (prenatal death) and deaths at birth. Vorhees (1987) observed similar effects using oral gavage to administer VPA doses (0, 150, 200, 300, 400, or 600 mg/kg) to pregnant rats between GD 7 and 18 (middle to the end of a rat's pregnancy). The highest dose (600 mg/kg) was lethal for two of the four dams, and it was lethal to all pups, while the 400 mg/kg resulted in high rates of pup mortality with more than half of the pups being resorbed or dead at birth. In contrast, the lower VPA doses (150–300 mg/kg) resulted in better survival rates but increased incidence of visible birth defects, including limb, cardiovascular, and skeletal abnormalities (see Table 1 for comparison of studies, doses, and the main findings).

In addition to physical malformations, Vorhees (1987) observed that lower VPA doses (150 and 200 mg/kg) from GD 7–18 led to decreased movement (locomotor activity), increased memory errors in a water maze, and reduced sudden noise response (startle response) compared to the 0 dose. These deficits, more pronounced at higher doses, indicate a dose-response relationship, where the increased VPA exposure was related to a worsening of the neurological and behavioural effects.

3.3 Discussion of Timing and Limitations

Administering VPA to pregnant (or recently given birth) animals has become a widely used approach in preclinical ASD research. Timing of gestational exposure is a critical factor. Several earlier studies have found that GD 11.5 or 12 (depending on whether vaginal plug detection is referred to as GD 0.5 or 1) has the strongest influence on behaviours typically associated with ASD (Bringas et al. 2013; Kim et al. 2011; Olde Loohuis et al. 2017; Schneider and Przewłocki 2005; Stanton et al. 2007), although similar effects have been described for GD 9 (Dufour-Rainfray et al. 2010; Tsujino et al. 2007) or GD 13 (Kataoka et al. 2013; Wagner et al. 2006; see

Table 1 Key studies on VPA-induced ASD rat and mouse models focused on single administration studies

Study	Animal	Dose (mg/kg)	Administration method	Day of exposure (GD or PND)	Main findings
Schneider and Przewłocki (2005)	Rats Wistar	600	IP	GD 12	↓ Pain sensitivity ↓ Prepulse inhibition (i.e., ↓information processing or attention) ↑ Locomotor activity ↑ Repetitive behaviours ↓Social interaction & initiation
Wagner et al. (2006)	Mice BALB/c	600	SC	GD 13	↓ Impaired motor activity = water maze escape improvement over 5 days (i.e., impaired memory) ↑ Locomotor activity
Stanton et al. (2007)	Rats Long-Evans	600	IP	GD 12	↑ Initial learning of eye-blink conditioning but ↓ reversal learning
Tsujino et al. (2007)	Rats Wistar	800	Orally—syringe fed	GD 9	↑ Locomotor activity in novel environment, disturbance in circadian rhythm (i.e. ↑ light cycle activity ↓ light cycle feeding) ↓ Body weight
Markram et al. (2008)	Rats Wistar-Han	500	IP	GD 12	↓ Social interaction ↑ Repetitive behaviours ↓ Sensitivity to pain
Yochum et al. (2008)	Mice BALB/c	400	SC	GD 14	↓ Social Interaction ↓ Play behaviour ↓ Locomotor activity
Dufour-Rainfray et al. (2010)	Rats Wistar	600	IP	GD 9	↓ Social interaction ↑ Locomotor activity
Narita et al. (2010)	Rats Wistar	800	Orally—syringe fed	GD 9	= No significant differences in the social interaction tests or memory test (water maze performance) ↑ Locomotor activity in a novel environment
Bambini-Junior et al. (2011)	Rats Wistar	600	IP	GD 12	↓ Learning in the conditioned place preference ↓ Preference for social novelty
Kim et al. (2011)	Rats Sprague-Dawley	400	SC	GD 7, 9, 12, 15[a]	↓ Sociability for rats exposed to VPA on GD12 ↓ Social preference for rats

(continued)

Table 1 (continued)

Study	Animal	Dose (mg/kg)	Administration method	Day of exposure (GD or PND)	Main findings
					exposed to VPA on GD12 ↑ Sensitivity to electric shocks for GD 12 exposed rats compared to rats exposed on other GD ↓ Live births for VPA exposure on GD 9
Bringas et al. (2013)	Rats Sprague-Dawley	500	IP	GD 12	↓ Head-dips (i.e., less exploration) ↑ Duration of head-dips (i.e., fixation)
Cohen et al. (2013)	Rats Long-Evans	350	IP	GD 12	↑ Social investigation ↑ Play fighting and behaviour = No social preference or change in locomotor activity
Kataoka et al. (2013)	Mice CD1	500	IP	GD 9, 12, 14[a]	↓ Social Interaction for VPA exposure on GD12 not 9 or 14 ↓ Sniffing behaviour for VPA exposure on GD12 not 9 or 14
Oguchi-Katayama et al. (2013)	Rats Wistar Hannover/Rcc	400	SC	PND 14	↑ Anxiety-like behaviour (i.e., increased time in closed arms of elevated plus maze) in males but not females
Pragnya et al. (2014)	Mice BALB/c	400	SC	PND 14	↓ Locomotor activity ↓ Pain sensitivity
Štefánik et al. (2015)	Rats Wistar	600	IP	GD 13	↑ Social investigation (i.e., increased time with novel conspecific[b] in three-chamber sociability test) ↑ Social initiation with conspecific
Olde Loohuis et al. (2015)	Rats Wistar	495	IP	GD 12	↓ Play behaviour ↑ Non-social behaviour with conspecific[b] ↓ Prepulse inhibition (i.e., decreased information processing or attention)
Al Sagheer et al. (2018)	Mice C57BL/6 J	450	IP	GD 12	↑ Time till eyes opened and ↑ righting reflex (i.e., developmental deficit)

(continued)

Table 1 (continued)

Study	Animal	Dose (mg/kg)	Administration method	Day of exposure (GD or PND)	Main findings
					↓ Sociability in males but not female ↓ Motor coordination ↓ Gait in males (and to a lesser extent in females)
Dai et al. (2018)	Rats Wistar	600	IP	GD 12	↓ Vocalizations (i.e., decreased communication) ↓ Social preference and no effect of social discrimination or preference of familiar compared with novel conspecifics (stranger rats)
Hou et al. (2018)	Rats Sprague-Dawley	600	IP	GD 12	↑ Righting reflex ↓ Motor coordination ↑ Anxiety-like behaviours ↓ Locomotor activity and exploration behaviour ↓ Social interaction (i.e., decreased time with conspecific in three-chamber sociability test) ↓ Spatial learning and learning (i.e., increased time to learn location of escape in water maze)
Schiavi et al. (2019)	Rats Wistar	500	IP	GD 12	= No difference between VPA exposed and control rats in conditioned place preference, amphetamine induced locomotor activity and sucrose preference ↓ Play response ↓ Sociability
Norton et al. (2020)	Mice C57BL/6	400	SC	PND 14	↑ Aggression ↓ Anxiety sensitivity ↓ Learning in adult offspring
Guerra et al. (2023)	Mice C57BL/6	500	IP	GD 10	= No difference in locomotor activity = No social preference (i.e., no difference between time with conspecific vs object in three chamber sociability test) ↓ Social novelty

(continued)

Table 1 (continued)

Study	Animal	Dose (mg/kg)	Administration method	Day of exposure (GD or PND)	Main findings
					↑ Repetitive behaviours (i.e., increased self-grooming)
Turpin et al. (2023)	Mice C57BL/6	450	IP	GD 12	= No difference in grooming behaviour, ↑ Frequency & Duration of rearing behaviour

GD Gestational day, *PND* Postnatal day, *IP* Intraperitoneal injection, *SC* Subcutaneous injection, *no* difference; ↑ = increased; ↓ = decreased
[a]Comparison of signal VPA doses on each of the GD
[b]Conspecific refer to an individual of the same species, not a litter or playmate

Table 1 for comparison of doses, timing, and related findings). However, when comparing different time points of VPA exposure during gestation, Kim et al. (2011) showed that the most effective time point was GD 12 for mice and rats when inducing ASD-like behaviours. While VPA exposure on GD 12 more consistently shows ASD-like behaviours, other gestational days (GD 9, 10, and GD 13) also produce a variety of ASD-like outcomes, reinforcing the view that VPA's effect on brain development can result in ASD-like behaviours when administered prenatally around GD 12. Beyond that, VPA exposure postnatally perhaps leads to a different behavioural profile, with rats and mice exhibiting increased aggression and reduced anxiety (Norton et al. 2020; Oguchi-Katayama et al. 2013; Pragnya et al. 2014).

It is important to note that while animal models provide valuable insights, there are limitations in extrapolating these findings to humans. The differences in gestational periods and brain development between species can complicate the direct application of these results to human condition (Mehra et al. 2022). Furthermore, as researchers we will never reproduce a complex psychiatric disorder in animals; as such, animal model research focuses on symptoms that are similar to those seen in the psychiatric disorder when drawing comparisons. Despite these challenges, the VPA model remains a critical tool for studying ASD due to its reproducibility and the robustness of its effects on neurodevelopment.

4 Exposure to VPA as an Animal Model for ASD

Beyond the described early VPA studies, there have been decades of research examining VPA within the context of the VPA-induced animal model of ASD. Unlike other pharmaceuticals, VPA demonstrates a remarkable specificity in increasing the risk of ASD. This specificity is highlighted by studies on the prenatal exposure to valpromide, which is pharmaceutically like VPA but lacks the HDAC inhibition. Research by Kataoka et al. (2013) found that valpromide does not induce ASD-like behaviours, unlike VPA (see the pharmacodynamics section on VPA later in the chapter). As shown in Table 1, VPA exposure has been consistently found to induce ASD-like behaviours in mice (Guerra et al. 2023; Turpin et al. 2023) and rats (Chomiak et al. 2013; Gąssowska-Dobrowolska et al. 2020; Hou et al. 2018), and while not covered in this chapter, it has also been found in non-human primates (Watanabe et al. 2021; Zhao et al. 2019) and zebrafish (Chen et al. 2018; Wang et al. 2024). For a review, see Ornoy et al. (2024).

Building on established evidence linking maternal VPA exposure to an increased risk of ASD in children, researchers have extensively used VPA-induced animal models to explore ASD-like behaviours and underlying neural mechanisms of those behaviours (Christensen et al. 2013; Ranger and Ellenbroek 2016; Rasalam et al. 2005). Earlier studies, such as those by Rodier et al. (1996, 1997) and Schneider and Przewłocki (2005), provided key insights for using VPA in preclinical research, such that rats and mice exhibit ASD-like behaviours when VPA is administered at specific pre- and postnatal stages. Subsequent research has expanded these findings, ultimately situating the timing of VPA exposure as a key variable in ASD model. As we will discuss in the remainder of this chapter, the timing (before or after neural tube closure) and the dosage of VPA play salient roles in inducing ASD-like behaviours, particularly in relation to social behaviour exhibited by rats and mice.

5 VPA Exposure: Single Administration Studies

The majority of research on prenatal VPA exposure as an animal model for ASD, has used a single-dose approach. In a landmark paper, Schneider and Przewłocki (2005) demonstrated that adolescent rats exposed prenatally to a single dose of 600 mg/kg of VPA on GD 12 exhibited reduced play and social behaviour, which persisted into adulthood with decreased social interaction. These findings were also found by Olde Loohuis et al. (2015). Similarly, deficits in social initiation were observed in adolescent rats exposed to VPA on GD 9 (Dufour-Rainfray et al. 2010), with similar social deficits noted in those exposed on GD 12 (Bambini-Junior et al. 2011; Kim et al. 2011; Markram et al. 2008; Al Sagheer et al. 2018). A systematic review by Chaliha et al. (2020) reiterated that the typical induction of ASD-like behaviours involves administering 400–600 mg/kg of VPA with an injection (interperitoneally or subcutaneously) between GD 12 and GD 13. Their analysis of 150 studies revealed that prenatal VPA exposure led to disruptions (often reductions or deficits) in rat and mouse pups' social behaviours (e.g., vocalisations, play, interaction, approach/avoidance), as well as increased repetitive behaviours (e.g., open field or marble burying behaviours) and cognitive deficits (e.g., delayed learning and decreased alterations in Y or T mazes; Chaliha et al. (2020)). These studies show that the single dose of VPA during gestation can be successful in inducing ASD-like behaviour in rats and mice.

Despite the widespread use of the VPA-induced animal model of ASD, the observed ASD-like behaviours in such animals are not always consistent. Some studies have reported increased rather than decreased social and play behaviours (Cohen et al. 2013; Štefánik et al. 2015), while others have found no significant differences compared to non-VPA-exposed animals (Narita et al. 2010). Discrepancies in findings may stem from variations in administration methods, such as the timing and dosage of VPA exposure. For instance, Narita et al. (2010) administered 800 mg/kg of VPA via oral gavage on GD 9 and found no significant differences in social interaction in adult offspring. Conversely, Cohen et al. (2013) and Štefánik et al. (2015) reported increased social investigation and play fighting following administration of 350 mg/kg and 600 mg/kg of VPA (respectively) via IP injection at GD 13 and GD 12, respectively. Some studies found minimal or no effects of prenatal exposure to VPA related to the offspring's behaviour (e.g., Dai et al. 2018; Larner et al. 2021; Schiavi et al. 2019; Tartaglione et al. 2019). While these administration methods (i.e., IP injection) were similar to those used by Schneider and Przewłocki (2005), the conflicting findings persist, suggesting that additional factors may contribute to the variability in outcomes. It is important to emphasise that also in humans, VPA induces ASD in only a small subpopulation of children. Nonetheless, Larner et al. (2021) emphasised the need for standardisation in VPA model protocols and behavioural tests to enhance the reliability and reproducibility of the findings.

While prenatal VPA exposure has been extensively studied, it is important to note rats and mice, unlike humans, are born at an earlier stage of brain development. This means that in rats and mice, early postnatal stages represent developmental periods comparable to late prenatal or early postnatal stages in humans (Ellenbroek and Youn 2016). Therefore, studying postnatal VPA exposure in these rodents allows researchers to gain additional insight into how drug exposure and disruptions during these crucial developmental windows may contribute to long-term neurobehavioral outcomes, mirroring vulnerabilities in late foetal human neurodevelopment.

To explore the effects of postnatal VPA exposure, studies have consistently identified behavioural deficits in sociability, locomotor activity, and sensory processing. Yochum et al. (2008) administered 400 mg/kg VPA to mice on PND 14, resulting in reduced interaction and play behaviour, as well as decreased locomotor activity. Norton et al. (2020) administered 400 mg/kg of VPA on PND 14 to mice, finding heightened aggression, reduced anxiety, and learning difficulties in adult offspring, suggesting a potential disinhibited or impulsive behaviour pattern. Similarly, Pragnya et al. (2014) observed decreased locomotor activity following PND 14 exposure to VPA, while Oguchi-Katayama et al. (2013) reported anxiety-like impairments. As such, the significance of timing in the VPA exposure profoundly influences the development of behaviours, and the postnatal exposure may also lead to a different behavioural profile.

Across pre- and postnatal single-dose studies, inconsistencies in behaviour raise significant concerns regarding the validity of VPA administration methods. Fundamental to this model is the administration of VPA, which is evaluated based on the construct, face, and predictive validity. Despite variability demonstrated in many studies, it is possible to broadly say that prenatal VPA exposure induces ASD-like behaviour (e.g., deficits in communication or social behaviour and abnormal cognitive performance), supporting the face validity of this model. However, the model's construct validity is weaker due to its deviation from human VPA exposure, which typically involves women taking VPA daily throughout pregnancy, whereas the current animal model typically uses a single dose, and this is notwithstanding the neglect of genetics as a factor in the development of ASD in humans and in this ASD animal model. The current single-dose approach, while effective in inducing ASD-like behaviours, may not fully replicate the cumulative and multifactorial influences seen in human ASD development. Therefore, enhancing the model's validity requires improvements in administration methods and consideration of the cumulative doses of VPA, such as incorporating multiple doses of VPA.

6 VPA Exposure: Multiple Administration Studies

Attempts to enhance the construct validity of the VPA-induced animal model for ASD, through repeated exposure of pregnant rodents to VPA have been made to better reflect the prolonged and cumulative nature of VPA exposure when a woman is taking it throughout pregnancy. Yet there is scarcity of studies on repeated or chronic administrations. Of the existing studies, the administration methods and the results are varied (see Table 2). Multiple administration studies provide insights into the cumulative effects of VPA exposure over time, shedding light on potential dose-response relationships and the long-term neurodevelopmental consequences of VPA exposure.

The earliest study we found using multiple VPA doses and examining offspring behaviour was conducted by Chapman and Cutler (1984). In their research, they administered 160–180 mg/kg/day of VPA in drinking water to mice throughout gestation and lactation periods. Interestingly, there were no observed behavioural changes in the offspring until adulthood, when social investigation increased. More recently, Bass et al. (2020) investigated the naturalistic (e.g., home cage) behaviours of mice exposed to VPA throughout pregnancy, ceasing VPA exposure at birth. They dissolved VPA in sweetened drinking water, determining a consumption range of 500–700 mg/kg/day based on parent mice weights and a lickometer. Their study revealed that prenatal VPA exposure led to delayed litter viability (e.g., extended time for a dam to produce pups without impact on litter size), and the offspring exhibited alterations in locomotor activity (e.g., more frequent but shorter bouts of locomotor activity, increased sheltering behaviours, and increased nocturnal wheel running), as well as increased feeding behaviour and increased social interaction (i.e., measured based on proximity to other sex and VPA exposure-matched mice, specifically conspecifics but not littermates). The findings from Chapman and Cutler (1984) and Bass et al. (2020), contrast with much of the behavioural findings from a single-administration studies, although, as mentioned above, increased social behaviour has been reported in some single-administration studies (see Table 1).

Table 2 The findings of studies which used multiple prenatal VPA exposure in rodents including VPA doses and administration routes

Study	Animal	Dose (mg/kg)	Administration method	Days of exposure (GD or PND)	Main findings
Chapman and Cutler (1984)	Mice Outbred	160–180	Drinking water	Conception to birth (& lactation to weaning)	↑ Social investigation in adulthood
Wagner et al. (2006)	Mice BALB/c	200	SC	GD 12–17	↑ Activity and self-injurious behaviours only when challenged with stimulant drug
Frisch et al. (2009)	Rats Wistars	825	Drinking water	Conception to birth	↓ Water maze spatial memory ↓ Rotarod motor coordination
Dendrinos et al. (2011)	Rats Sprague Dawley	200	IP	GD 11–13	No social behaviour changes
Reynolds et al. (2012)	Rats Long-Evans	150	IP	PND 6–12	↓ Auditory response ↓ Locomotor performance
Sabers et al. (2014)	Rats Wistar	20 or 100	IP	GD 12–21 (& lactation to weaning)	↑ Neocortical cell ↑ Cortical thickness
Wellmann et al. (2014)	Rats Sprague-Dawley	200	IP	GD 12–13[a]	↓ Number of vocalisations ↑ Locomotor activity ↑ Play behaviour
Juliandi et al. (2015)	Mice Unspecified	300	Oral[b]	GD 12–14	↓ Learning and memory ↓ Contextual and fear association ↑ HDAC inhibition
Fujimura et al. (2016)	Mice CD1	Not specified[c]	Drinking water	Conception to birth	↑ Epigenetic changes ↓ Neuronal proliferation
Mony et al. (2016)	Rats Sprague-Dawley	300	SC	PND 2 and 4 ±	↑ Locomotor activity ↑ Anxiety-like behaviour (i.e., time in and more entries into closed arms of elevated plus maze)

(continued)

Table 2 (continued)

Study	Animal	Dosc (mg/kg)	Administration method	Days of exposure (GD or PND)	Main findings
Bertelsen et al. (2017)	Rats Wistar	20 or 100	IP	PND 12–21	↓ Play behaviour for pre-pubescent males exposed to 20 mg/kg/day
Semmler et al. (2017)	Rats Wistar	500 or 825	Drinking water	Conception to birth	↑ Hippocampal neurons ↓ Astrocytes
Kinjo et al. (2019)	Rats Wistar	100 or 200	IP	GD 12 to birth	↑ Locomotor activity ↓ Anxiety-like behaviour (more time in open arms of elevated plus maze) ↑ Hippocampal proliferation
Bass et al. (2020)	Mice C57BL/6 J	500-600	Drinking water	PND 4–11	↑ Time to produce litter ↑ Nocturnal wheel running
Elnahas et al. (2021)	Rats Wistar	300	SC	PND 2 and 4 ±	↓ Play behaviour ↓ Pain sensitivity ↑ Repetitive behaviours
Vakili Shahrbabaki et al. (2022)	Rats Wistar	400–600	IP or SC	PND 2 and 4 ±	↓ Social interaction (i.e., reduced exploration of stranger conspecific) ↓ Social behaviour

GD Gestational day(s), *PND* Postnatal day(s), *IP* Intraperitoneal injection, *SC* Subcutaneous injection; = no difference; ± = VPA administered twice on PND2 and 3, and once on PND 4; ↑ = increased; ↓ = decreased

[a]Wellmann et al. (2014) administered an additional dose on GD 12 (pm)

[b]Juliandi et al. (2015) did not specify the exact oral administration methods

[c]Fujimura et al. (2016) administered VPA at a concentration of 0.4% in the drinking water, achieving VPA plasma level of 10–60 μg/ml in the dams

Interestingly, the repeated administration of VPA in the drinking water has been associated with epigenetic changes and disruptions in neuronal function (Fujimura et al. 2016; Semmler et al. 2017), which are findings that parallel similar connectivity impairments in those living with ASD (Gogolla et al. 2009). Frisch et al. (2009) administrated VPA in the drinking water for female rats 7 days before mating and again from 48 h after mating until delivery and observed varied effects on rat offspring based on the VPA exposure, such that medium doses (~470 mg/kg/day) led to increased locomotor activity and improved cognition, while very high doses

(~720 mg/kg/day) resulted in deficits in the same measures. Using similar dosing methods, Semmler et al. (2017) administered VPA in drinking water, leading to daily medium (500 mg/kg/day) and high (825 mg/kg/day) doses. VPA administration was paused after 7 days for mating and resumed 48 h later, continuing until delivery. The authors found VPA dose-dependent effects on neuronal measures (e.g., increased neuron counts and decreased astrocyte counts in the hippocampus), aligning with evidence from birth registries data in humans that show that increased exposure to VPA is associated with increased severity of behavioural outcomes (Tomson et al. 2015). Together these studies provide evidence that oral administration in drinking water can be an effective and relatively stress-free method for prolonged VPA administration to pregnant rodents that induces a range of neurodevelopmental changes, mirroring the dose-dependent relationships seen in human ASD populations.

Various other (more solid) ingestion methods have been employed to administer multiple doses of VPA during pregnancy. Main and Kulesza (2017), for instance, used peanut butter mixed with a high dose of VPA (800 mg/kg) to deliver doses to pregnant rats on GD 10 and GD 12. Their investigation revealed that VPA-exposed rats weighed less as adolescents, with smaller brains, and exhibited significant deficits in motor coordination tasks (e.g., rotarod, beam walking, and gait patterns). Main and Kulesza suggested that disrupted cerebellar development could underlie these deficits in coordination. Notably, examination of the cerebellum revealed significantly diminished Purkinje cells, which are critically involved in motor coordination. Work from our group (Pettie 2020) shows that VPA dissolved in gelatine pellets to administer 300–600 mg/kg/day to rats leads to ASD-like behaviours (e.g., decreased maternal separation communication, increased social behaviour, and social aversion) in male and female offspring. Similarly, despite not specifying their exact oral administration methods, Juliandi et al. (2015) exposed mice to multiple oral VPA doses between GD 12 and 14 and found neurogenesis deficits and abnormal hippocampal morphology; however, the observed behavioural differences in anxiety and cognitive measures were minimal compared to control mice. In a recent study, Jazayeri et al. (2020) administered 20 g VPA/kg in the rat's food throughout pregnancy, which they suggest is approximately equivalent to 1,000 mg/kg/day of VPA. Their experiment showed that rats fed diets with VPA concentrations up to this level consumed similar amounts of food as standard (untreated) chow-fed rats, indicating that 20 g/kg was palatable and bioavailable, yielding blood levels around 220 μmol/L. This continuous VPA exposure during pregnancy led to shorter rat pup lengths and altered vertebrae. These studies shed light on the various neurodevelopmental consequences of prolonged prenatal VPA exposure, mirroring neurodevelopmental changes observed in children prenatally exposed to VPA.

Repeated administration of VPA through intraperitoneal injections (IP) has also been explored. While this is likely a significantly more stressful procedure for the rat or mouse, it reduces the variability in exposure levels typically seen with voluntary intake. Sabers et al. (2014) administered 20 mg/kg/day or 100 mg/kg/day of VPA to pregnant rats from GD 12.5 to GD 21, continuing treatment during lactation until

PND 23 (typically a point when pups are removed from the dam). Offspring exposed to VPA prenatally and postnatally exhibited increased neocortical cells and cortical thickness, indicating significant disruptions in cortical growth and organisation, along with cerebellar disturbances. Similarly, Bertelsen et al. (2017) administered 20 or 100 mg/kg/day of VPA via IP from GD 12 to birth, observing reduced play behaviours in prepubescent male rats exposed to 20 mg/kg/day of VPA, while no differences were noted with 100 mg/kg/day of VPA. The authors suggest that the lack of behavioural changes at the higher dose may be related to a series of biochemical changes that vary depending on the prenatal dose and that these biochemical changes are distinct from those induced at a lower dose. The behavioural changes may be attributed to histone deacetylation, DNA alterations, and increased GABA activity. Additionally, research from Kinjo et al. (2019) and Ito et al. (2021) repeatedly administered 100 mg/kg/day or 200 mg/kg/day of VPA using IP injections from GD 12.5 to GD 21.5 (covering the last 9–12 days of pregnancy), which led to increased locomotor activity and reduced anxiety sensitivity in the offspring, alongside enhanced hippocampal neurogenesis (increased number of hippocampal cells). However, Dendrinos et al. (2011) found no sociability changes in rats following VPA administration via IP between GD 11 and 13. Collectively, these studies suggest that multiple VPA exposures during gestation can lead to diverse behavioural and neurodevelopmental effects, and the effects are at odds with the behavioural and neurodevelopmental outcomes of a single VPA administration during gestation. Overall, although far less studied, multiple or prolonged VPA exposures (i.e., using drinking water, food, or injections) seem to lead to different (and sometimes even opposite) effects compared to a single dose of VPA (see Table 2).

The evidence regarding multiple exposures to VPA only during the early postnatal period is limited, but the few available studies suggest profound effects on offspring motor activity, coordination, and later-life ASD-like behaviours. For instance, Mony et al. (2016) exposed rats to 300 mg/kg/day of VPA on PND 2 and 3 (twice daily) and again on PND 4 (once), resulting in hyperactive locomotor activity, increased anxiety, and reduced sociability and social preference. Elnahas et al. (2021) and Vakili Shahrbabaki et al. (2022) found similar behavioural alterations in rats using the same dosing schedule. However, the distinction between behavioural profiles from postnatal VPA exposure and multiple or single prenatal VPA exposure remains unclear and, as such, presents a critical gap in the current research.

A study by Wagner et al. (2006) provides insight into this research gap, investigating various VPA doses and exposure timings in mice. They examined four experimental groups: (a) single dose of 600 mg/kg VPA on GD 13, (b) multiple doses of 200 mg/kg, GD 12–17, (c) single dose of VPA 400 mg/kg on PND 14 subcutaneously (for male mice only), and (d) single dose of VPA 50 mg/kg on PND 14 to control for the VPA sedation effects during behavioural testing (which revealed no behavioural disruptions). Results showed that the single dose on GD 13 led to motor coordination deficits, while the multiple prenatal doses (GD 13–17) resulted in minor deficits. Male mice treated with 400 mg/kg of VPA on PND

14 showed increased activity and self-injurious (i.e., repetitive) behaviours only when challenged with d-amphetamine, indicating behavioural sensitivity to stimulants. The findings from Wagner et al. (2006) are indicative of differential behavioural profiles of VPA exposure depending on the VPA dose and the developmental stage of the offspring.

However, not all studies that exposed rats and mice to VPA during the postnatal period have led to the same behavioural outcomes. For example, Reynolds et al. (2012) found that 150 mg/kg/day of VPA administered once per day from PND 6 to PND 12 in rats led to sensory and motor impairments (i.e., reduced response to auditory stimuli and impaired locomotor activity), which may resemble features often seen in human ASD populations. Whereas, Bass et al. (2020) found that mice exposed to 100 mg/kg/day or 200 mg/kg/day of VPA once daily from PND 4–11 exhibited no change in locomotor activity or anxiety behaviours that were statistically different from control mice. As such, the observations from prolonged postnatal VPA administration are indicative of further nuances in the VPA-induced animal model of ASD, and more work is needed to understand the brain development changes induced by postnatal VPA administration.

Given the combined evidence from multiple VPA administration studies, there is considerable variability in the approaches, clearly demonstrating the importance of specific VPA doses and the developmental stage the dose is given for shaping the offspring's behavioural outcomes. Moreover, it is possible that multiple VPA exposures during the prenatal and/or postnatal period may present differential behavioural profiles, which could be indicative of inducing various developmental disorders and not just ASD (Lotufo Denucci et al. 2021). However, much more research is needed to understand and compare the various combinations of dose (i.e., mg/kg), timing (i.e., prenatal, postnatal, or both), and duration (i.e., number of doses) to create distinct behavioural profiles, which then may be examined considering ASD and/or other psychiatric disorders. As we identified earlier, Larner et al. (2021) emphasised the need for standardised behavioural tests (e.g., social behaviour, three-chamber test) and the subsequent analysis to allow for direct comparisons of the behaviour and neuronal outcomes following the exposure to VPA during critical developmental periods, thereby enhancing the reliability and reproducibility of the findings. Lastly, we would like to note the inherent limitations of ASD animal models, in that there is not one model that replicates all core behaviours of the condition (Ruhela et al. 2015); rather they usually model specific behavioural or disease aspects that may apply to more than one condition.

Current and future researchers who use (or intend to use) the single VPA administration method might like to consider the choice between a single-dose method compared to a multiple-dose approach when seeking to induce ASD-like behaviours in animal models. From a construct validity perspective, single VPA administration may be limited in accurately reflecting the cumulative exposure to VPA that a developing human foetus is exposed to when women take VPA throughout pregnancy. However, as it is often necessary in preclinical research, simplifying factors, like opting for a single dose of VPA, can facilitate standardisation, make the studies and individual experiments more feasible, and perhaps yield more consistent results, as we have seen in relation to behavioural outcomes following VPA exposure on GD 12. Furthermore, researchers may be overlooking translational relevance by opting for a single administration method. Consideration of alternative dosing regimens or methods that could better address research questions and objectives as well as the intended impact of their research. As such, researchers may like to reflect on the reviewed studies and findings to consider which approach may provide the desired insights into VPA exposure and its effects on neurodevelopment and behaviour.

7 Pharmacodynamics of Valproate

The description of the perinatal effects of VPA on the development of brain and behaviour after perinatal administration warrants the question of how a relatively simple molecule like VPA can have such complex and widespread effects. As briefly touched upon earlier, VPA affects a multitude of different biochemical and molecular processes within the cell. Pharmacodynamically, VPA increases the availability of gamma-aminobutyric acid (GABA) and inhibits GABA's metabolism (Löscher 2002; Owens and Nemeroff 2003), thereby enhancing GABAergic activity in the brain. In addition, VPA regulates neurotransmission by blocking voltage-dependent sodium channels, inhibiting the completion of action potentials (Bollino et al. 2015), and decreasing the conduction of potassium channels at the same time (VanDongen et al. 1986; Chateauvieux et al. 2010). Additionally, VPA activates extracellular signal-regulated kinase, promoting cell survival and preventing apoptosis, or cell death (Michaelis et al. 2006). VPA use increases the expression of brain-derived neurotrophic factor (BDNF), which is involved in neurogenesis and the facilitation of cortex development (Almeida et al. 2014; Fagiolini et al. 2022). These combined effects contribute to VPA's antiepileptic actions by reducing nervous system activity through increased GABA concentrations. Furthermore, VPA's pharmacodynamic profile extends to the treatment of various clinical symptoms, including pain, affective disorders, and cancer (Schäfer and Brandl 2020; Tomson et al. 2016). These intricate pharmacodynamic mechanisms emphasise the multifaceted effects of VPA on neural function and development.

As VPA crosses the placental barrier, it exerts teratogenic effects on a developing foetus (Dickinson et al. 1979; Nau et al. 1981a; Ranger and Ellenbroek 2016). If VPA is used while the neural tube is open, typically between gestational days 21–23 for humans and GD 10–12 for rats and mice (Ergaz et al. 2016), then VPA's impact is more severe (Phiel et al. 2001). Its teratogenic mechanism involves HDAC inhibition, leading to epigenetic alterations in the foetus (Göttlicher et al. 2001; Gurvich et al. 2005; Phiel et al. 2001; Ranger and Ellenbroek 2016). VPA binds to HDAC enzymes, preventing their function of removing acetyl groups from histones. Acetylation of histones involves the addition of acetyl groups to histone proteins, is a critical regulator of gene transcription, and affects cell function, differentiation, and proliferation (Göttlicher et al. 2001). As histone acetylation leads to an opening up of the 3-dimensional structure of DNA, HDAC inhibition increases gene expression and slows cell growth and its processes that may slow the development of cancer cells (Aroosa et al. 2023; Brodie and Brandes 2014; Tomson et al. 2016). Conversely, histone deacetylation that is inhibited by VPA leads to epigenetic alterations and, therefore, disruptions in typical neuronal development, particularly during critical developmental windows (Kataoka et al. 2013). The combination of VPA's placental transport and HDAC inhibition during the first trimester of pregnancy leads to harmful effects, including the increased risk of ASD (as we have discussed in this chapter; Kataoka et al. 2013; Phiel et al. 2001).

VPA's pharmacodynamics are complex and involve interaction with various factors. In addition to the already mentioned effect, VPA enhances the production of reactive oxygen species (ROS), endogenous molecules that can cause damage to DNA, proteins, and cellular communication (Defoort et al. 2006; Na et al. 2003; Tung and Winn 2011). During foetal development, the immature antioxidant systems fail to adequately detoxify ROS, particularly during organogenesis, which makes the foetus more vulnerable to damage caused by ROS (Tung and Winn 2011). The mechanisms of VPA actions are multifaceted, potentially leading to an interactive effect, where HDAC inhibition increases ROS levels, resulting in increased cell damage (Carew et al. 2008). Elevated ROS levels have been linked with various adverse effects, including neural tube defects, embryonic growth reduction, and apoptosis (Defoort et al. 2006; Tung and Winn 2011). Studies in mice have also demonstrated that VPA-induced germ cell toxicity is associated with increased oxidative stress (Khan et al. 2011). And so, there is an interplay between VPA's HDAC inhibition and increased ROS levels that results in various outcomes, ranging from cellular damage and neural tube defects to embryonic growth reduction and apoptosis. Taken together, it is apparent that prenatal VPA can have detrimental effects on the neural development of the unborn foetus.

8 Conclusions and Perspectives

As we have highlighted above, the impact of drug exposure on brain development is multifaceted. The interplay between exposure to a teratogen during the highly sensitive neurodevelopmental period can have detrimental effects on the neural function and behaviours of both animals and humans. Our in-depth analysis of VPA as a pharmacological model for studying neurodevelopmental conditions, ASD in particular, emphasised the complex interplay between the timing of drug exposure, dosage, and duration of exposure, as well as how this can affect the behavioural outcomes. Especially while the neural tube is open, VPA's teratogenic effects are responsible for a heightened risk that the offspring will develop ASD-like behaviours. Attention should also be paid to the teratogenic mechanism HDAC inhibition, as the subsequent epigenetic changes may lead to unforeseeable long-term effects that may further impact the physical or neurological development, so more work is needed to explore this avenue. These insights enhance our understanding of the neurobiological underpinnings of ASD and the translational relevance of VPA-induced ASD animal models and will aid the research community in our quest to continue to develop targeted interventions.

References

Adab N, Kini U, Vinten J, Ayres J, Baker G, Clayton-Smith J, Coyle H, Fryer A, Gorry J, Gregg J, Mawer G, Nicolaides P, Pickering L, Tunnicliffe L, Chadwick DW (2004) The longer term outcome of children born to mothers with epilepsy. J Neurol Neurosurg Psychiatry 75(11): 1575–1583. https://doi.org/10.1136/jnnp.2003.029132

Akimova D, Wlodarczyk BJ, Lin Y, Ross ME, Finnell RH, Chen Q, Gross SS (2017) Metabolite profiling of whole murine embryos reveals metabolic perturbations associated with maternal valproate-induced neural tube closure defects. Birth Defects Res 109:106–119

Al Sagheer T, Haida O, Balbous A, Francheteau M, Matas E, Fernagut P-O, Jaber M (2018) Motor impairments correlate with social deficits and restricted neuronal loss in an environmental model of autism. Int J Neuropsychopharmacol 21(9):871–882. https://doi.org/10.1093/ijnp/pyy043

Almeida LEF, Roby CD, Krueger BK (2014) Increased BDNF expression in fetal brain in the valproic acid model of autism. Mol Cell Neurosci 59:57–62. https://doi.org/10.1016/j.mcn.2014.01.007

Alwan S, Chambers CD (2015) Identifying human teratogens: an update. J Pediatr Genet 4:39–41

Aroosa M, Malik JA, Ahmed S, Bender O, Ahemad N, Anwar S (2023) The evidence for repurposing anti-epileptic drugs to target cancer. Mol Biol Rep 50(9):7667–7680. https://www.ncbi.nlm.nih.gov/pmc/articles/PMC10460753/pdf/11033_2023_Article_8568.pdf

Bambini-Junior V, Rodrigues L, Behr GA, Moreira JCF, Riesgo R, Gottfried C (2011) Animal model of autism induced by prenatal exposure to valproate: behavioral changes and liver parameters. Brain Res 1408:8–16. https://doi.org/10.1016/j.brainres.2011.06.015

Bass JS, Tuo AH, Ton LT, Jankovic MJ, Kapadia PK, Schirmer C, Krishnan V (2020) On the digital psychopharmacology of valproic acid in mice [Original Research]. Front Neurosci 14. https://doi.org/10.3389/fnins.2020.594612

Bertelsen F, Folloni D, Møller A, Landau AM, Scheel-Krüger J, Winterdahl M (2017) Suppressed play behaviour and decreased oxytocin receptor binding in the amygdala after prenatal exposure to low-dose valproic acid. Behav Pharmacol 28(6):450. https://doi.org/10.1097/FBP.0000000000000316

Blotière P-O, Raguideau F, Weill A, Elefant E, Perthus I, Goulet V, Rouget F, Zureik M, Coste J, Dray-Spira R (2019) Risks of 23 specific malformations associated with prenatal exposure to 10 antiepileptic drugs. Neurology 93(2):e167–e180. https://doi.org/10.1212/WNL.0000000000007696

Bollino D, Balan I, Aurelian L (2015) Valproic acid induces neuronal cell death through a novel calpain-dependent necroptosis pathway. J Neurochem 133(2):174–186. https://doi.org/10.1111/jnc.13029

Bringas ME, Carvajal-Flores FN, López-Ramírez TA, Atzori M, Flores G (2013) Rearrangement of the dendritic morphology in limbic regions and altered exploratory behavior in a rat model of autism spectrum disorder. Neuroscience 241:170–187. https://doi.org/10.1016/j.neuroscience.2013.03.030

Bromley R, Mawer G, Clayton-Smith J, Baker G, Liverpool & Group MN (2008) Autism spectrum disorders following in utero exposure to antiepileptic drugs. Neurology 71:1923–1924

Brodie SA, Brandes JC (2014) Could valproic acid be an effective anticancer agent? The evidence so far. Expert Rev Anticancer Ther 14(10):1097–1100. https://doi.org/10.1586/14737140.2014.940329

Carew JS, Giles FJ, Nawrocki ST (2008) Histone deacetylase inhibitors: Mechanisms of cell death and promise in combination cancer therapy. Cancer Lett 269(1):7–17. https://doi.org/10.1016/j.canlet.2008.03.037
Chaliha D, Albrecht M, Vaccarezza M, Takechi R, Lam V, Al-Salami H, Mamo J (2020) A systematic review of the valproic-acid-induced rodent model of autism. Dev Neurosci 42(1): 12–48. https://doi.org/10.1159/000509109
Chapman JB, Cutler MG (1984) Sodium valproate: effects on social behaviour and physical development in the mouse. Psychopharmacology 83:390–396. https://link.springer.com/article/10.1007/BF00428553
Chateauvieux S, Morceau F, Dicato M, Diederich M (2010) Molecular and therapeutic potential and toxicity of valproic acid. Biomed Res Int 2010(1):479364. https://doi.org/10.1155/2010/479364
Chen J, Lei L, Tian L, Hou F, Roper C, Ge X, Zhao Y, Chen Y, Dong Q, Tanguay RL, Huang C (2018) Developmental and behavioral alterations in zebrafish embryonically exposed to valproic acid (VPA): An aquatic model for autism. Neurotoxicol Teratol 66:8–16. https://doi.org/10.1016/j.ntt.2018.01.002
Chomiak T, Turner N, Hu B (2013) What we have learned about autism spectrum disorder from valproic acid. Pathol Res Int 2013(1):712758. https://doi.org/10.1155/2013/712758
Christensen J, Grønborg TK, Sørensen MJ, Schendel D, Parner ET, Pedersen LH, Vestergaard M (2013) Prenatal valproate exposure and risk of autism spectrum disorders and childhood autism. JAMA 309(16):1696–1703. https://doi.org/10.1001/jama.2013.2270
Christianson AL, Chester N, Kromberg JGR (1994) Fetal valproate syndrome: clinical and neuro-developmental features in two sibling pairs. Dev Med Child Neurol 36(4):361–369. https://doi.org/10.1111/j.1469-8749.1994.tb11858.x
Cohen OS, Varlinskaya EI, Wilson CA, Glatt SJ, Mooney SM (2013) Acute prenatal exposure to a moderate dose of valproic acid increases social behavior and alters gene expression in rats. Int J Dev Neurosci 31(8):740–750. https://doi.org/10.1016/j.ijdevneu.2013.09.002
Dai Y-C, Zhang H-F, Schön M, Böckers TM, Han S-P, Han J-S, Zhang R (2018) Neonatal oxytocin treatment ameliorates autistic-like behaviors and oxytocin deficiency in valproic acid-induced rat model of autism. Front Cell Neurosci 12:355. https://doi.org/10.3389/fncel.2018.00355
Dean JC, Hailey H, Moore SJ, Lloyd DJ, Turnpenny PD, Little J (2002) Long term health and neurodevelopment in children exposed to antiepileptic drugs before birth. J Med Genet 39(4): 251–259. https://doi.org/10.1136/jmg.39.4.251
Defoort E, Kim P, Winn L (2006) Valproic acid increases conservative homologous recombination frequency and reactive oxygen species formation: a potential mechanism for valproic acid-induced neural tube defects. Mol Pharmacol 69:1304–1310. https://doi.org/10.1124/mol.105.017855
Delage C, Palayer M, Etain B, Hagenimana M, Blaise N, Smati J, Chouchana M, Bloch V, Besson VC (2023) Valproate, divalproex, valpromide: are the differences in indications justified? Biomed Pharmacother 158:114051. https://doi.org/10.1016/j.biopha.2022.114051
Demirtaş MS (2020) The pathogenesis of congenital anomalies: roles of teratogens and infections. In: Congenital anomalies in newborn infants—clinical and etiopathological perspectives. IntechOpen. https://doi.org/10.5772/intechopen.92580
Dendrinos G, Hemelt M, Keller A (2011) Prenatal VPA exposure and changes in sensory processing by the superior colliculus. Front Integr Neurosci 5. https://doi.org/10.3389/fnint.2011.00068
Dickinson RG, Harland RC, Lynn RK, Smith WB, Gerber N (1979) Transmission of valproic acid (Depakene) across theplacenta: half-life of the drug in mother and baby. J Pediatr 94(5): 832–835. https://www.jpeds.com/article/S0022-3476(79)80172-9/abstract
Dufour-Rainfray D, Vourc'h P, Le Guisquet A-M, Garreau L, Ternant D, Bodard S, Jaumain E, Gulhan Z, Belzung C, Andres CR, Chalon S, Guilloteau D (2010) Behavior and serotonergic disorders in rats exposed prenatally to valproate: a model for autism. Neurosci Lett 470(1): 55–59. https://doi.org/10.1016/j.neulet.2009.12.054

Ellenbroek BA, Youn J (2016) Gene – environment interaction in psychiatry. Nature, nurture, neuroscience. Elsevier, p 366

Elnahas EM, Abuelezz SA, Mohamad MI, Nabil MM, Abdelraouf SM, Bahaa N, Hassan GAM, Ibrahim EA, Ahmed AI, Aboul-Fotouh S (2021) Validation of prenatal versus postnatal valproic acid rat models of autism: a behavioral and neurobiological study. Prog Neuro-Psychopharmacol Biol Psychiatry 108:110185. https://doi.org/10.1016/j.pnpbp.2020.110185

Ergaz Z, Weinstein-Fudim L, Ornoy A (2016) Genetic and non-genetic animal models for autism spectrum disorders (ASD). Reprod Toxicol 64:116–140. https://doi.org/10.1016/j.reprotox.2016.04.024. (44th Annual Conference of the European Teratology Society)

Fagiolini A, Cuomo A, McIntyre RS (2022) Valproate. In: Fagiolini A, Cuomo A, McIntyre RS (eds) Pocket guide to practical psychopharmacology: lithium and anticonvulsants in psychiatric practice. Springer International Publishing, pp 23–34. https://doi.org/10.1007/978-3-030-98060-3_2

Frisch C, Hüsch K, Angenstein F, Kudin A, Kunz W, Elger CE, Helmstaedter C (2009) Dose-dependent memory effects and cerebral volume changes after in utero exposure to valproate in the rat. Epilepsia 50(6):1432–1441. https://doi.org/10.1111/j.1528-1167.2008.01943.x

Fujimura K, Mitsuhashi T, Shibata S, Shimozato S, Takahashi T (2016) In utero exposure to valproic acid induces neocortical dysgenesis via dysregulation of neural progenitor cell proliferation/differentiation. J Neurosci 36(42):10908–10919. https://doi.org/10.1523/JNEUROSCI.0229-16.2016

Gąssowska-Dobrowolska M, Cieślik M, Czapski GA, Jęśko H, Frontczak-Baniewicz M, Gewartowska M, Dominiak A, Polowy R, Filipkowski RK, Babiec L, Adamczyk A (2020) Prenatal exposure to valproic acid affects microglia and synaptic ultrastructure in a brain-region-specific manner in young-adult male rats: Relevance to autism spectrum disorders. Int J Mol Sci 21(10):3576. https://doi.org/10.3390/ijms21103576

Genton P, Semah F, Trinka E (2006) Valproic acid in epilepsy. Drug Saf 29(1):1–21. https://doi.org/10.2165/00002018-200629010-00001

Gill D, Derry S, Wiffen PJ, Moore RA (2011) Valproic acid and sodium valproate for neuropathic pain and fibromyalgia in adults. Cochrane Database Syst Rev 2011(10):CD009183. https://doi.org/10.1002/14651858.CD009183.pub2

Gogolla N, LeBlanc JJ, Quast KB, Südhof TC, Fagiolini M, Hensch TK (2009) Common circuit defect of excitatory-inhibitory balance in mouse models of autism. J Neurodev Disord 1:172

Göttlicher M, Minucci S, Zhu P, Krämer OH, Schimpf A, Giavara S, Sleeman JP, Lo Coco F, Nervi C, Pelicci PG, Heinzel T (2001) Valproic acid defines a novel class of HDAC inhibitors inducing differentiation of transformed cells. EMBO J 20(24):6969–6978. https://doi.org/10.1093/emboj/20.24.6969

Guerra M, Medici V, Weatheritt R, Corvino V, Palacios D, Geloso MC, Farini D, Sette C (2023) Fetal exposure to valproic acid dysregulates the expression of autism-linked genes in the developing cerebellum. Transl Psychiatry 13(1):114. https://doi.org/10.1038/s41398-023-02391-9

Gurvich N, Berman M, Wittner B, Gentleman R, Klein P, Green J (2005) Association of valproate-induced teratogenesis with histone deacetylase inhibition in vivo. FASEB J 19(9):1166–1168. https://doi.org/10.1096/fj.04-3425fje

Harden CL (2013) In utero valproate exposure and autism: long suspected, finally proven: In utero valproate exposure and autism. Epilepsy Curr 13(6):282–284. https://doi.org/10.5698/1535-7597-13.6.282

Hou Q, Wang Y, Li Y, Chen D, Yang F, Wang S (2018) A developmental study of abnormal behaviors and altered gabaergic signaling in the vpa-treated rat model of autism [Original Research]. Front Behav Neurosci 12. https://doi.org/10.3389/fnbeh.2018.00182

Ito M, Kinjo T, Seki T, Horie J, Suzuki T (2021) The long-term prognosis of hippocampal neurogenesis and behavioral changes of offspring from rats exposed to valproic acid during pregnancy. Neuropsychopharmacol Rep 41(2):260–264. https://doi.org/10.1002/npr2.12181

Jazayeri D, Braine E, McDonald S, Dworkin S, Powell KL, Griggs K, Vajda FJE, O'Brien TJ, Jones NC (2020) A rat model of valproate teratogenicity from chronic oral treatment during pregnancy. Epilepsia 61(6):1291–1300. https://doi.org/10.1111/epi.16536

Johannessen CU (2000) Mechanisms of action of valproate: a commentatory. Neurochem Int 37(2): 103–110. https://doi.org/10.1016/S0197-0186(00)00013-9

Juliandi B, Tanemura K, Igarashi K, Tominaga T, Furukawa Y, Otsuka M, Moriyama N, Ikegami D, Abematsu M, Sanosaka T, Tsujimura K, Narita M, Kanno J, Nakashima K (2015) Reduced adult hippocampal neurogenesis and cognitive impairments following prenatal treatment of the antiepileptic drug valproic acid. Stem Cell Rep 5(6):996–1009. https://doi.org/10.1016/j.stemcr.2015.10.012

Kataoka S, Takuma K, Hara Y, Maeda Y, Ago Y, Matsuda T (2013) Autism-like behaviours with transient histone hyperacetylation in mice treated prenatally with valproic acid. Int J Neuropsychopharmacol 16(1):91–103. https://doi.org/10.1017/s1461145711001714

Khan S, Ahmad T, Parekh CV, Trivedi PP, Kushwaha S, Jena G (2011) Investigation on sodium valproate induced germ cell damage, oxidative stress and genotoxicity in male Swiss mice. Reprod Toxicol 32(4):385–394. https://doi.org/10.1016/j.reprotox.2011.09.007

Kim KC, Kim P, Go HS, Choi CS, Yang S-I, Cheong JH, Shin CY, Ko KH (2011) The critical period of valproate exposure to induce autistic symptoms in Sprague–Dawley rats. Toxicol Lett 201(2):137–142. https://doi.org/10.1016/j.toxlet.2010.12.018

Kinjo T, Ito M, Seki T, Fukuhara T, Bolati K, Arai H, Suzuki T (2019) Prenatal exposure to valproic acid is associated with altered neurocognitive function and neurogenesis in the dentate gyrus of male offspring rats. Brain Res 1723:146403. https://doi.org/10.1016/j.brainres.2019.146403

Larner O, Roberts J, Twiss J, Freeman L (2021) A need for consistency in behavioral phenotyping for asd: Analysis of the valproic acid model. Autism Res Treat 2021:8863256. https://doi.org/10.1155/2021/8863256

Lloyd KA (2013) A scientific review: mechanisms of valproate-mediated teratogenesis. Biosci Horiz Int J Student Res 6. https://doi.org/10.1093/biohorizons/hzt003

López-Muñoz F, Shen WW, D'Ocon P, Romero A, Álamo C (2018) A history of the pharmacological treatment of bipolar disorder. Int J Mol Sci 19(7):2143. https://doi.org/10.3390/ijms19072143

Löscher W (2002) Basic pharmacology of valproate. CNS Drugs 16(10):669–694. https://doi.org/10.2165/00023210-200216100-00003

Lotufo Denucci B, Silva de Lima L, Ferreira Lima Mota L, Rocha Madureira Azevedo J, Germino Veras L, Montenegro Luzardo Bicca JV, de Miranda Santana B, Beserra Pinheiro G, Gonçalves Coelho G, Mortari MR (2021) Current knowledge, challenges, new perspectives of the study, and treatments of Autism Spectrum Disorder. Reprod Toxicol 106:82–93. https://doi.org/10.1016/j.reprotox.2021.10.010

Main SL, Kulesza RJ (2017) Repeated prenatal exposure to valproic acid results in cerebellar hypoplasia and ataxia. Neuroscience 340:34–47. https://doi.org/10.1016/j.neuroscience.2016.10.052

Markram K, Rinaldi T, Mendola DL, Sandi C, Markram H (2008) Abnormal fear conditioning and amygdala processing in an animal model of autism. Neuropsychopharmacology 33(4):901–912. https://doi.org/10.1038/sj.npp.1301453

Mehra S, Ul Ahsan A, Seth E, Chopra M (2022) Critical evaluation of valproic acid-induced rodent models of autism: Current and future perspectives. J Mol Neurosci 72(6):1259–1273. https://doi.org/10.1007/s12031-022-02033-7

Michaelis M, Suhan T, Michaelis UR, Beek K, Rothweiler F, Tausch L, Werz O, Eikel D, Zörnig M, Nau H, Fleming I, Doerr HW, Cinatl J (2006) Valproic acid induces extracellular signal-regulated kinase 1/2 activation and inhibits apoptosis in endothelial cells. Cell Death Differ 13(3):446–453. https://doi.org/10.1038/sj.cdd.4401759

Mony TJ, Lee JW, Dreyfus C, DiCicco-Bloom E, Lee HJ (2016) Valproic acid exposure during early postnatal gliogenesis leads to autistic-like behaviors in rats. Clin Psychopharmacol Neurosci 14(4):338–344. https://doi.org/10.9758/cpn.2016.14.4.338

Moore SJ, Turnpenny P, Quinn A, Glover S, Lloyd DJ, Montgomery T, Dean JC (2000) A clinical study of 57 children with fetal anticonvulsant syndromes. J Med Genet 37(7):489–497. https://doi.org/10.1136/jmg.37.7.489

Na L, Wartenberg M, Nau H, Hescheler J, Sauer H (2003) Anticonvulsant valproic acid inhibits cardiomyocyte differentiation of embryonic stem cells by increasing intracellular levels of reactive oxygen species. Birth Defects Res A Clin Mol Teratol 67(3):174–180. https://doi.org/10.1002/bdra.10030

Narita M, Oyabu A, Imura Y, Kamada N, Yokoyama T, Tano K, Uchida A, Narita N (2010) Nonexploratory movement and behavioral alterations in a thalidomide or valproic acid-induced autism model rat. Neurosci Res 66(1):2–6. https://doi.org/10.1016/j.neures.2009.09.001

Nau H, Koch S, Häuser I, Helge H (1981a) Valproic acid and its metabolites: placental transfer, neonatal pharmacokinetics, transfer via mother's milk and clinical status in neonates of epileptic mothers. J Pharmacol Exp Ther 219(3):768–777. https://jpet.aspetjournals.org/content/219/3/768.long

Nau H, Zierer R, Spielmann H, Neubert D, Gansau C (1981b) A new model for embryotoxicity testing: teratogenicity and pharmacokinetics of valproic acid following constant-rate administration in the mouse using human therapeutic drug and metabolite concentrations. Life Sci 29(26):2803–2813

Norton SA, Gifford JJ, Pawlak AP, Derbaly A, Sherman SL, Zhang H, Wagner GC, Kusnecov AW (2020) Long-lasting behavioral and neuroanatomical effects of postnatal valproic acid treatment. Neuroscience 434:8–21. https://doi.org/10.1016/j.neuroscience.2020.02.029

Oguchi-Katayama A, Monma A, Sekino Y, Moriguchi T, Sato K (2013) Comparative gene expression analysis of the amygdala in autistic rat models produced by pre- and post-natal exposures to valproic acid. J Toxicol Sci 38(3):391–402. https://doi.org/10.2131/jts.38.391

Olde Loohuis N, Kole K, Glennon J, Karel P, Van der Borg G, Van Gemert Y, Shahabipour F (2015) Elevated microRNA-181c and microRNA-30d levels in the enlarged amygdala of the valproic acid rat model of autism. Neurobiol Dis 80:42–53

Olde Loohuis NFM, Martens GJM, van Bokhoven H, Kaplan BB, Homberg JR, Aschrafi A (2017) Altered expression of circadian rhythm and extracellular matrix genes in the medial prefrontal cortex of a valproic acid rat model of autism. Prog Neuro-Psychopharmacol Biol Psychiatry 77: 128–132. https://doi.org/10.1016/j.pnpbp.2017.04.009

Ornoy A, Echefu B, Becker M (2024) Valproic acid in pregnancy revisited: neurobehavioral, biochemical and molecular changes affecting the embryo and fetus in humans and in animals: A narrative review. Int J Mol Sci 25(1):390. https://doi.org/10.3390/ijms25010390

Owens MJ, Nemeroff CB (2003) Pharmacology of valproate. Psychopharmacol Bull 37(Suppl 2): 17–24. http://www.ncbi.nlm.nih.gov/pubmed/14624230

Pettie M (2020) When genetic and environmental factors meet in modelling autism

Phiel CJ, Zhang F, Huang EY, Guenther MG, Lazar MA, Klein PS (2001) Histone deacetylase is a direct target of valproic acid, a potent anticonvulsant, mood stabilizer, and teratogen *. J Biol Chem 276(39):36734–36741. https://doi.org/10.1074/jbc.M101287200

Pragnya B, Kameshwari JSL, Veeresh B (2014) Ameliorating effect of piperine on behavioral abnormalities and oxidative markers in sodium valproate induced autism in BALB/C mice. Behav Brain Res 270:86–94. https://doi.org/10.1016/j.bbr.2014.04.045

Ranger P, Ellenbroek BA (2016) Perinatal influences of valproate on brain and behaviour: an animal model for autism. In: Kostrzewa RM, Archer T (eds) Neurotoxin modeling of brain disorders—life-long outcomes in behavioral teratology, vol 29. Springer International Publishing, pp 363–386. https://doi.org/10.1007/7854_2015_404

Rasalam A, Hailey H, Williams JHG, Moore SJ, Turnpenny PD, Lloyd DJ, Dean JCS (2005) Characteristics of fetal anticonvulsant syndrome associated autistic disorder. Dev Med Child Neurol 47(8):551–555. https://doi.org/10.1111/j.1469-8749.2005.tb01190.x

Reynolds S, Millette A, Devine DP (2012) Sensory and motor characterization in the postnatal valproate rat model of autism. Dev Neurosci 34(2–3):258–267. https://doi.org/10.1159/000336646

Richens A, Ahmad S (1975) Controlled trial of sodium valproate in severe epilepsy. BMJ 4(5991): 255–256. https://doi.org/10.1136/bmj.4.5991.255

Rodier PM, Ingram JL, Tisdale B, Nelson S, Romano J (1996) Embryological origin for autism: Developmental anomalies of the cranial nerve motor nuclei. J Comp Neurol 370(2):247–261. https://doi.org/10.1002/(SICI)1096-9861(19960624)370:2<247::AID-CNE8>3.0.CO;2-2

Rodier PM, Ingram JL, Tisdale B, Croog VJ (1997) Linking etiologies in humans and animal models: Studies of autism. Reprod Toxicol 11(2–3):417–422. https://doi.org/10.1016/S0890-6238(97)80001-U

Roullet FI, Lai JKY, Foster JA (2013) In utero exposure to valproic acid and autism – a current review of clinical and animal studies. Neurotoxicol Teratol 36:47–56. https://doi.org/10.1016/j.ntt.2013.01.004. (Special Issue: Environmental Influences and Emerging Mechanisms in the Etiology of Autism)

Ruhela RK, Prakash A, Medhi B (2015) An urgent need for experimental animal model of autism in drug development. Ann Neurol 22:44–49

Sabers A, Bertelsen FCB, Scheel-Krüger J, Nyengaard JR, Møller A (2014) Long-term valproic acid exposure increases the number of neocortical neurons in the developing rat brain. A possible new animal model of autism. Neurosci Lett 580:12–16. https://doi.org/10.1016/j.neulet.2014.07.036

Schäfer M, Brandl EJ (2020) Mood stabilizers: Valproate. In: Riederer P, Laux G, Mulsant B, Le W, Nagatsu T (eds) NeuroPsychopharmacotherapy. Springer International Publishing, pp 1–9. https://doi.org/10.1007/978-3-319-56015-1_49-1

Schiavi S, Iezzi D, Manduca A, Leone S, Melancia F, Carbone C, Petrella M, Mannaioni G, Masi A, Trezza V (2019) Reward-related behavioral, neurochemical and electrophysiological changes in a rat model of autism based on prenatal exposure to valproic acid. Front Cell Neurosci 13:479. https://doi.org/10.3389/fncel.2019.00479

Schneider T, Przewłocki R (2005) Behavioral alterations in rats prenatally exposed to valproic acid: animal model of autism. Neuropsychopharmacology 30(1):80–89. https://doi.org/10.1038/sj.npp.1300518

Semmler A, Frisch C, Bleul C, Smith D, Bigler L, Prost J-C, Blom H, Linnebank M (2017) Intrauterine valproate exposure is associated with alterations in hippocampal cell numbers and folate metabolism in a rat model of valproate teratogenicity. Seizure 46:7–12. https://doi.org/10.1016/j.seizure.2017.01.003

Smith V, Brown N (2014) Prenatal valproate exposure and risk of autism spectrum disorders and childhood autism. Arch Dis Child Educ Pract Ed 99(5):198. https://doi.org/10.1136/archdischild-2013-305636

Stanton ME, Peloso E, Brown KL, Rodier P (2007) Discrimination learning and reversal of the conditioned eyeblink reflex in a rodent model of autism. Behav Brain Res 176(1):133–140. https://doi.org/10.1016/j.bbr.2006.10.022. (Animal Models for Autism)

Štefánik P, Olexová L, Kršková L (2015) Increased sociability and gene expression of oxytocin and its receptor in the brains of rats affected prenatally by valproic acid. Pharmacol Biochem Behav 131:42–50. https://doi.org/10.1016/j.pbb.2015.01.021

Tartaglione AM, Schiavi S, Calamandrei G, Trezza V (2019) Prenatal valproate in rodents as a tool to understand the neural underpinnings of social dysfunctions in autism spectrum disorder. Neuropharmacology 159:107477. https://doi.org/10.1016/j.neuropharm.2018.12.024. (The Neuropharmacology of Social Behavior: From Bench to Bedside)

Tomson T, Battino D, Perucca E (2016) Valproic acid after five decades of use in epilepsy: time to reconsider the indications of a time-honoured drug. Lancet Neurol 15(2):210–218. https://doi.org/10.1016/S1474-4422(15)00314-2

Tomson T, Marson A, Boon P, Canevini MP, Conavis A, Gaily E, Kaiviainen R, Trinka E (2015) Valproate in the treatment of epilepsy in girls and women of childbearing potential. Epilepsia 56:1006–1019

Tsujino N, Nakatani Y, Seki Y, Nakasato A, Nakamura M, Sugawara M, Arita H (2007) Abnormality of circadian rhythm accompanied by an increase in frontal cortex serotonin in animal model of autism. Neurosci Res 57(2):289–295. https://doi.org/10.1016/j.neures.2006.10.018

Tung EWY, Winn LM (2011) Valproic acid increases formation of reactive oxygen species and induces apoptosis in postimplantation embryos: a role for oxidative stress in valproic acid-induced neural tube defects. Mol Pharmacol 80(6):979–987. https://doi.org/10.1124/mol.111.072314

Turpin V, Schaffhauser M, Thabault M, Aubert A, Joffre C, Balado E, Longueville J-E, Francheteau M, Burucoa C, Pichon M, Layé S, Jaber M (2023) Mice prenatally exposed to valproic acid do not show autism-related disorders when fed with polyunsaturated fatty acid-enriched diets. Sci Rep 13(1):11235. https://doi.org/10.1038/s41598-023-38423-z

Vajda FJE, Graham JE, Hitchcock AA, Lander CM, O'Brien TJ, Eadie MJ (2019) Antiepileptic drugs and foetal malformation: analysis of 20 years of data in a pregnancy register. Seizure 65: 6–11. https://doi.org/10.1016/j.seizure.2018.12.006

Vakili Shahrbabaki SS, Jonaidi H, Sheibani V, Bashiri H (2022) Early postnatal handling alters social behavior, learning, and memory of pre- and postnatal VPA-induced rat models of autism in a context-based manner. Physiol Behav 249:113739. https://doi.org/10.1016/j.physbeh.2022.113739

VanDongen AMJ, VanErp MG, Voskuyl RA (1986) Valproate reduces excitability by blockage of sodium and potassium conductance. Epilepsia 27(3):177–182. https://www.ncbi.nlm.nih.gov/pubmed/3084227

Viinikainen K, Eriksson K, Mönkkönen A, Äikiä M, Nieminen P, Heinonen S, Kälviäinen R (2006) The effects of valproate exposure in utero on behavior and the need for educational support in school-aged children. Epilepsy Behav 9(4):636–640. https://doi.org/10.1016/j.yebeh.2006.08.007

Vorhees CV (1987) Teratogenicity and developmental toxicity of valproic acid in rats. Teratology 35(2):195–202. https://onlinelibrary.wiley.com/doi/10.1002/tera.1420350205

Wagner GC, Reuhl KR, Cheh M, McRae P, Halladay AK (2006) A new neurobehavioral model of autism in mice: pre- and postnatal exposure to sodium valproate. J Autism Dev Disord 36(6): 779–793. https://doi.org/10.1007/s10803-006-0117-y

Wang J, Zou L, Jiang P, Yao M, Xu Q, Hong Q, Zhu J, Chi X (2024) Vitamin A ameliorates valproic acid-induced autism-like symptoms in developing zebrafish larvae by attenuating oxidative stress and apoptosis. Neurotoxicology 101:93–101. https://doi.org/10.1016/j.neuro.2023.12.015

Watanabe S, Kurotani T, Oga T, Noguchi J, Isoda R, Nakagami A, Sakai K, Nakagaki K, Sumida K, Hoshino K, Saito K, Miyawaki I, Sekiguchi M, Wada K, Minamimoto T, Ichinohe N (2021) Functional and molecular characterization of a non-human primate model of autism spectrum disorder shows similarity with the human disease. Nat Commun 12(1):5388. https://doi.org/10.1038/s41467-021-25487-6

Wellmann KA, Varlinskaya EI, Mooney SM (2014) D-Cycloserine ameliorates social alterations that result from prenatal exposure to valproic acid. Brain Res Bull 108:1–9

Williams PG, Hersh JH (1997) A male with fetal valproate syndrome and autism. Dev Med Child Neurol 39(9):632–634. https://doi.org/10.1111/j.1469-8749.1997.tb07500.x

Yochum CL, Dowling P, Reuhl KR, Wagner GC, Ming X (2008) VPA-induced apoptosis and behavioral deficits in neonatal mice. Brain Res 1203:126–132. https://doi.org/10.1016/j.brainres.2008.01.055

Zhao H, Wang Q, Yan T, Zhang Y, Xu H-J, Yu H-P, Tu Z, Guo X, Jiang Y-H, Li X-J, Zhou H, Zhang YQ (2019) Maternal valproic acid exposure leads to neurogenesis defects and autism-like behaviors in non-human primates. Transl Psychiatry 9(1):267. https://doi.org/10.1038/s41398-019-0608-1

How Early-Life Antipsychotic Drug Administration Modifies Behavioral and Brain Function in Adulthood

Mark E. Bardgett

This chapter is dedicated to John G. Csernansky, M.D.

Contents

Abstract The use of antipsychotic drugs in children has greatly expanded over the past 30 years. This increase occurred despite a lack of research addressing how these drugs affect brain development. This review summarizes and synthesizes preclinical studies that have ascertained the short- and long-term consequences of prenatal and early postnatal antipsychotic drug administration. Antipsychotic drugs are well-known for their ability to block D_2-type dopamine receptors and this action features in two main themes that emerge from the literature. First, prenatal antipsychotic drug exposure generally leads to a long-term hypodopaminergic state while early postnatal antipsychotic drug exposure produces a hyperdopaminergic one. Second, the effects of postnatal antipsychotic drug exposure appear roughly similar regardless of

M. E. Bardgett (✉)
Department of Psychological Science, Northern Kentucky University, Highland Heights, KY, USA
e-mail: bardgettm@nku.edu

Curr Topics Behav Neurosci (2025) 75: 69–97
https://doi.org/10.1007/7854_2024_567
Published Online: 07 March 2025

age, but drug administration any time prior to puberty (pre or early postnatal) leaves a lasting imprint on neural and behavioral function that persists long after treatment cessation. The enduring brain and behavioral changes seen after early-life antipsychotic administration in animal models provide support for initiatives aimed at reducing the number of children treated with antipsychotics and limiting the dose and length of treatment for those that need them.

Keywords Atypical · Children · D_2 receptor · Dopamine · Forebrain · Haloperidol · Olanzapine · Postnatal · Prenatal · Rat · Risperidone · Typical

1 Introduction

This review describes how early-life administration of antipsychotic drugs might lead to persistent biochemical and behavioral anomalies later in life. To better appreciate if and how such changes occur, it is worth going back to when the first generation of antipsychotic drugs was introduced and consider how they evolved into the present generation of drugs more widely used today in children.

In adults, antipsychotic drugs are the predominant pharmacological treatment for schizophrenia and related psychotic disorders. Their efficacy was discovered in the early 1950s when chlorpromazine and, soon thereafter, more potent drugs such as haloperidol were demonstrated to profoundly reduce psychosis (Filippino and Csernansky 2004; Freedman 2005) via the blockade of dopamine D_2 receptors. Yet due to their antagonism of D_2 receptors, first-generation, "typical" antipsychotics produced parkinsonism, tardive dyskinesia, and excessive plasma prolactin levels (Filippino and Csernansky 2004) to such an aversive degree that treatment discontinuation by patients was common (Lieberman et al. 2005).

Not long after the introduction of chlorpromazine, another antipsychotic drug, clozapine, was synthesized and used in Europe. It was remarkably devoid of motor side-effects and alleviated symptoms in individuals resistant to the therapeutic effects of typical antipsychotics (Filippino and Csernansky 2004). Unfortunately, clozapine greatly diminished white cell counts (i.e., caused agranulocytosis) in some patients and was withdrawn from clinical use. It was reintroduced into practice in the late 1980s following guidelines aimed at reducing agranulocytosis (Kane et al. 1988; see Filippino and Csernansky 2004 for review).

Clozapine's "atypical" clinical profile stems from a unique pharmacological profile – it antagonizes both D_2 and serotonin $5HT_{2A}$ receptors. Given clozapine's efficacy and lack of motor side-effects, much drug development was aimed at identifying a clozapine-like antipsychotic that blocked D_2 and $5HT_{2A}$ receptors without eliciting agranulocytosis. This line of work led to a second generation of antipsychotic drugs, including olanzapine, quetiapine, ziprasidone, and risperidone, that were approved for use in the 1990s (Londino et al. 2004). While none were associated with agranulocytosis, these second-generation drugs did not completely emulate clozapine's clinical profile since some caused motor impairment and excessive weight gain at higher doses. Nonetheless, they have been portrayed as safer than first-generation antipsychotic drugs. Perhaps because of this impression, their clinical use expanded significantly beyond schizophrenia and into the treatment of affective and nonpsychotic disorders (Londino et al. 2004). Another group in which second-generation, atypical antipsychotic drugs became more widely used was children, mainly to manage the behavioral symptoms associated with autism, ADHD (attention-deficit hyperactivity disorder), and disruptive behavior disorders.

2 Increased Use of Antipsychotic Drugs in Children

In the late 1990s and early 2000s, the number of children and adolescents treated with an antipsychotic drug increased sixfold in the United States (Olfson et al. 2006). Similar trends were observed around the same time in other Western countries, such as the Netherlands (Kalverdijk et al. 2008) and Canada (Murphy et al. 2013; Ronsley et al. 2013). Even for children younger than seven years age, prescribing rates for antipsychotics doubled beginning around 2000 (Zito et al. 2007; Olfson et al. 2010; Lohr et al. 2015) with many of these children maintained on antipsychotics for lengthy periods of time (Kalverdijk et al. 2008; Constantine et al. 2012; Lohr et al. 2021). Over the past decade, antipsychotic use in children has continued to increase in Europe but leveled off in the USA, although the number of antipsychotic prescriptions in the USA remains relatively high compared to many other countries (Kalverdijk et al. 2017; Bushnell et al. 2021).

Most antipsychotics prescribed to kids are second-generation drugs; of these, risperidone is the most widely used (Kalverdijk et al. 2017). In youths, boys are much more likely to receive antipsychotics than girls (Kalverdijk et al. 2017; Bushnell et al. 2021). In the United States, the Food and Drug Administration (FDA) has approved risperidone and aripiprazole for the treatment of autism in children between the ages of 5 and 17 and quetiapine for the treatment of bipolar disorder in children between the ages of 10 and 17. However, antipsychotics are predominantly used in children for off-label purposes, as such as the treatment of attention-deficit hyperactivity disorder (ADHD) and disruptive behavior disorder, and mood disorders (Kalverdijk et al. 2017; Sultan et al. 2019; Bushnell et al. 2021; Lohr et al. 2021). In addition, many children receiving antipsychotic drugs are also prescribed other psychotherapeutic medications, such as stimulants and antidepressants (Olfson et al. 2015; Lohr et al. 2021).

Most clinical studies of antipsychotic drug use in children have focused on their impact on aggression, conduct problems, and irritability. This work has demonstrated that risperidone or aripiprazole significantly reduces these symptoms in autism spectrum disorders (Pandina et al. 2007; Vitiello et al. 2009; Ching and

Pringsheim 2012; Aman et al. 2015), ADHD, and disruptive behavior disorders (Loy et al. 2017; Breaux et al. 2023). In children with autism, symptom reduction can be achieved without dramatic cognitive impairment (Pandina et al. 2007; Aman et al. 2015; Farmer et al. 2017), a concern linked to the sedation produced by higher doses of antipsychotic drugs (Kent et al. 2013).

Despite their clinical benefits, the side-effect profile of second-generation antipsychotics in children is worrisome. Some children may possess a heightened sensitivity to such side-effects (e.g., Hammerman et al. 2008), and antipsychotic treatment in children may disrupt the development of important metabolic pathways in a way that persists long after treatment cessation. The most frequently noted adverse outcome is weight gain (Correll et al. 2009; Ching and Pringsheim 2012; Aman et al. 2015; Loy et al. 2017; Pillay et al. 2018), which has been associated in some but not all (Aman et al. 2015) studies with metabolic disruptions like changes in low-density lipoprotein and triglycerides levels (Correll et al. 2009; Pillay et al. 2018), reduced insulin resistance (Nicol et al. 2018), and a greater prevalence of type-2 diabetes (Galling et al. 2016). Both risperidone and aripiprazole are also more likely to produce extra-pyramidal motor symptoms, sedation, and somnolence in children (Pringsheim et al. 2011; Sharma and Shaw 2012; Ching and Pringsheim 2012; Pillay et al. 2018; Coustals et al. 2021). Most troubling of all, youths treated with antipsychotic drugs experience a higher rate of unexpected deaths (Ray et al. 2019). In response to these risks, several US states have implemented peer-reviewed, prior authorization panels to oversee antipsychotic prescriptions to children receiving publicly assisted health insurance (Zito et al. 2018). This approach has effectively stabilized or reduced the use of antipsychotics among youths in these states (Zito et al. 2018; Akincigil et al. 2020; Candon et al. 2021).

While research has addressed the clinical effects of antipsychotics in children, relatively less empirical effort has been directed at determining whether antipsychotic drug exposure early in life directly alters brain development. This can be partially attributed to the unregulated and complicated nature of antipsychotic use in children. The heterogeneity of disorders treated with antipsychotics and the various drugs, doses, treatment durations, and ages (i.e., the very young (below the age of six), school-age youths (6–12), and early adolescents) makes it difficult to ascertain how a given antipsychotic impacts brain maturation and behavioral and cognitive function during adulthood.

3 Animal Models of Brain Development and Antipsychotic Drug Action

In this regard, rodent models of human brain development offer a powerful tool for discerning how antipsychotic drugs may alter brain maturation. First, the effects of antipsychotics on behavior and brain physiology in adult rats have been extensively characterized. This literature has been useful in predicting the clinical effects of antipsychotic drugs in adult humans (see Bardgett 2004 for review) and serves as a substantive reference for comparable studies in developing rats aimed at predicting the long-term consequences of antipsychotic use in children (e.g., see studies by Moran-Gates et al. 2007 and Choi et al. 2010 where adult reference data are explicitly used in such comparisons). Second, rat brains are remarkably similar in structure and physiology to human brains, and the order of brain development in both species follows the same relative sequence (Andersen 2003; Brenhouse and Andersen 2011). Thus, data gained from studies of early-life antipsychotic drug exposure in rats should validly illustrate how such exposure may affect outcomes in adult humans exposed to antipsychotics as children.

As a first step towards such an understanding, it is important to appreciate the neural targets of antipsychotic drugs. Nearly all antipsychotic drugs antagonize D_2 receptors or, in the case of aripiprazole, act as partial agonists at these sites (Stahl and Shayegan 2003; Farah 2005). The blockade of D_2 receptors in the mesolimbic dopamine pathway extending from the dopaminergic cell bodies in the ventral tegmental area to their terminal regions in the nucleus accumbens and frontal cortex has been implicated in the antipsychotic efficacy of these drugs. Antagonism of D_2 receptors in the nigrostriatal pathway comprised of dopamine neurons in the substantia nigra and their terminal fields in the caudate putamen appears to contribute to the motor side-effects associated with many antipsychotic drugs. Risperdone and aripiprazole also possess appreciable affinity for $5HT_{2A}$, $5HT_{2C}$, and α_1 receptors, with activity at the latter receptor associated with a decreased risk of motor impairment relative to typical antipsychotic drugs (Stahl and Shayegan 2003).

To anticipate the types of behavioral changes one could expect to see after prolonged antipsychotic drug treatment early in life, it is worth noting that after cessation of chronic, daily antipsychotic administration to adult rodents, a host of behavioral changes can be seen including increases in baseline and dopamine-agonist elevated locomotor activity, stereotyped behavior, self-stimulation, and drug reward after treatment cessation (Seeger and Gardner 1979; Spear et al. 1980; Waddington and Gamble 1980; Boye and Rompré 2000; Samaha et al. 2008; Tadokoro et al. 2012). It is generally assumed that such behavioral changes reflect compensatory elevations in D_2 receptor sensitivity or density produced in response to the prolonged receptor blockade caused by chronic antipsychotic administration (Prosser et al. 1989; Tarazi et al. 2001; Moran-Gates et al. 2006; Choi et al. 2017; although see Tadokoro et al. 2012 regarding aripiprazole). A review by Muller and Seeman (1978) suggested that the compensatory changes observed in adult animals after treatment cessation are transient and linger for a period equivalent to the duration of daily drug administration. As described below, many studies of prenatal and early postnatal antipsychotic drug administration in animals reveal more enduring neural and behavioral changes that depend on the timing of exposure.

4 Effects of Prenatal Antipsychotic Drug Administration

4.1 Prenatal Haloperidol and Other Typical Antipsychotics

Initial concerns about the effects of antipsychotic drugs on brain development stemmed from their use during pregnancy or lactation. Much of this research predates the advent of second-generation antipsychotics and extends back to when the typical drug, haloperidol, was the most prescribed antipsychotic. At a gross anatomical level, haloperidol administration during gestation reduces body and brain weight by ~10% in rat offspring in a manner that persists into adulthood (Williams et al. 1992). Assessments of neurotransmitter receptors have revealed that prenatal haloperidol decreases D_2 receptor binding in the caudate and nucleus accumbens (Rosengarten and Friedhoff 1979; Scalzo et al. 1989a; although see Moon 1984), and D_1 (Scalzo et al. 1989a) and μ-opioid receptors in the caudate (Moon 1984). (See Table 1 for a summary of results from these and other cited studies of chronic antipsychotic administration on forebrain D_1 and D_2 receptor densities.) These decreases in receptor density appear to be mainly confined to postsynaptic sites (Scalzo et al. 1989b) and occur in the absence of changes in striatal dopamine content or turnover (Scalzo et al. 1989a), although an earlier study found that

Table 1 Effects of chronic prenatal, early postnatal, and adult antipsychotics on forebrain dopamine D_1 and D_2 receptor density

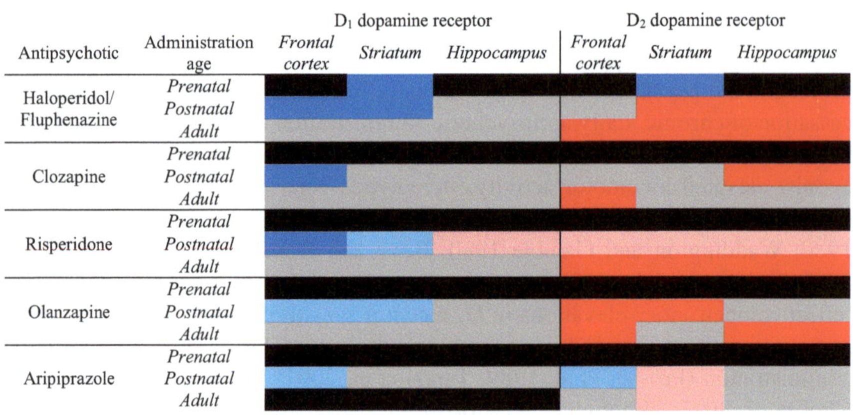

Colors represent increases (red), decreases (blue), or no change (gray) in receptor density based on previous reports. The lighter red and light blue boxes indicate areas where most but not all reports support respective increases or decreases in receptor density. Black boxes indicate areas where no data exist. Data based on Bardgett et al. (2024), Choi et al. (2017, 2019); Cuomo et al. (1981, 1983a); De Santis et al. (2016); Lian et al. (2016); Milstein et al. (2013); Moon (1984); Moran-Gates et al. (2006, 2007); Rosengarten and Friedhoff (1979): Scalzo et al. (1989a); Tadokoro et al. (2012); Varela et al. (2014); Vinish et al. (2013)

combined prenatal and postnatal thioproperazine administration in rats reduces striatal dopamine release (Velley et al. 1975).

Forebrain dopamine function after prenatal haloperidol has been assessed mainly by testing the ability of the D_1/D_2 receptor agonist, apomorphine, to elicit stereotyped behavior (e.g., licking, sniffing, biting) and modify locomotor activity. Apomorphine produces a bi-phasic, dose-dependent effect on ongoing behavior. Immediately after drug administration, locomotion is decreased, presumably due to the blunting of dopamine release by the stimulation of high-affinity D_2 autoreceptors on dopamine terminals. Thereafter, activity levels increase followed by bouts of stereotypy – these effects may reflect delayed occupation of lower affinity, postsynaptic dopamine receptors (Cuomo et al. 1985; McDougall et al. 2013). Rats exposed to haloperidol prior to birth are less sensitive to apomorphine-induced stereotypy, which aligns with the decreased number of postsynaptic receptors found in these animals (Rosengarten and Friedhoff 1979: Cuomo et al. 1985). Pituitary and plasma prolactin levels, another marker of dopamine function, are increased after weaning and near puberty in rats exposed to prenatal haloperidol (Iñiguez et al. 1990). Since prolactin production and release are inhibited by D_2 receptors (Iñiguez et al. 1990), this effect, like the decreased sensitivity to apomorphine, may stem from fewer D_2 receptors in the offspring of pregnant dams administered haloperidol.

Prenatal haloperidol has been found to alter other aspects of behavior and neuronal function in several species. Mice, rats, and even zebra fish exposed to haloperidol during embryonic or prenatal stages of development are more activity during adulthood (Dallemagne and Weiss 1982; Scalzo et al. 1989a; Oliveri and Levin 2019), although a period of hypoactivity may be seen earlier in postnatal development (Scalzo et al. 1989a) or in different phases of the circadian light cycle (Oliveri and Levin 2019). Increased anxiety has also been observed in studies of adult rats (Singh and Singh 2002), but not adult mice (Wang et al. 2019), exposed to gestational haloperidol. Suggestive of a more nuanced emotional effect, Cagiano et al. (1988) found that prenatal haloperidol leads to more vigorous and prolonged ultrasonic vocalizations during sexual behavior in rat offspring. Cognitive studies have shown that prenatal haloperidol does not affect fixed ratio operant responding (Dallemagne and Weiss 1982) but impairs object recognition memory (Wang et al. 2019) in mice, and delays the acquisition of an impulse control task (differential reinforcement of low rates responding) in rats (Cuomo et al. 1981). The object recognition deficit was accompanied by neuronal changes such as decreased proliferation and maturation in adult-born granule cells and reduced dendritic length and branching in granule cells within the hippocampus (Wang et al. 2019), a region crucial to this form of memory. Since dendritic growth seen within dopamine circuits early in development can depend on D_2 receptors (Levitt et al. 1997), the D_2 antagonism invoked by haloperidol at this time may limit dendritic elaboration.

4.2 Prenatal Second-Generation Antipsychotics

With the introduction of second-generation, atypical antipsychotics in the 1990s, later studies determined if and how maternal treatment with these drugs affected brain and behavioral development. Like haloperidol, prenatal risperidone reduces fetal and postnatal body and brain weights (Zuo et al. 2008; Singh et al. 2016) with specific decreases in cortical and hippocampal thickness and striatal volume reported at birth (Singh et al. 2016; Singh and Singh 2017). Depending on the dose and the type of behavioral assay, adult rats exposed to risperidone before birth appear more active and less anxious than controls (Zuo et al. 2008; Singh and Singh 2017), and exhibit deficiencies in spatial learning and active/passive avoidance (Singh et al. 2016). Prenatal risperidone in mice leads to similar outcomes in adulthood, such as decrements in recognition memory, hippocampal neurogenesis, and dendritic length and spine density in hippocampal granule cells, but does not affect locomotor activity, anxiety, or hippocampal N-methyl-D-aspartate (NMDA) receptor expression (Wang et al. 2019).

Singh and Tripathi (2015) found that prenatal quetiapine produces many of the same effects of prenatal risperidone: reduced postnatal body weight gain, impairments in water maze learning, decreases in fetal hippocampal thickness with high doses (although low doses were associated with increased thickness), and fewer hippocampal pyramidal cells. Notably, offspring exposed to a high dose of quetiapine (100 mg/kg) during pregnancy failed to thrive after birth and died prior to weaning (Singh and Tripathi 2015). A similar study (Singh and Tripathi 2014) of prenatal aripiprazole found lower fetal and postnatal body weights in treated offspring. To date, there are no published reports on the impact of prenatal risperidone, aripiprazole, or quetiapine on indices of dopamine neurotransmission or function.

It should be noted that recent large-scale, population-based studies in humans (Wang et al. 2021; Huybrechts et al. 2023) have not found dramatic evidence for a greater likelihood of preterm birth or congenital malformations, reduced birth weight, or altered incidences of ADHD or autism in offspring gestationally exposed to first- or second-generation antipsychotics. Given that prenatal antipsychotics can reduce body weights or other gross measures in animals, the above data from humans suggest important species and dosing differences. However, even though antipsychotic treatment during pregnancy appears to be relatively safe in human offspring, it would be worth investigating whether it is associated with any later-life changes in brain and cognitive function like those revealed in animal models.

5 Effects of Postnatal Antipsychotic Administration

5.1 Postnatal Haloperidol and Other Typical Antipsychotics

Initial research on the effects of antipsychotic drug exposure during early postnatal life was principally driven by concerns about drug exposure via breast milk. In rat models, these concerns were addressed by providing nursing dams with antipsychotics either through drinking water or via direct injection, or by subcutaneous injections of the pups. Early research in this area was primarily focused on the effects of haloperidol.

Rat studies conducted by Spear and colleagues found that a combination of gestational and lactational haloperidol altered many behavioral and biochemical outcomes in offspring. Soon after weaning on postnatal days 23–24, such rats were hyperactive and more sensitive to the cataleptic effects of haloperidol but less sensitive to the locomotor stimulant effects of the dopamine/norepinephrine agonist, d-amphetamine (Spear et al. 1980). These effects disappeared by early adolescence (postnatal days 35–36) but reemerged in late adolescence/early adulthood (postnatal days 47–48). Further work (Shalaby and Spear 1980) revealed that adult offspring similarly exposed to haloperidol displayed less sensitivity to the hypolocomotive effects of a low dose of apomorphine, suggesting that early-life haloperidol exposure decreases D_2 autoreceptor function. In support of the former claim, Scalzo and Spear (1985) also found that adult rats exposed to haloperidol from postnatal days 1 to 20 were insensitive to the dopamine-elevating effects of gamma-butyrolactone, which increases striatal dopamine content by suppressing autoreceptor function. In contrast to these pre/postnatal haloperidol effects, chronic haloperidol administration during adulthood resulted in greater locomotor sensitivity to amphetamine, less catalepsy after acute haloperidol, and enhanced dopamine autoreceptor function (Spear et al. 1980), suggesting that neuronal responses to chronic haloperidol administration spanning pre- to early postnatal ages are distinct from those seen during adulthood.

Around the same time, others began showing contrasting effects of prenatal versus early postnatal exposure to haloperidol. As mentioned previously, rat offspring exposed to haloperidol during gestation were found to possess fewer striatal D_2 receptors and less behavioral sensitivity to apomorphine (Rosengarten and Friedhoff 1979; Cuomo et al. 1985). On the other hand, those exposed to haloperidol after birth either via lactation (Rosengarten and Friedhoff 1979) or direct injections (Cuomo et al. 1981, 1983c) displayed higher densities of striatal D_2 receptors (Table 1). Rats administered fluphenazine from postnatal days 21 to 42 also showed more D_2 and D_4 receptors in the nucleus accumbens and dorsal striatum, more D_2 receptors in the hippocampus, and fewer D_1 receptors in medial prefrontal and dorsolateral cortex and nucleus accumbens (Moran-Gates et al. 2006). Early postnatal haloperidol exposure has also been associated with greater sensitivity to the respective stimulating and suppressing effects of apomorphine on stereotypy and locomotion (Rosengarten and Friedhoff 1979; Cuomo et al. 1981, 1983b). On the other hand, dopamine turnover evoked by a later haloperidol challenge is reduced by such exposure suggesting that the changes in D_2 receptor sensitivity seen after early-postnatal haloperidol may not be uniform (Cuomo et al. 1981).

Since combined pre- and postnatal haloperidol (Shalaby and Spear 1980) produces a dopaminergic phenotype more aligned with the effects of prenatal haloperidol (Rosengarten and Friedhoff 1979; Cuomo et al. 1985), prenatal treatment must impact the development of dopaminergic circuitry (see Levitt et al. 1997; Andersen

2003; Brenhouse and Andersen 2011 for reviews) in a manner that is impervious to later reversal by postnatal treatment. The outcome of postnatal haloperidol better matches that of haloperidol administration during adulthood (although see Spear et al. 1980), but the effects of early postnatal treatment appear more persistent than the transitory ones seen after adult treatment (see Cuomo et al. 1983c for review). Finally, not all consequences of pre- and postnatal haloperidol differ since either treatment can delay the acquisition of an impulse control task (differential reinforcement of low rates (DRL) responding) in adulthood (Cuomo et al. 1981, 1985).

5.2 *Postnatal Clozapine*

While much of the early focus on the neurodevelopmental effects of antipsychotics centered on haloperidol, at least one study (Cuomo et al. 1983a) from the same time investigated the effects of early-life clozapine on later behavior. Rat pups administered daily clozapine from postnatal days 1 through 20 demonstrated typical stereotypy and locomotor responses to apomorphine at postnatal day 60. However, these rats exhibited greater basal levels of the dopamine metabolite, homovanillic acid, and were delayed in their acquisition of a DRL task in a manner similar to that produced by early-life haloperidol (Cuomo et al. 1981, 1985). Later work (Moran-Gates et al. 2006) showed that daily clozapine administration from postnatal days 21 to 42 lowered D_1 receptor density in the medial prefrontal and dorsal lateral cortex, increased D_2 receptor density in the hippocampus, and increased D_4 receptor density in the nucleus accumbens and dorsal striatum when assessed one day after treatment cessation (Table 1). The effects of clozapine on D_1 and D_2 receptors were specific to neurodevelopment since they were not seen in adult rats one day after the cessation of chronic clozapine administration (Moran-Gates et al. 2006). The decreased D_1 receptor density in the prefrontal cortex may explain the DRL delays reported by Cuomo et al. (1983a) since dopamine depletion in the frontal cortex also disrupts DRL performance (Sokolowski and Salamone 1994).

5.3 *Postnatal Second-Generation Antipsychotics*

As mentioned above, when second-generation antipsychotic drugs were introduced in the 1990s, they were perceived to be safer drugs due to their unique pharmacology and decreased side-effect liability. Because the use of these drugs in children increased dramatically through the 2000s, animal models of early-life antipsychotic drug treatment pivoted from exposure of nursing dams to first-generation drugs to direct administration of second-generation antipsychotics to rat offspring at ages analogous to human preschool years through early adolescence.

5.3.1 Effects on Dopamine Receptors and Neurotransmission

Since risperidone is the most widely prescribed antipsychotic in children (Kalverdijk et al. 2017), most preclinical research has assessed its impact on brain development. Tarazi and colleagues (Moran-Gates et al. 2007) were among the first to characterize the effects of early-life risperidone on dopamine receptor density in young adult, male rats (Table 1). In their studies, three risperidone doses were administered daily to male rats from postnatal days 21 to 42 and receptor density assessed 24 h after the last injection. Overall, early-life risperidone elevated dopamine receptor density in various forebrain regions. Specifically, the two higher risperidone doses increased D_1 receptors in the nucleus accumbens and dorsal striatum and D_2 receptors in the nucleus accumbens, while the highest dose alone increased D_2 receptors in the dorsal striatum. All three doses increased D_2 receptor density in the medial prefrontal cortex and hippocampus, and D_4 receptor density in nucleus accumbens, dorsal striatum, and hippocampus. Each effect was also seen after chronic administration of the highest risperidone dose to adult rats except for the elevation of forebrain D_1 receptors.

Deng and colleagues (De Santis et al. 2016; Lian et al. 2016) have measured dopamine receptor density at two or 60 days after the cessation of early-life risperidone administration to female and male rats. In both studies, a single risperidone dose (0.3 mg/kg) was provided three times a day from either postnatal days 22 to 50 (De Santis et al. 2016) or 23 to 43 (Lian et al. 2016) via an oral route that may have best approximated the 1.0 mg/kg dose used once a day by Moran-Gates et al. (2007). Two days after treatment cessation, D_1 receptor density was decreased in the medial prefrontal and cingulate cortex of male rats administered risperidone, with no effects on D_1 or D_2 receptors or dopamine transporters observed elsewhere in the forebrain (Lian et al. 2016). These results differ (e.g., no change in D_2 receptor density, decreased frontal cortical D_1 receptors) from those reported by Moran-Gates et al. (2007) perhaps owing to the larger, single doses of risperidone used in the latter study as well as differences in the route of administration, the timing of tissue collection (one day vs. two days) after the last treatment, or the inclusion of female rats.

When assessed approximately two months after cessation of early-life risperidone (De Santis et al. 2016), rats administered risperidone exhibited more D_1 receptors in the hippocampus (females only) but fewer D_1 receptors in the nucleus accumbens (males only). There was no effect of early-life risperidone on forebrain D_2 receptor density. Levels of phosphorylated tyrosine hydroxylase were elevated in the prefrontal cortex and hippocampus of male rats previously administered risperidone. Dopamine transporters were increased in the ventral tegmental area of the risperidone rats but decreased in the striatum. We have also found fewer striatal dopamine transporters near postnatal day 60 in rats administered risperidone from postnatal days 14 to 42 (Bardgett et al. 2020). However, we have also seen more striatal D_2 receptors in such animals (Bardgett et al. 2024) unlike the lack of an effect reported by De Santis et al. (2016). This difference may be due to the higher risperidone dose

(3.0 mg/kg) and the shorter posttreatment time-period (two weeks vs. two months) used in our study.

Other work has assessed the effects of early-life olanzapine administration on forebrain dopamine function in adulthood. Daily administration of 5 mg/kg of olanzapine to male rats from postnatal days 21 to 42 decreases D_1 receptor density in the medial prefrontal and dorsolateral cortex when measured 24 h after the last injection (Moran-Gates et al. 2006). An identical regimen of olanzapine administration to adult rats did not elicit such changes. Olanzapine administration to either young or adult rats increased D_2 and D_4 receptor densities in the nucleus accumbens and medial and lateral striatum, whereas adult administration led exclusively to more D_2 receptors in the medial prefrontal cortex. There was no effect of olanzapine at either age on forebrain D3 receptors.

A similar study (Lian et al. 2016) assessed dopamine receptors in female and male rats administered 3 mg/kg of oral olanzapine daily from postnatal days 23 to 43. When measured 48 h after the last olanzapine injection, male rats exhibited more D_1 receptors in the cingulate cortex and more D_2 receptors in the nucleus accumbens, whereas female rats had fewer D_2 receptors in the nucleus accumbens. When assessed eight weeks after the end of a similar regimen of olanzapine, forebrain D_1 receptor density was unaffected but D_2 receptors in the prefrontal cortex and nucleus accumbens were elevated (De Santis et al. 2016). Two additional studies assessing adult male rats administered oral olanzapine (7.5 mg/day) from postnatal days 28 to 49 (Milstein et al. 2013; Vinish et al. 2013) also found more D_2 receptors in the prefrontal cortex and nucleus accumbens at 180 days of age, but fewer D_1 receptors in these same regions plus the orbitofrontal cortex. Additionally, Vinish et al. (2013) reported that evoked dopamine release in the nucleus accumbens was reduced in adult rats administered olanzapine early in life. Notwithstanding the decrease in D_2 receptors seen soon after treatment cessation in female rats (Lian et al. 2016), early-life olanzapine appears establish a long-lasting increase in D_2 receptors and a more variable diminution of D_1 receptors in the nucleus accumbens and prefrontal cortex. If the elevations in D_2 receptors represent greater autoreceptor function, then they may explain the reduced dopamine release (Vinish et al. 2013) seen in rats administered olanzapine early in life.

The impact of early postnatal aripiprazole administration on dopaminergic neurotransmission has been investigated in several studies. Striatal D_2 receptors are increased, and D_2-mediated suppression of locomotor activity is more pronounced in male and female rats up to eight days after the cessation of daily aripiprazole (or haloperidol) injections from postnatal days 10 to 20 (Varela et al. 2014). Oral aripiprazole administration from postnatal days 23 to 43 also leads to more D_2 receptors in the nucleus accumbens and dorsal striatum of female rats but fewer D_1 receptors in the prefrontal cortex of male rats when assessed 48 h posttreatment (Lian et al. 2016). However, when assessed eight weeks after treatment cessation, there are no dramatic, long-lasting effects of aripiprazole on dopamine receptors or transporters, although D_1 receptors in the ventral tegmental area and phosphorylated tyrosine hydroxylase in the prefrontal cortex are decreased in female rats, and tyrosine hydroxylase in the ventral tegmental area of female and male rats is reduced

(De Santis et al. 2016). The effects of adult aripiprazole administration on D_2 receptor density are mixed. Chronic daily administration of aripiprazole just after puberty (from postnatal days 36 to 56) leads to fewer D_2 receptors in the prefrontal cortex and hippocampus when measured approximately four weeks after the last injection (Choi et al. 2019). Chronic administration later in adulthood increases cortical, striatal, and hippocampal D_2 receptors when measured one day after treatment cessation (Choi et al. 2017) but does not affect striatal D_2 receptor levels one week after treatment (Tadokoro et al. 2012). These studies suggest D_2 receptor density is temporarily elevated upon cessation of aripiprazole regardless of age (Varela et al. 2014; Lian et al. 2016; Choi et al. 2017). With time, these receptors then return to basal levels in rats administered aripiprazole early in postnatal life (De Santis et al. 2016) or return to baseline levels (Tadokoro et al. 2012) or decrease (Choi et al. 2019) in a region-dependent manner in rats administered aripiprazole during adulthood.

5.3.2 Effects on Dopamine-Related Behaviors

The research described above suggests that early-life exposure to some second-generation, atypical antipsychotics modifies forebrain dopamine function in a persistent manner. These changes raise an obvious question as to whether behaviors and cognitive processes linked to dopamine are similarly affected. One of the key behavioral measures used to answer this question has been spontaneous locomotion, a construct comprised of ambulatory movements, rearing, and sniffing observed in novel and familiar environments. This amalgamation of behaviors is highly sensitive to dopamine receptor modulation. In our own studies (Bardgett et al. 2013, 2019, 2024; Stevens et al. 2016; Stubbeman et al. 2017), we have consistently observed increased locomotor activity in adult female and male rats that received daily injections of risperidone (either 1.0 or 3.0 mg/kg) from postnatal days 14 to 42. This locomotor hyperactivity has been observed as late as nine months of age (Bardgett et al. 2013), during both dark and light phases of the circadian cycle (Stubbeman et al. 2017), and in the absence of changes in stereotypy (Stubbeman et al. 2017).

Elevated locomotor activity can also be seen in rats administered daily, chronic risperidone during adulthood (Stevens et al. 2016). However, the emergence and persistence of hyperlocomotion past the end of risperidone administration differs as a function of early-life or adult exposure. Adult rats administered risperidone daily from postnatal days 74 to 102 display increased (compensatory?) activity just 23 h after the first risperidone injection and continue to demonstrate this delayed hyperactivity throughout the injection period. Developing rats administered risperidone daily from postnatal days 14 to 42 do not exhibit this pattern of delayed hyperactivity until the third week (postnatal day 36) of injections. Since D_2 receptors are functional in early postnatal life (Varela et al. 2014), our data suggest that compensatory behaviors linked to receptor blockade either take a longer time (three weeks vs. one day) to emerge from maturing brains or can only be established after puberty.

Regardless, this same study showed that at one-week post-risperidone administration, only rats administered risperidone early in life continued to show increased activity. This finding provides further evidence that early-life antipsychotics induce enduring changes in brain function unlike the transient ones seen after treatment in adulthood.

The increased activity observed in adult rats administered risperidone early in life mirrors the reported effects of early postnatal haloperidol (Spear et al. 1980) but not those of clozapine (Cuomo et al. 1983a), suggesting a more typical antipsychotic profile for risperidone early in life. Others have also found that early-life risperidone leads to greater activity during adulthood (De Santis et al. 2016), although this effect was not observed in adult rats chronically administered risperidone just after puberty (Choi et al. 2019), perhaps suggesting that prepubertal risperidone exposure has a more enduring effect on behavior. In a preliminary study from our lab (Bardgett et al. 2013), rats that received risperidone from postnatal days 14 to 28 or 14 to 42 were more active between two and five months of age than those that received risperidone from days 28 to 42, suggesting a sensitive period prior to puberty. Single dose studies of early-life olanzapine and aripiprazole (Milstein et al. 2013; De Santis et al. 2016) have not found enduring locomotor effects in adult rats, indicating a more clozapine-like profile for these drugs (Cuomo et al. 1983a), although the initiation of treatment in the third week of life may have missed an earlier sensitive period.

The elevated locomotor activity observed in adult rats administered risperidone early in life is consistent with a shift in dopaminergic tone (e.g., increased D_2 receptors (Moran-Gates et al. 2007; Bardgett et al. 2024)) but does not serve as direct evidence of it. To better assay the impact of early-life risperidone on dopamine function during adulthood, we have assessed the effects of apomorphine and the D_2/D_3 agonist, quinpirole, on locomotor activity and stereotypy. Risperidone rats were more sensitive to the locomotor-elevating effects of apomorphine and quinpirole and the stereotypy-inducing effects of quinpirole (Bardgett et al. 2024). These results are consistent with the effects of early postnatal haloperidol administration (Cuomo et al. 1981). The ability of early-life risperidone and haloperidol, but not clozapine, to elevate forebrain D_2 receptor density (Moran-Gates et al. 2006, 2007; Bardgett et al. 2024) could explain the enhanced sensitivity to apomorphine and quinpirole observed in our study (Bardgett et al. 2024).

Adult rats administered risperidone early in life are also more sensitive to the locomotor-activating effects of amphetamine (Stubbeman et al. 2017). Similar outcomes have been reported for rats administered haloperidol or aripiprazole from postnatal days 10 to 20, such that the effects of large amphetamine doses on activity were more pronounced between postnatal days 20 and 28 (Varela et al. 2014). Second, amphetamine conditioned place preference is enhanced by early-life risperidone (Bardgett et al. 2020) or olanzapine (Vinish et al. 2013). These enhancements may rely on shifts in forebrain dopamine tone since they were accompanied by increases in nucleus accumbens D_2 receptors (Vinish et al. 2013), decreases in nucleus accumbens dopamine release and D_1 receptors (Vinish et al. 2013) and fewer striatal dopamine transporters (Bardgett et al. 2020). However, not every behavioral effect of amphetamine is permanently modified by early-life antipsychotics since amphetamine-induced stereotypy and working memory deficits are not affected by early-life risperidone (Bardgett et al. 2019, 2020).

5.3.3 Effects on Non-Dopaminergic Receptors and Dendritic Morphology

Given the antagonism of other non-dopaminergic receptors by many atypical antipsychotics and the essential interactions between dopamine and other neurotransmitters, several studies have considered how early-life antipsychotic drug administration affects other neurotransmitter systems besides dopamine. Some of this work has focused on serotonin receptors, given that many antipsychotics possess higher affinities for $5HT_{2A}$ receptors than for D_2 receptors (Richelson and Souder 2000). Choi et al. (2010) administered daily injections of three risperidone doses to male rats from postnatal days 21 to 42 and assayed the density of different forebrain serotonin receptors on postnatal day 43. These treatments reduced $5HT_{2A}$ receptor densities in the prefrontal and dorsolateral cortex, as did a similar risperidone regimen administered during adulthood. On the other hand, early-life risperidone increased $5HT_{1A}$ receptors in the same areas plus the hippocampus, with the latter effect not seen after adult risperidone administration. A similar study by Lian et al. (2016) also found fewer $5HT_{2A}$ receptors in the prefrontal cortex of adult rats administered risperidone, olanzapine, or aripiprazole early in life, with additional reductions observed in the cingulate cortex, nucleus accumbens, and dorsal striatum of the olanzapine rats. Early-life olanzapine also reduced $5HT_{2C}$ receptors in the same brain regions. None of the drug treatments in this latter study altered forebrain serotonin transporter expression. To date, no studies have assessed the behavioral pharmacology of serotonin agonists or antagonists in animals chronically administered antipsychotics early in life.

Others have investigated the effects of early-life antipsychotic drug administration on forebrain glutamate and GABA receptors. Using a design identical to that employed in their study of serotonin receptors (Choi et al. 2010 – see last paragraph), Choi et al. (2009) found that chronic risperidone decreased NMDA receptor density throughout the striatum when administered to rats during development or adulthood. The latter group also demonstrated fewer NMDA receptors in the hippocampus. On the other hand, developmental risperidone alone increased AMPA receptor density in prefrontal cortex, whereas developmental and adult risperidone elevated AMPA receptors in the dorsal striatum. Conversely, Lian et al. (2016) reported that developmental risperidone or aripiprazole, but not olanzapine, elevated striatal NMDA receptor binding at two days posttreatment. The discrepant data regarding glutamate receptors may reflect drug- and dose-dependent effects, differences in the timing, duration and route of administration, and post-administration time-period.

Three weeks of daily aripiprazole (but not risperidone or olanzapine) administered between postnatal days 23 and 44 increases $GABA_A$ receptor density in the prefrontal cortex and striatum of rats when assessed two days after the end of treatment (Lian and Deng 2019). When assessed seven months posttreatment, a similar regimen of early-life olanzapine can also increase $GABA_A$ receptor density in the prefrontal cortex (Milstein et al. 2013). Further research into the effects of early-life antipsychotics on glutamate and GABA receptors and related behavioral function is warranted given their crucial role in synaptic physiology throughout the brain.

Not only can early-life antipsychotics alter the density of various neurotransmitter receptors, but it can modify neuronal morphology and dendritic spine density. Frost et al. (2010) examined dendritic spines and arbor in six-month-old mice that received twice daily injections of haloperidol or olanzapine from postnatal days 3 to 21. Dendritic length and branching, as well as the density of apical and basal spines, were decreased in the nucleus accumbens and the prefrontal, orbitofrontal, and parietal cortices of mice administered either drug. In rats, later administration of olanzapine from postnatal days 28 to 49 also reduces spine density in the prefrontal cortex when assessed immediately after treatment cessation (Milstein et al. 2013). When assessed eight months after treatment, there are fewer spines in the dentate gyrus but more spines in the prefrontal and orbitofrontal cortices of rats administered olanzapine (Milstein et al. 2013). Given that dendritic spines play a crucial role in synaptic connectivity, these data suggest that the long-term receptor blockade effected by early-life antipsychotic administration can modify such connectivity or overall neuronal arbor in a species-, age-, or brain region-dependent manner. Recent studies of psychedelic drugs that work through $5HT_{2A}$ receptors to promote dendritic plasticity (Vargas et al. 2023, see Roth and Gumpper 2023 for review) suggest a mechanism for olanzapine's effects on cortical spine density. Perhaps by blocking $5HT_{2A}$ receptors, early-life olanzapine leads to short-term decreases in cortical spine density followed by compensatory upregulation seen several months after the end of treatment.

5.3.4 Effects on Social, Emotional, and Cognitive Processes

Because antipsychotics target brain regions associated with cognition, emotion, and sociability, behavioral assays have been performed to assess the impact of early-life antipsychotic administration on each process. Social interactions between pups and dams and between conspecifics are altered during a course of early-life risperidone administration (Gannon et al. 2015). For several hours after drug injections, pups are profoundly sedated and make few contacts with their dams. But even at 23 h postinjection, pups administered risperidone continue to make less contact their dam and control litter mates, which raises concerns about disrupted dam-pup attachment leading to disordered social interactions later in life. For example, after weaning, rats administered risperidone are more active and make more nonsocial contacts with their cage-mates but exhibit less social grooming. Studies of social interaction during adulthood are needed to determine if these changes in social interaction associated with early-life risperidone are permanent. Such work is important given the approved use of risperidone for managing symptoms of autism, a disorder marked by reduced sociability.

The effects of early-life antipsychotic administration on affect and emotional behaviors during adulthood are mixed. Rats administered risperidone, aripiprazole, and olanzapine for three weeks beginning near the fourth or fifth week of life show typical patterns of exploratory behavior in the elevated plus maze, a common test of anxiety (Milstein et al. 2013; Choi et al. 2019). However, a similar study performed by De Santis et al. (2016) found that daily oral administration of aripiprazole and risperidone, but not olanzapine, from postnatal days 22 to 50 led to less anxiety-like behavior in the plus maze in male rats during adulthood. Adult male rats administered olanzapine early in life were also less likely to exhibit depression-like behavior (as assessed in the forced swim test), whereas adult female rats administered olanzapine or risperidone early in life displayed more depression-like behavior. These data illustrate the importance of considering the influence of sex when assessing responses to antipsychotic drugs and suggest sex-dependent effects of early-life antipsychotics on affective tone.

Several studies have determined whether early-life antipsychotic drug administration modifies learning and memory during adulthood. This interest is based, in part, on the idea that many of the brain regions that regulate cognitive control, working memory and spatial memory, such as the frontal cortex, hippocampus, and striatum, possess high densities of dopaminergic, serotonergic, and adrenergic receptors that are blocked by antipsychotics. Clinical research has shown that risperidone treatment for up to a year does not impair cognitive processing or recall memory in children with disruptive behavior disorder and intellectual impairment (Pandina et al. 2007, 2009) or with ADHD marked by severe aggression (Farmer et al. 2017). Further research assessing the impact of longer treatment durations or outcomes after cessation of chronic treatments would be informative.

Animal studies have provided evidence of small, yet enduring impairments in working memory after early-life antipsychotic administration. Male pig-tailed macaques administered risperidone or quetiapine during their second year of life demonstrate working memory-like perseveration deficits in a learning set task after drug discontinuation (Mandell et al. 2011). We have also seen working memory deficits in female and male adult rats administered risperidone daily from postnatal days 14 to 42 (Bardgett et al. 2019). Chronic oral olanzapine administration to male rats from postnatal days 28 to 49 also impairs working memory and delays the extinction of fear conditioning during adulthood but does not affect water maze-based spatial memory (Milstein et al. 2013). Conversely, chronic administration of aripiprazole, but not risperidone, later in puberty (postnatal days 36–56) has been linked to improved working memory during adulthood in male rats (Choi et al. 2019). Thus, it appears that chronic antipsychotic drug administration prior to puberty creates a lasting memory impairment after the cessation of treatment in drug- and domain-specific ways. The mechanism underlying these effects could be related to changes in glutamate receptors (Choi et al. 2009; Lian and Deng 2019), modified dendritic spine densities (Milstein et al. 2013), or other forms of neural changes induced by second-generation antipsychotics.

6 Limitations of Animal Models of Early-Life Antipsychotic Drug Treatment

Much remains to be discovered about the long-term impact of early-life antipsychotic drug treatment on neural, behavioral, and cognitive function. However, as research in this area progresses, it would be helpful to find better consensus across studies regarding several issues. For example, a perpetually moving target in preclinical psychopharmacology that is particularly problematic in animal models of antipsychotic drug action is choosing an appropriate dose. Several approaches have been adopted to select the most clinically relevant dose in animals, including choosing a dose that: (1) produces an equivalent, clinically effective plasma and tissue drug level or mimics the dose per kg of body weight in humans, (2) elicits a 50% change in some behavioral measure (e.g., the ED_{50} for reversing amphetamine-induced hyperactivity (Arnt 1995) or IC_{50} for reducing conditioned avoidance responding (Olsen et al. 2008)), (3) achieves 60–80% D_2 receptor occupancy (e.g., Csernansky et al. 1993; Kapur et al. 2003; Olsen et al. 2008), or (4) has been used in previous studies. Given these choices, one is hard-pressed to reach agreement on the best measure to use in adult studies. Accordingly, choosing a clinically relevant dose in an animal model of development is even more fraught with uncertainty and controversy.

In studies of early-life antipsychotic drug action, approaches to dosing, as well as the route and timing of drug administration, have varied widely. Clinical studies often assess children that receive oral antipsychotic doses once or twice a day and, in some cases, involve a progressive increase from a low dose to a desired maintenance dose (Günther et al. 2006; Pandina et al. 2007, 2009). Many studies of early-life antipsychotic drug action in animals have utilized a single daily injection of a specific drug dose (e.g., Rosengarten and Friedhoff 1979; Cuomo et al. 1981; Moran-Gates et al. 2007; Choi et al. 2009; Bardgett et al. 2013). However, this approach is questionable since, even in adult rats, antipsychotics possess relatively rapid half-lives (~2–4 h) (e.g., Aravagiri et al. 1999; Aravagiri and Marder 2002) which might leave lengthy periods of low or non-detectable plasma antipsychotic levels between injections.

In order to achieve a more consistent plasma and tissue antipsychotic concentration, some animal studies have used a twice-daily injection regimen (Spear et al. 1980; Moran-Gates et al. 2006; Frost et al. 2010), provided the drug through a cookie dough vehicle offered three times a day (De Santis et al. 2016; Lian et al. 2016), or administered the drug through the drinking water (Milstein et al. 2013; Vinish et al. 2013). While these approaches address the circadian variability in antipsychotic concentrations associated with single daily injections, they may be limited in other ways. Drug delivery through drinking water or food adulteration requires that animals be single-housed or that gang-housed animals all receive the same treatment, which can lead to isolation stress or confounds of drug effects with cage effects. A more ideal approach may be to deliver drugs via a mini pump (e.g., Kapur et al. 2003; Samaha et al. 2008) but this route is impractical for developing rats and mice given their small initial size and rapid daily growth. Moreover, pumps need to be loaded with extremely high doses to produce clinically relevant behavioral or biochemical activity (e.g., 60–80% D_2 receptor occupancy, see Kapur et al. 2003). Finally, continuous delivery of antipsychotics can lead to tolerance or more limited drug effects on behavioral and biochemical outcomes in comparison to the effects of intermittent daily injections (Samaha et al. 2008).

It seems that many of these issues could be circumvented by a simple design modification – study multiple doses. Most of the research cited above focused on the effects of a single antipsychotic dose, perhaps because including multiple dose groups is time and labor-intensive and the effective dose range may be limited. Fewer studies (Dallemagne and Weiss 1982; Iñiguez et al. 2008; Soiza-Reilly and Azcurra 2009; Moran-Gates et al. 2007; Choi et al. 2009, 2010; Bardgett et al. 2013; Gannon et al. 2015; Stevens et al. 2016; Stubbeman et al. 2017) have included multiple doses. Despite the added time and expense, this approach enables one to include doses that straddle the various measures (e.g., behavioral ED_{50} or D_2 occupancy) and allows researchers to cover the wide range of dosing used in children (Ray et al. 2019). Clinical guidelines recommend doses in children that are 50–75% less than those used in adults – as such, it makes sense that dosing in young rats should be reduced accordingly. However, notwithstanding the potential for extremely high rates of receptor occupancy or catalepsy, an emphasis on lower doses ignores the reality that some children are prescribed antipsychotic doses within the range of those recommended for adults (see the "Second Generation Antipsychotic Treatment Effectiveness and Tolerability in Youth Study (SATIETY)" Sikich 2013). From this perspective, it may be better to first assay a reasonably high antipsychotic dose to initially establish whether any dose can significantly impact later behavioral or biochemical function. If a high dose is active, then one can discern the effective dose by subsequently investigating lower doses or by simply performing a dose-response study at the outset.

Another variable that routinely differs across studies of early-life antipsychotic drug action is the timing and duration of drug administration. Many early papers focused on antipsychotic exposure during gestation and nursing incorporated drug injections to pregnant and nursing dams (e.g., Spear et al. 1980) or direct injections to pups from the first few days of postnatal life until weaning (e.g., Cuomo et al. 1981; Frost et al. 2010). Others have focused on drug regimens that begin near puberty (e.g., postnatal days 28–35) and continue into a period analogous to mid- to late adolescence (Milstein et al. 2013; De Santis et al. 2016; Lian et al. 2016; Vinish et al. 2013). Since clinical studies have reported antipsychotic treatment in very young children (Zito et al. 2007; Olfson et al. 2010; Lohr et al. 2015) and treatment durations that can span years (Kalverdijk et al. 2017; Lohr et al. 2021), our approach (e.g., Bardgett et al. 2013, 2019, 2020) has been to examine a dosing regimen from postnatal days 14 to 42 – ages intended to mimic early childhood years through early adolescence. Others (Moran-Gates et al. 2006, 2007; Choi et al. 2009, 2010) have taken a similar approach that begins one week later but ends at the same time. An argument that can be made that any animal model of childhood antipsychotic use should include ages (postnatal days 14–28) considered analogous to juvenile or prepubertal children. But a more informative (yet time-intensive) approach (e.g., Bardgett et al. 2013) would compare groups of rats administered antipsychotics across different age periods (e.g., prepubertal vs. postpubertal) to determine if sensitive periods of exposure exist that significantly impact later brain or behavioral function.

A related factor to consider in the evaluation of animal models of early-life antipsychotic drug treatment is whether outcomes observed during adulthood represent developmental effects or simply the effects of chronic drug administration independent of age. This can be addressed by including an adult control group that receives the same drug dose and administration duration beginning sometime in adulthood. Such studies (Moran-Gates et al. 2006, 2007; Choi et al. 2009, 2010; Stevens et al. 2016; Bardgett et al. 2020) have revealed changes in behavior and neurotransmitter receptors that are specific to developmental or adult exposure, or nonspecific to age. For many other studies reviewed above (including several of our own (Bardgett et al. 2013; Gannon et al. 2015; Stubbeman et al. 2017; Bardgett et al. 2019)) that do not incorporate an adult control group, it is impossible to determine whether early-life antipsychotic administration interfered with a specific maturational process. However, the outcomes of these studies remain relevant since they show that pre- or early postnatal antipsychotic administration leads to lingering neural, cognitive, and behavioral changes.

7 Conclusions

While concerns about antipsychotic drug use in children became more widespread in the 1990s and 2000s due to escalations in pediatric prescriptions, preclinical work on antipsychotic drug exposure early in life extends back into the 1970s and 1980s. What emerges from this long arc of research is the same conclusion that can be applied to many areas of research: more work needs to be done. This is especially true regarding the effects of antipsychotics on brain development since despite efforts to monitor and limit the number of prescriptions in children (Medicaid.gov 2016; Akincigil et al. 2020; Candon et al. 2021), the rate of antipsychotic use in European and American children remains relatively high (Kalverdijk et al. 2017).

The decades of preclinical research on early-life antipsychotic drug exposure have yielded findings that coalesce around several themes (Table 1 and Fig. 1). First, prenatal antipsychotic drug administration appears to exert a qualitatively different outcome than early postnatal drug administration. The former scenario seems to lead to a hypodopaminergic state in the forebrain (e.g., fewer dopamine receptors, less sensitivity to dopamine agonists during adulthood) in a way that suggests that antipsychotics disrupt the fundamental establishment of dopamine pathways in the brain. In contrast, early postnatal antipsychotic drug administration produces a hyperdopaminergic tone later in life. This outcome may reflect a kind of long-term drug rebound effect: a compensatory plasticity that is unleashed in still maturing dopamine circuits after drug-induced receptor blockade and yields a sustained

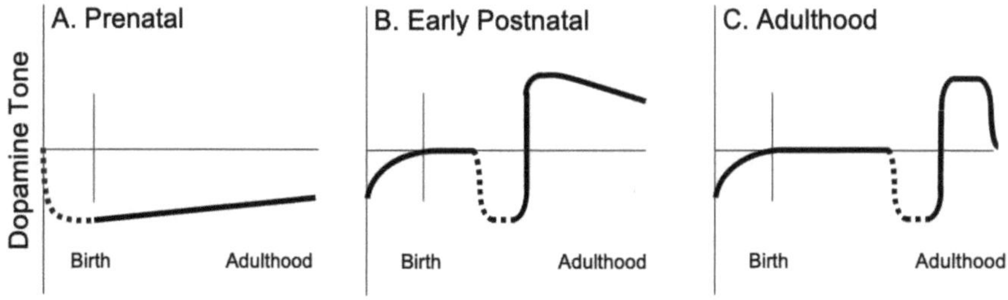

Fig. 1 General impact of chronic prenatal, early postnatal, or adult antipsychotic administration on dopaminergic tone (e.g., receptors, release and metabolites, behavioral sensitivity). The dotted portions of the line represent tone during active daily drug administration. Early treatment (**a**) leads to reduced dopamine function later in life while postnatal treatments (**b** & **c**) lead to compensatory increases in dopamine function with more enduring increases elicited by early postnatal treatment (**b**)

increase in the "dopaminergic set-point" of the forebrain. This change hints at a second theme that emerges from the work on early postnatal antipsychotic administration. Most of biochemical and behavioral alterations observed later in adulthood are similar to those seen after antipsychotic administration during adulthood; however, the effects of early postnatal administration appear to last longer. Finally, while early work suggested that a typical antipsychotic drug like haloperidol has a more significant impact on later neural and behavioral function than the prototypical atypical drug clozapine, later work with second-generation, atypical drugs like risperidone and olanzapine suggests that their impact has been more "typical" than "atypical". Some aspects of their "typical" profile could be related to timing and dose as opposed to true categorical distinctions, analogous to clinical research suggesting more quantitative versus qualitative differences between the effects of many antipsychotic drugs (Lieberman et al. 2005).

Future translational research should consider whether co-administration of psychostimulants or antidepressant drugs modifies the impact of antipsychotics on later behavioral and brain function. Many children prescribed antipsychotics receive other psychotherapeutic medications – most often drugs to treat ADHD or an antidepressant (Olfson et al. 2015; Lohr et al. 2021). It will be important to discern whether these drugs mitigate, exacerbate, or have no effect on the impact of antipsychotic drugs on brain development. A second issue deserving attention is how early-life antipsychotics affect outcomes in animal models of the disorders (e.g., ADHD, autism) for which they are most often used (although it is important to note that many children prescribed antipsychotics do not receive a formal diagnosis (Lohr et al. 2015)). While this strategy may be confounded by an inability to tease out changes inherent in the animal model versus those imposed by drug alone, it is a more clinically valid approach to some cases of childhood antipsychotic exposure. Moreover, it is possible that drug actions seen as problematic in a "neurotypical" animal may serve corrective purposes in an animal model of ADHD or autism. Finally, as most antipsychotics have actions at multiple receptor sites, it would be beneficial to investigate the neurodevelopmental effects of such drugs in genetic models where single receptors have been knocked out or silenced temporarily. These studies could help to isolate the molecular mechanisms responsible for the long-term impact of early-life antipsychotics.

So how should a clinician interpret the preclinical or translational data presented here? In one sense, the overall effect sizes of early-life antipsychotic exposure on later behavioral and brain function tend to be small and consistent with effects previously reported after adult dosing. This would suggest that antipsychotics should remain in the clinician's toolbox when crafting treatment plans for pediatric patients. On the other hand, when coupled with the oft-reported, worrisome side-effect profile reported in the clinical literature, the significant and persistent effects of early antipsychotic administration seen in animal models should serve as an additional warning sign regarding the overuse of these drugs in pediatric psychiatric populations. Consistent with calls for more caution (e.g., Medicaid.gov 2016; Akincigil et al. 2020; Candon et al. 2021; Lohr et al. 2021), the preclinical literature offers further support for initiatives aimed at decreasing drug doses and treatment durations, and reducing the number of children treated with antipsychotics. Moreover, the many long-term effects seen in animal models of early-life antipsychotic use should compel clinical researchers to design and implement longitudinal studies that track children treated with antipsychotics into adulthood and ascertain whether any detrimental effects on later brain and behavioral function occur. With a better understanding of such consequences and with a more judicious and optimized treatment approach, it is hoped that clinicians can better enable kids with serious mental health issues to realize their full potential and experience fulfilling lives.

Acknowledgments This work was supported by the National Institute of Health (grant numbers P20GM103436 and R16149498). This funding source had no other role in the research other than funding. The author declares no conflict of interest.

References

Akincigil A, Mackie TI, Cook S et al (2020) Effectiveness of mandatory peer review to reduce antipsychotic prescriptions for Medicaid-insured children. Health Serv Res 55:596–603. https://doi.org/10.1111/1475-6773.13297

Aman M, Rettiganti M, Nagaraja HN et al (2015) Tolerability, safety, and benefits of risperidone in children and adolescents with autism: 21-month follow-up after 8-week placebo-controlled trial. J Child Adolesc Psychopharmacol 25:482–493. https://doi.org/10.1089/cap.2015.0005

Andersen SL (2003) Trajectories of brain development: point of vulnerability or window of opportunity? Neurosci Biobehav Rev 27:3–18. https://doi.org/10.1016/s0149-7634(03)00005-8

Aravagiri M, Marder SR (2002) Brain, plasma and tissue pharmacokinetics of risperidone and 9-hydroxyrisperidone after separate oral administration to rats. Psychopharmacology 159:424–431. https://doi.org/10.1007/s00213-001-0933-x

Aravagiri M, Teper Y, Marder SR (1999) Pharmacokinetics and tissue distribution of olanzapine in rats. Biopharm Drug Dispos 20:369–377. https://doi.org/10.1002/1099-081x(199911)20:8<369::aid-bdd200>3.0.co;2-6

Arnt J (1995) Differential effects of classical and newer antipsychotics on the hypermotility induced by two dose levels of D-amphetamine. Eur J Pharmacol 283:55–62. https://doi.org/10.1016/0014-2999(95)00292-s

Bardgett ME (2004) Behavioral models of atypical antipsychotic drug action in rodents. In: Csernansky JG, Lauriello J (eds) Atypical antipsychotics: from bench to bedside. Marcel Dekker, New York, pp 61–92

Bardgett ME, Franks-Henry JM, Colemire KR, Juneau KR, Stevens RM, Marczinski CA, Griffith MS (2013) Adult rats treated with risperidone during development are hyperactive. Exp Clin Psychopharmacol 21:259–267. https://doi.org/10.1037/a0031972

Bardgett ME, Crane C, Baltes Thompson EC, Cox B, Downnen T (2019) The effects of amphetamine on working memory and locomotor activity in adult rats administered risperidone early in life. Behav Brain Res 362:64–70. https://doi.org/10.1016/j.bbr.2018.12.044

Bardgett ME, Downnen T, Crane C, Baltes Thompson EC, Muncie B, Steffen SA, Yates JR, Pauly JR (2020) Chronic risperidone administration leads to greater amphetamine-induced conditioned place preference. Neuropharmacology 179:108276. https://doi.org/10.1016/j.neuropharm.2020.108276

Bardgett ME, Griffith MS, Robinson KR, Stevens RM, Gannon MA, Knuth MD, Hawk GS, Pauly JR (2024) Early-life risperidone alters locomotor responses to apomorphine and quinpirole in adulthood. Behav Brain Res 473:115171. https://doi.org/10.1016/j.bbr.2024.115171

Boye SM, Rompré PP (2000) Behavioral evidence of depolarization block of dopamine neurons after chronic treatment with haloperidol and clozapine. J Neurosci 20:1229–1239. https://doi.org/10.1523/JNEUROSCI.20-03-01229.2000

Breaux R, Baweja R, Eadeh HM, Shroff DM, Cash AR, Swanson CS, Knehans A, Waxmonsky JG (2023) Systematic review and meta-analysis: pharmacological and nonpharmacological interventions for persistent nonepisodic irritability. J Am Acad Child Adolesc Psychiatry 62:318–334. https://doi.org/10.1016/j.jaac.2022.05.012

Brenhouse HC, Andersen SL (2011) Developmental trajectories during adolescence in males and females: a cross-species understanding of underlying brain changes. Neurosci Biobehav Rev 35:1687–1703. https://doi.org/10.1016/j.neubiorev.2011.04.013

Bushnell GA, Crystal S, Olfson M (2021) Trends in antipsychotic medication use in young privately insured children. J Am Acad Child Adolesc Psychiatry 60:877–886. https://doi.org/10.1016/j.jaac.2020.09.023

Cagiano R, Barfield RJ, White NR, Pleim ET, Weinstein M, Cuomo V (1988) Subtle behavioural changes produced in rat pups by in utero exposure to haloperidol. Eur J Pharmacol 157:45–50. https://doi.org/10.1016/0014-2999(88)90469-4

Candon M, Shen S, Fadeyibi O, Smith JL, Rothbard A (2021) Trends in antipsychotic prescribing for approved and unapproved indications to Medicaid-enrolled youth in Philadelphia, Pennsylvania between 2014 and 2018. BMC Psychiatry 21:524. https://doi.org/10.1186/s12888-021-03533-3

Ching H, Pringsheim T (2012) Aripiprazole for autism spectrum disorders (ASD). Cochrane Database Syst Rev 16:CD009043. https://doi.org/10.1002/14651858.CD009043.pub2

Choi YK, Gardner MP, Tarazi FI (2009) Effects of risperidone on glutamate receptor subtypes in developing rat brain. Eur Neuropsychopharmacol 19:77–84. https://doi.org/10.1016/j.euroneuro.2008.08.010

Choi YK, Moran-Gates T, Gardner MP, Tarazi FI (2010) Effects of repeated risperidone exposure on serotonin receptor subtypes in developing rats. Eur Neuropsychopharmacol 20:187–194. https://doi.org/10.1016/j.euroneuro.2009.09.002

Choi YK, Adham N, Kiss B, Gyertyán I, Tarazi FI (2017) Long-term effects of aripiprazole exposure on monoaminergic and glutamatergic receptor subtypes: comparison with cariprazine. CNS Spectr 22:484–494. https://doi.org/10.1017/S1092852916000894

Choi HJ, Im SJ, Park HR, Park S, Kim CE, Ryu S (2019) Long-term effects of aripiprazole treatment during adolescence on cognitive function and dopamine D2 receptor expression in neurodevelopmentally normal rats. Clin Psychopharmacol Neurosci 17:400–408. https://doi.org/10.9758/cpn.2019.17.3.400

Constantine RJ, Jentz S, Bengtson M, McPherson M, Andel R, Jones MB (2012) Exposure to antipsychotic medications over a 4-year period among children who initiated antipsychotic treatment before their sixth birthday. Pharmacoepidemiol Drug Saf 21:152–160. https://doi.org/10.1002/pds.2189

Correll CU, Manu P, Olshanskiy V, Napolitano B, Kane JM, Malhotra AK (2009) Cardiometabolic risk of second-generation antipsychotic medications during first-time use in children and adolescents. JAMA 302:1765–1773. https://doi.org/10.1001/jama.2009.1549

Coustals N, Ménard ML, Cohen D (2021) Aripiprazole in children and adolescents. J Child Adolesc Psychopharmacol 31:4–32. https://doi.org/10.1089/cap.2020.0014

Csernansky JG, Wrona CT, Bardgett ME, Early TS, Newcomer JW (1993) Subcortical dopamine and serotonin turnover during acute and subchronic administration of typical and atypical neuroleptics. Psychopharmacology 110:145–151. https://doi.org/10.1007/BF02246964

Cuomo V, Cagiano R, Coen E, Mocchetti I, Cattabeni F, Racagni G (1981) Enduring behavioural and biochemical effects in the adult rat after prolonged postnatal administration of haloperidol. Psychopharmacology 74:166–169. https://doi.org/10.1007/BF00432686

Cuomo V, Cagiano R, Mocchetti I, Coen E, Cattabeni F, Racagni G (1983a) Behavioural and biochemical effects in the adult rat after prolonged postnatal administration of clozapine. Psychopharmacology 81:239–243. https://doi.org/10.1007/BF00427270

Cuomo V, Cattabeni F, Racagni G (1983b) The influence of early postnatal treatment with haloperidol on the effects induced by small and large doses of apomorphine on locomotion of adult rats. Neuropharmacology 22:1137–1139. https://doi.org/10.1016/0028-3908(83)90050-3

Cuomo V, Cagiano R, Mocchetti I, Coen E, Cattabeni F, Racagni G (1983c) Biochemical and behavioural effects after early postnatal administration of neuroleptics in rats. In: Zbinden G, Cuomo V, Racagni G, Weiss B (eds) Application of behavioral pharmacology in toxicology. Raven Press, New York, pp 173–185

Cuomo V, Cagiano R, Renna G, Serinelli A, Brunello N, Racagni G (1985) Comparative evaluation of the behavioural consequences of prenatal and early postnatal exposure to haloperidol in rats. Neurobehav Toxicol Teratol 7:489–492

Dallemagne G, Weiss B (1982) Altered adult behavior of mice following postnatal treatment with haloperidol. Pharmacol Biochem Behav 16:761–767. https://doi.org/10.1016/0091-3057(82)90232-5

De Santis M, Lian J, Huang XF, Deng C (2016) Early antipsychotic treatment in juvenile rats elicits long-term alterations to the dopamine neurotransmitter system. Int J Mol Sci 17:1944. https://doi.org/10.3390/ijms17111944

Farah A (2005) Atypicality of atypical antipsychotics. Prim Care Companion J Clin Psychiatry 7: 268–274. https://doi.org/10.4088/pcc.v07n0602

Farmer CA, Epstein JN, Findling RL, Gadow KD, Arnold LE, Kipp H, Kolko DJ, Butter E, Schneider J, Bukstein OG, McNamara NK, Molina BS, Aman MG (2017) Risperidone added to psychostimulant in children with severe aggression and attention-deficit/hyperactivity disorder: lack of effect on attention and short-term memory. J Child Adolesc Psychopharmacol 27: 117–124. https://doi.org/10.1089/cap.2016.0040

Filippino S, Csernansky JG (2004) Atypical antipsychotic drugs: theories of mechanism of action. In: Csernansky JG, Lauriello J (eds) Atypical antipsychotics: from bench to bedside. Marcel Dekker, New York, pp 1–18

Freedman R (2005) The choice of antipsychotic drugs for schizophrenia. N Engl J Med 353:1286–1288. https://doi.org/10.1056/NEJMe058200

Frost DO, Page SC, Carroll C, Kolb B (2010) Early exposure to haloperidol or olanzapine induces long-term alterations of dendritic form. Synapse 64:191–199. https://doi.org/10.1002/syn.20715

Galling B, Roldán A, Nielsen RE, Nielsen J, Gerhard T, Carbon M, Stubbs B, Vancampfort D, De Hert M, Olfson M, Kahl KG, Martin A, Guo JJ, Lane HY, Sung FC, Liao CH, Arango C, Correll CU (2016) Type 2 diabetes mellitus in youth exposed to antipsychotics: a systematic review and meta-analysis. JAMA Psychiatry 73:247–259. https://doi.org/10.1001/jamapsychiatry.2015.2923

Gannon MA, Brown CJ, Stevens RM, Griffith MS, Marczinski CA, Bardgett ME (2015) Early-life risperidone administration alters maternal-offspring interactions and juvenile play fighting. Pharmacol Biochem Behav 130:90–96. https://doi.org/10.1016/j.pbb.2015.01.007

Günther T, Herpertz-Dahlmann B, Jolles J, Konrad K (2006) The influence of risperidone on attentional functions in children and adolescents with attention-deficit/hyperactivity disorder and co-morbid disruptive behavior disorder. J Child Adolesc Psychopharmacol 16:725–735. https://doi.org/10.1089/cap.2006.16.725

Hammerman A, Dreiher J, Klang SH, Munitz H, Cohen AD, Goldfracht M (2008) Antipsychotics and diabetes: an age-related association. Ann Pharmacother 42:1316–1322. https://doi.org/10.1345/aph.1L015

Huybrechts KF, Straub L, Karlsson P, Pazzagli L, Furu K, Gissler M, Hernandez-Diaz S, Nørgaard M, Zoega H, Bateman BT, Cesta CE, Cohen JM, Leinonen MK, Reutfors J, Selmer RM, Suarez EA, Ulrichsen SP, Kieler H (2023) Association of in utero antipsychotic medication exposure with risk of congenital malformations in nordic countries and the US. JAMA Psychiatry 80:156–166. https://doi.org/10.1001/jamapsychiatry.2022.4109

Iñiguez C, Tamayo P, Romero E, Gayoso MJ, De Juan J (1990) Alterations in prolactin secretion during the 1st postnatal month following perinatal dopaminergic blockade with haloperidol. Neuroendocrinology 51:700–704. https://doi.org/10.1159/000125413

Iñiguez SD, Cortez AM, Crawford CA, McDougall SA (2008) Effects of aripiprazole and terguride on dopamine synthesis in the dorsal striatum and medial prefrontal cortex of preweanling rats. J Neural Transm 115:97–106. https://doi.org/10.1007/s00702-007-0820-7

Kalverdijk LJ, Tobi H, van den Berg PB, Buiskool J, Wagenaar L, Minderaa RB, de Jong-van den Berg LT (2008) Use of antipsychotic drugs among Dutch youths between 1997 and 2005. Psychiatr Serv 59:554–560. https://doi.org/10.1176/appi.ps.59.5.554

Kalverdijk LJ, Bachmann CJ, Aagaard L, Burcu M, Glaeske G, Hoffmann F, Petersen I, Schuiling-Veninga CCM, Wijlaars LP, Zito JM (2017) A multi-national comparison of antipsychotic drug use in children and adolescents, 2005-2012. Child Adolesc Psychiatry Ment Health 11:55. https://doi.org/10.1186/s13034-017-0192-1

Kane J, Honigfeld G, Singer J, Meltzer H (1988) Clozapine for the treatment-resistant schizophrenic. A double-blind comparison with chlorpromazine. Arch Gen Psychiatry 45:789–796. https://doi.org/10.1001/archpsyc.1988.01800330013001

Kapur S, Vander Spek SC, Brownlee BA, Nobrega JN (2003) Antipsychotic dosing in preclinical models is often unrepresentative of the clinical condition: a suggested solution based on in vivo occupancy. J Pharmacol Exp Ther 305:625–631. https://doi.org/10.1124/jpet.102.046987

Kent JM, Kushner S, Ning X, Karcher K, Ness S, Aman M, Singh J, Hough D (2013) Risperidone dosing in children and adolescents with autistic disorder: a double-blind, placebo-controlled study. J Autism Dev Disord 43:1773–1783. https://doi.org/10.1007/s10803-012-1723-5

Levitt P, Harvey JA, Friedman E, Simansky K, Murphy EH (1997) New evidence for neurotransmitter influences on brain development. Trends Neurosci 20:269–274. https://doi.org/10.1016/s0166-2236(96)01028-4

Lian J, Deng C (2019) Early antipsychotic exposure affects NMDA and GABAA receptor binding in the brains of juvenile rats. Psychiatry Res 273:739–745. https://doi.org/10.1016/j.psychres.2019.02.001

Lian J, Pan B, Deng C (2016) Early antipsychotic exposure affects serotonin and dopamine receptor binding density differently in selected brain loci of male and female juvenile rats. Pharmacol Rep 68:1028–1035. https://doi.org/10.1016/j.pharep.2016.06.003

Lieberman JA, Stroup TS, McEvoy JP et al (2005) Effectiveness of antipsychotic drugs in patients with chronic schizophrenia. N Engl J Med 353:1209–1223. https://doi.org/10.1056/NEJMoa051688

Lohr WD, Chowning RT, Stevenson MD, Williams PG (2015) Trends in atypical antipsychotics prescribed to children six years of age or less on Medicaid in Kentucky. J Child Adol Psychopharmacol 25:440–443. https://doi.org/10.1089/cap.2014.0057

Lohr WD, Jawad K, Feygin Y, Le J, Creel L, Pasquenza N, Williams PG, Jones VF, Myers J, Davis DW (2021) Antipsychotic medications for low-income preschoolers: long duration and psychotropic medication polypharmacy. Psychiatr Serv 73:510–517. https://doi.org/10.1176/appi.ps.202000673

Londino DL, Wiggins L, Dallas J, Stirewalt E, Buckley P (2004) Evolving use of atypical antipsychotic medication in childhood psychiatric conditions. In: Csernansky JG, Lauriello J (eds) Atypical antipsychotics: from bench to bedside. Marcel Dekker, New York, pp 239–273

Loy JH, Merry SN, Hetrick SE, Stasiak K (2017) Atypical antipsychotics for disruptive behaviour disorders in children and youths. Cochrane Database Syst Rev 8:CD008559. https://doi.org/10.1002/14651858.CD008559.pub3

Mandell DJ, Unis A, Sackett GP (2011) Post-drug consequences of chronic atypical antipsychotic drug administration on the ability to adjust behavior based on feedback in young monkeys. Psychopharmacology 215:345–352. https://doi.org/10.1007/s00213-010-2147-6

McDougall SA, Mohd-Yusof A, Kaplan GJ, Abdulla ZI, Lee RJ, Crawford CA (2013) Postnatal manganese exposure does not alter dopamine autoreceptor sensitivity in adult and adolescent male rats. Eur J Pharmacol 706:4–10. https://doi.org/10.1016/j.ejphar.2013.02.030

Medicaid.gov (2016) A review of state medicaid approaches on child antipsychotic monitoring programs. https://www.medicaid.gov/medicaid-chip-program-information/by-topics/prescription-drugs/downloads/state-medicaid-dur-summaries.pdf. Accessed 12 July 2023

Milstein JA, Elnabawi A, Vinish M, Swanson T, Enos JK, Bailey AM, Kolb B, Frost DO (2013) Olanzapine treatment of adolescent rats causes enduring specific memory impairments and alters cortical development and function. PLoS One 8:e57308. https://doi.org/10.1371/journal.pone.0057308

Moon SL (1984) Prenatal haloperidol alters striatal dopamine and opiate receptors. Brain Res 323:109–113. https://doi.org/10.1016/0006-8993(84)90270-1

Moran-Gates T, Gan L, Park YS, Zhang K, Baldessarini RJ, Tarazi FI (2006) Repeated antipsychotic drug exposure in developing rats: dopamine receptor effects. Synapse 59:92–100. https://doi.org/10.1002/syn.20220

Moran-Gates T, Grady C, Shik Park Y, Baldessarini RJ, Tarazi FI (2007) Effects of risperidone on dopamine receptor subtypes in developing rat brain. Eur Neuropsychopharmacol 17:448–455. https://doi.org/10.1016/j.euroneuro.2006.10.004

Muller P, Seeman P (1978) Dopaminergic supersensitivity after neuroleptics: time-course and specificity. Psychopharmacology 60:1–11. https://doi.org/10.1007/BF00429171

Murphy AL, Gardner DM, Cooke C, Kisely S, Hughes J, Kutcher SP (2013) Prescribing trends of antipsychotics in youth receiving income assistance: results from a retrospective population database study. BMC Psychiatry 13:198. https://doi.org/10.1186/1471-244X-13-198

Nicol GE, Yingling MD, Flavin KS, Schweiger JA, Patterson BW, Schechtman KB, Newcomer JW (2018) Metabolic effects of antipsychotics on adiposity and insulin sensitivity in youths: a randomized clinical trial. JAMA Psychiatry 75:788–796. https://doi.org/10.1001/jamapsychiatry.2018.1088

Olfson M, Blanco C, Liu L, Moreno C, Laje G (2006) National trends in the outpatient treatment of children and adolescents with antipsychotic drugs. Arch Gen Psychiatry 63:679–685. https://doi.org/10.1001/archpsyc.63.6.679

Olfson M, Crystal S, Huang C, Gerhard T (2010) Trends in antipsychotic drug use by very young, privately insured children. J Am Acad Child Adolesc Psychiatry 49(1):13–23. https://doi.org/10.1097/00004583-201001000-00005

Olfson M, King M, Schoenbaum M (2015) Treatment of young people with antipsychotic medications in the United States. JAMA Psychiatry 72:867–874. https://doi.org/10.1001/jamapsychiatry.2015.0500

Oliveri AN, Levin ED (2019) Dopamine D_1 and D_2 receptor antagonism during development alters later behavior in zebrafish. Behav Brain Res 356:250–256. https://doi.org/10.1016/j.bbr.2018.08.028

Olsen CK, Brennum LT, Kreilgaard M (2008) Using pharmacokinetic-pharmacodynamic modelling as a tool for prediction of therapeutic effective plasma levels of antipsychotics. Eur J Pharmacol 584:318–327. https://doi.org/10.1016/j.ejphar.2008.02.005

Pandina GJ, Bilder R, Harvey PD, Keefe RS, Aman MG, Gharabawi G (2007) Risperidone and cognitive function in children with disruptive behavior disorders. Biol Psychiatry 62:226–234. https://doi.org/10.1016/j.biopsych.2006.09.036

Pandina GJ, Zhu Y, Cornblatt B (2009) Cognitive function with long-term risperidone in children and adolescents with disruptive behavior disorder. J Child Adolesc Psychopharmacol 19:749–756. https://doi.org/10.1089/cap.2008.0159

Pillay J, Boylan K, Newton A, Hartling L, Vandermeer B, Nuspl M, MacGregor T, Featherstone R, Carrey N (2018) Harms of antipsychotics in children and young adults: a systematic review update. Can J Psychiatr 63:661–678. https://doi.org/10.1177/0706743718779950

Pringsheim T, Panagiotopoulos C, Davidson J, Ho J, Canadian Alliance for Monitoring Effectiveness and Safety of Antipsychotics in Children (CAMESA) Guideline Group (2011) Evidence-based recommendations for monitoring safety of second-generation antipsychotics in children and youth. Paediatr Child Health 16:581–589

Prosser ES, Pruthi R, Csernansky JG (1989) Differences in the time course of dopaminergic supersensitivity following chronic administration of haloperidol, molindone, or sulpiride. Psychopharmacology 99:109–116. https://doi.org/10.1007/BF00634463

Ray WA, Stein CM, Murray KT, Fuchs DC, Patrick SW, Daugherty J, Hall K, Cooper WO (2019) Association of antipsychotic treatment with risk of unexpected death among children and youths. JAMA Psychiatry 76:162–171. https://doi.org/10.1001/jamapsychiatry.2018.3421

Richelson E, Souder T (2000) Binding of antipsychotic drugs to human brain receptors focus on newer generation compounds. Life Sci 68:29–39. https://doi.org/10.1016/s0024-3205(00)00911-5

Ronsley R, Scott D, Warburton WP, Hamdi RD, Louie DC, Davidson J, Panagiotopoulos C (2013) A population-based study of antipsychotic prescription trends in children and adolescents in British Columbia, from 1996 to 2011. Can J Psychiatr 58:361–369. https://doi.org/10.1177/070674371305800608

Rosengarten H, Friedhoff AJ (1979) Enduring changes in dopamine receptor cells of pups from drug administration to pregnant and nursing rats. Science 203:1133–1135. https://doi.org/10.1126/science.570724

Roth BL, Gumpper RH (2023) Psychedelics as transformative therapeutics. Am J Psychiatry 180:340–347. https://doi.org/10.1176/appi.ajp.20230172

Samaha AN, Reckless GE, Seeman P, Diwan M, Nobrega JN, Kapur S (2008) Less is more: antipsychotic drug effects are greater with transient rather than continuous delivery. Biol Psychiatry 64:145–152. https://doi.org/10.1016/j.biopsych.2008.01.010

Scalzo FM, Spear LP (1985) Chronic haloperidol during development attenuates dopamine autoreceptor function in striatal and mesolimbic brain regions of young and older adult rats. Psychopharmacology 85:271–276. https://doi.org/10.1007/BF00428186

Scalzo FM, Holson RR, Gough BJ, Ali SF (1989a) Neurochemical effects of prenatal haloperidol exposure. Pharmacol Biochem Behav 34:721–725. https://doi.org/10.1016/0091-3057(89)90265-7

Scalzo FM, Newport GD, Gough BJ, Ali SF, Holson RR (1989b) Chronic prenatal haloperidol exposure: lack of effect on presynaptic dopamine autoreceptors. Neurotoxicology 10:485–490

Seeger TF, Gardner EL (1979) Enhancement of self-stimulation behavior in rats and monkeys after chronic neuroleptic treatment: evidence for mesolimbic supersensitivity. Brain Res 175:49–57. https://doi.org/10.1016/0006-8993(79)90513-4

Shalaby IA, Spear LP (1980) Chronic administration of haloperidol during development: later psychopharmacological responses to apomorphine and arecoline. Pharmacol Biochem Behav 13:685–690. https://doi.org/10.1016/0091-3057(80)90012-x

Sharma A, Shaw SR (2012) Efficacy of risperidone in managing maladaptive behaviors for children with autistic spectrum disorder: a meta-analysis. J Pediatr Health Care 26:291–299. https://doi.org/10.1016/j.pedhc.2011.02.008

Sikich LM (2013) Second-generation antipsychotic treatment indication effectiveness and tolerability in youth (SATIETY) study. National Library of Medicine. https://www.clinicaltrials.gov/study/NCT01269710?cond=second%20generation%20antipsychotic&rank=9. Accessed 12 July 2023

Singh KP, Singh M (2002) Effect of prenatal haloperidol exposure on behavioral alterations in rats. Neurotox Teratol 24:497–502. https://doi.org/10.1016/s0892-0362(02)00189-7

Singh KP, Singh MK (2017) In utero exposure to atypical antipsychotic drug, risperidone: effects on fetal neurotoxicity in hippocampal region and cognitive impairment in rat offspring. Prog Neuro-Psychopharmacol Biol Psychiatry 75:35–44. https://doi.org/10.1016/j.pnpbp.2016.12.006

Singh KP, Tripathi N (2014) Prenatal exposure of a novel antipsychotic aripiprazole: impact on maternal, fetal and postnatal body weight modulation in rats. Curr Drug Saf 9:43–48. https://doi.org/10.2174/1574886311308666006l

Singh KP, Tripathi N (2015) Prenatal exposure to a novel antipsychotic quetiapine: impact on neuro-architecture, apoptotic neurodegeneration in fetal hippocampus and cognitive impairment in young rats. Int J Dev Neurosci 42:59–67. https://doi.org/10.1016/j.ijdevneu.2015.02.011

Singh KP, Singh MK, Singh M (2016) Effects of prenatal exposure to antipsychotic risperidone on developmental neurotoxicity, apoptotic neurodegeneration and neurobehavioral sequelae in rat offspring. Int J Dev Neurosci 52:13–23. https://doi.org/10.1016/j.ijdevneu.2016.05.006

Soiza-Reilly M, Azcurra JM (2009) Developmental striatal critical period of activity-dependent plasticity is also a window of susceptibility for haloperidol induced adult motor alterations. Neurotoxicol Teratol 31:191–197. https://doi.org/10.1016/j.ntt.2009.03.001

Sokolowski JD, Salamone JD (1994) Effects of dopamine depletions in the medial prefrontal cortex on DRL performance and motor activity in the rat. Brain Res 642:20–28. https://doi.org/10.1016/0006-8993(94)90901-6

Spear LP, Shalaby IA, Brick J (1980) Chronic administration of haloperidol during development: behavioral and psychopharmacological effects. Psychopharmacology 70:47–58. https://doi.org/10.1007/BF00432369

Stahl SM, Shayegan DK (2003) The psychopharmacology of ziprasidone: receptor-binding properties and real-world psychiatric practice. J Clin Psychiatry 64:6–12

Stevens RM, Gannon MA, Griffith MS, Bardgett ME (2016) Delayed yet persistent effects of daily risperidone on activity in developing rats. Behav Pharmacol 27:460–469. https://doi.org/10.1097/FBP.0000000000000230

Stubbeman B, Brown CJ, Yates JR, Bardgett ME (2017) Early-life risperidone enhances locomotor responses to amphetamine during adulthood. Eur J Pharmacol 812:256–263. https://doi.org/10.1016/j.ejphar.2017.07.025

Sultan RS, Wang S, Crystal S, Olfson M (2019) Antipsychotic treatment among youths with attention-deficit/hyperactivity disorder. JAMA Netw Open 2:e197850. https://doi.org/10.1001/jamanetworkopen.2019.7850

Tadokoro S, Okamura N, Sekine Y, Kanahara N, Hashimoto K, Iyo M (2012) Chronic treatment with aripiprazole prevents development of dopamine supersensitivity and potentially supersensitivity psychosis. Schizophr Bull 38:1012–1020. https://doi.org/10.1093/schbul/sbr006

Tarazi FI, Zhang K, Baldessarini RJ (2001) Long-term effects of olanzapine, risperidone, and quetiapine on dopamine receptor types in regions of rat brain: implications for antipsychotic drug treatment. J Pharmacol Exp Ther 297:711–717. https://doi.org/10.1124/jpet.103.052597

Varela FA, Der-Ghazarian T, Lee RJ, Charntikov S, Crawford CA, McDougall SA (2014) Repeated aripiprazole treatment causes dopamine D2 receptor up-regulation and dopamine supersensitivity in young rats. J Psychopharmacol 28:376–386. https://doi.org/10.1177/0269881113504016

Vargas MV, Dunlap LE, Dong C, Carter SJ, Tombari RJ, Jami SA, Cameron LP, Patel SD, Hennessey JJ, Saeger HN, McCorvy JD, Gray JA, Tian L, Olson DE (2023) Psychedelics promote neuroplasticity through the activation of intracellular 5-HT2A receptors. Science 379: 700–706. https://doi.org/10.1126/science.adf0435

Velley L, Blanc G, Tassin JP, Thierry AM, Glowinski J (1975) Inhibition of striatal dopamine synthesis in rats injected chronically with neuroleptics in their early life. Naunyn Schmiedeberg's Arch Pharmacol 288:97–102. https://doi.org/10.1007/BF00501817

Vinish M, Elnabawi A, Milstein JA, Burke JS, Kallevang JK, Turek KC, Lansink CS, Merchenthaler I, Bailey AM, Kolb B, Cheer JF, Frost DO (2013) Olanzapine treatment of adolescent rats alters adult reward behaviour and nucleus accumbens function. Int J Neuropsychopharmacol 16:1599–1609. https://doi.org/10.1017/S1461145712001642

Vitiello B, Correll C, van Zwieten-Boot B, Zuddas A, Parellada M, Arango C (2009) Antipsychotics in children and adolescents: increasing use, evidence for efficacy and safety concerns. Eur Neuropsychopharmacol 19:629–635. https://doi.org/10.1016/j.euroneuro.2009.04.008

Waddington JL, Gamble SJ (1980) Neuroleptic treatment for a substantial proportion of adult life: behavioural sequelae of 9 months haloperidol administration. Eur J Pharmacol 67:363–369. https://doi.org/10.1016/0014-2999(80)90177-6

Wang H, Li JT, Zhang Y, Liu R, Wang XD, Si TM, Su YA (2019) Prenatal exposure to antipsychotics disrupts the plasticity of dentate neurons and memory in adult male mice. Int J Neuropsychopharmcol 22:71–82. https://doi.org/10.1093/ijnp/pyy073

Wang Z, Chan AYL, Coghill D, Ip P, Lau WCY, Simonoff E, Brauer R, Wei L, Wong ICK, Man KKC (2021) Association between prenatal exposure to antipsychotics and attention-deficit/hyperactivity disorder, autism spectrum disorder, preterm birth, and small for gestational age. JAMA Int Med 181:1332–1340. https://doi.org/10.1001/jamainternmed.2021.4571

Williams R, Ali SF, Scalzo FM, Soliman K, Holson RR (1992) Prenatal haloperidol exposure: effects on brain weights and caudate neurotransmitter levels in rats. Brain Res Bull 29:449–458. https://doi.org/10.1016/0361-9230(92)90082-9

Zito JM, Safer DJ, Valluri S, Gardner JF, Korelitz JJ, Mattison DR (2007) Psychotherapeutic medication prevalence in medicaid-insured preschoolers. J Child Adolesc Psychopharmacol 17: 195–203. https://doi.org/10.1089/cap.2007.0006

Zito JM, Burcu M, McKean S, Warnock R, Kelman J (2018) Pediatric use of antipsychotic medications before and after Medicaid peer review implementation. JAMA Psychiatry 75: 100–103. https://doi.org/10.1001/jamapsychiatry.2017.3493

Zuo J, Liu Z, Ouyang X, Liu H, Hao Y, Xu L, Lu XH (2008) Distinct neurobehavioral consequences of prenatal exposure to sulpiride (SUL) and risperidone (RIS) in rats. Prog Neuro-Psychopharmacol Biol Psychiatry 32:387–397. https://doi.org/10.1016/j.pnpbp.2007.09.005

Fetal Alcohol Spectrum Disorders

Riley J. Felicicchia, Christina R. Veziris, and Sarah N. Mattson

Contents

Abstract Prenatal alcohol exposure (PAE) can cause a wide range of physical, neurobehavioral, and neurocognitive impairments that impact developmental trajectories throughout the lifespan. Clinically, individuals who have been exposed to alcohol prenatally and who show physical or neurobehavioral difficulties may be classified as having a condition included in fetal alcohol spectrum disorders (FASDs). The aim of this chapter is to summarize the current knowledge of FASD including diagnostic criteria, neurobehavioral outcomes, lifespan considerations, and interventions. Individuals with PAE exhibit challenges in the cognitive domains of executive functioning, general intelligence, motor function, learning, and memory. Aggression, trouble with the law, and oppositional behavior are also commonly

R. J. Felicicchia, C. R. Veziris, and S. N. Mattson (✉)
Department of Psychology, Center for Behavioral Teratology, San Diego State University, San Diego, CA, USA
e-mail: sarah.mattson@sdsu.edu

Curr Topics Behav Neurosci (2025) 75: 99–116
https://doi.org/10.1007/7854_2024_569
Published Online: 07 March 2025

associated with individuals with FASD. The effects of PAE can be attributed to altered neural development such as smaller total brain volume and structural abnormalities. Prenatal exposure to alcohol increases risk for co-occurring neurodevelopmental conditions and psychiatric disorders. This chapter will also review the current literature on pre- and postnatal interventions to target the effects of PAE.

Keywords Fetal alcohol spectrum disorders (FASDs) · Prenatal alcohol exposure

1 Introduction

Alcohol is a well-documented teratogen that significantly impacts fetal development (Jones et al. 1973). Prenatal alcohol exposure (PAE) can cause a wide range of physical, neurobehavioral, and neurocognitive impairments that impact developmental trajectories throughout the lifespan (Mattson et al. 2019). Recent estimates suggest that one in ten women consume alcohol during pregnancy (Tan et al. 2015). Previous research indicates that the amount of alcohol consumed during pregnancy and the timing of exposure impacts the degree of impairment in neonatal outcomes (Coles et al. 2015). Additionally, the postnatal environment can also influence developmental trajectories in alcohol-exposed children (Streissguth et al. 1996, 2004). Clinically, individuals who have been exposed to alcohol prenatally and who show physical or neurobehavioral difficulties may be classified as having a condition included in fetal alcohol spectrum disorders (FASDs). This umbrella term encompasses several diagnoses of different degrees of severity, which are described below. While the exact prevalence of individuals with FASD remains unknown, recent studies suggest that an estimated 1.1–5.0% of the population within the United States are living with diagnosed or undiagnosed FASDs (May et al. 2018). Despite ongoing research and clinical efforts, diagnostic challenges lead to many individuals who are undiagnosed or misdiagnosed with other developmental disorders (Chasnoff et al. 2015). PAE has been identified as the most preventable cause of developmental impairments in children (Williams et al. 2015), and FASD is more common than other developmental disorders like autism spectrum disorder (Maenner et al. 2021). Thus, prenatal exposure to alcohol is a significant public health concern.

2 Diagnostic Criteria

Fetal alcohol spectrum disorder (FASD) includes a range of adverse outcomes that can include facial dysmorphology, pre- or postnatal growth restriction, and neurobehavioral deficits caused by PAE (Moore and Xia 2022). While FASD is caused by prenatal exposure to alcohol, outcomes vary in severity as there are many factors that influence developmental outcomes such as dose or timing of exposure (Coles et al. 2015). However, this chapter will focus on individuals with PAE who have an FASD diagnosis. Approximately 630,000 children are born with FASD each year (Moore and Xia 2022), and studies suggest that the prevalence of FASD ranges from 11.3 to 50 cases per 1000 children (May et al. 2018), though prevalence rates differ from country to country (Wozniak et al. 2019). In addition, many confounding variables, such as separation from their biological mother or unavailable or unreliable information about PAE, affect the ability to make a diagnosis (Wozniak et al. 2019).

The group of conditions included in FASD consists of fetal alcohol syndrome (FAS), partial FAS (PFAS), alcohol-related neurodevelopmental disorder (ARND), alcohol-related birth defects (ARBD), and neurobehavioral disorder associated with prenatal alcohol exposure (ND-PAE; see Table 1). FAS is the most severe of the disorders (Hoyme et al. 2016; Mattson and Riley 2011), and this diagnosis includes four criteria: facial dysmorphology; prenatal or postnatal growth impairment; abnormality in brain growth, morphogenesis, or neurophysiology; and neurobehavioral impairment (Hoyme et al. 2016). Facial dysmorphology includes short palpebral fissures, a thin vermillion border, and a smooth philtrum. PFAS also requires facial dysmorphology and neurobehavioral impairment with documented prenatal exposure to alcohol. If exposure is unknown or cannot be documented, PFAS additionally requires deficient somatic or brain growth (Hoyme et al. 2016). In contrast to FAS and PFAS, ARND, ARBD, and ND-PAE do not require the presence of facial features. ARND requires confirmed prenatal alcohol exposure and neurobehavioral impairment, which can present in cognitive impairment or behavioral impairment. All diagnoses require neurobehavioral impairment with the exception of ARBD, which requires documented prenatal alcohol exposure and the presence of specific major malformations (e.g., cardiac, skeletal; Hoyme et al. 2016). A diagnosis of ND-PAE can be given with or without a diagnosis of FAS or PFAS but requires confirmation of prenatal alcohol exposure. For ND-PAE, symptoms need to be present in three domains: neurocognitive functioning, self-regulation, and adaptive functioning (Doyle and Mattson 2015; Kable et al. 2016; Mattson et al. 2019). Currently, ND-PAE is listed under "Conditions for Further Study" in the *Diagnostic and Statistical Manual of Mental Disorders* (DSM; American Psychiatric Association 2013). While ARND and ND-PAE are similarly focused on neurobehavioral impairment, ARND is a medical diagnosis that can be made based on global cognitive deficits with or without the presence of behavioral issues and ND-PAE is a mental health diagnosis intended to be used in clinical settings (American Psychiatric Association 2013; Hoyme et al. 2016).

The severity of symptoms that arise from PAE is related to the amount and timing of the alcohol exposure, although multiple pre- and postnatal risk factors exist that can exacerbate or ameliorate the effects of PAE (Coles 1994). There is no specific amount of alcohol required for a diagnosis, and different diagnostic systems vary in their requirements. A large prevalence study defined PAE as six or more drinks per week for two or more weeks, three or more drinks per occasion at least twice, or

Table 1 Table of requirements for diagnoses included in FASD (American Psychiatric Association 2013; Hoyme et al. 2016)

	Documented prenatal exposure to alcohol	Facial dysmorphology	Growth deficiency	CNS structural abnormalities	Neurobehavioral impairment	Specific major malformations
FAS		✔	✔	✔	✔	
PFAS[a]	✔	✔			✔	
		✔	✔[b]		✔	
ARBD	✔					✔
ARND[c]	✔				✔	
ND-PAE[d]	✔				✔	

[a] Diagnosis can be made with or without documented exposure to alcohol
[b] One or more required if no documented exposure to alcohol
[c] Diagnosis cannot be made for children under 3 years old or when FAS or PFAS is present
[d] Diagnosis is independent of physical indicators of prenatal alcohol exposure and can be made in individuals with or without FAS or PFAS

documented social or legal problems related to alcohol around the time of the pregnancy (May et al. 2018). While this is lower than used elsewhere, there have been reports of alcohol-related effects at lower levels of exposure (Wozniak et al. 2019). Exposure during the first trimester, when many of the major organs form, can cause facial dysmorphia and neurological damage, and animal studies suggest that early exposure to alcohol is more damaging than later PAE (Coles 1994).

3 Impact on Neuropsychological Function

Prenatal exposure to alcohol can negatively impact neuropsychological functioning, including cognitive and behavioral outcomes (Mattson et al. 2019; Streissguth et al. 2004). The cognitive domains that are impacted by PAE include general intelligence, motor function, adaptive functioning, visual perception, learning, memory, and executive functioning (Mattson et al. 2019). Additionally, individuals with FASD also commonly exhibit behavioral challenges that significantly impact daily life (Streissguth et al. 2004).

3.1 Cognition

Individuals with FASD often have decreased intelligence quotient (IQ) scores, and performance on intelligence tests is negatively impacted (Mattson and Riley 1998). Average IQ scores typically fall between 70 and 90 for individuals with FASD, although the range of scores is broad (Mattson and Riley 1998). Motor function has also been shown to be impacted in children with FASD as evidenced by impaired balance, poor hand-eye coordination, and difficulties with fine motor control (Connor et al. 2006; Doney et al. 2014). Furthermore, individuals with FASD demonstrate impairments across domains of adaptive functioning including daily living skills, social skills, and communication (Carr et al. 2010). Individuals with FASD commonly display deficits in attention compared to non-exposed individuals (Lee et al. 2004; Mattson et al. 2006; Nanson and Hiscock 1990). In particular, children with FASD have shown difficulties with sustained attention and attending to visual stimuli (Rasmussen et al. 2013). In alcohol-exposed children, impairments in verbal learning skills and recall have been well-documented throughout the literature (Crocker et al. 2011). Challenges in verbal learning and memory persist even when controlling for IQ (Lewis et al. 2015), which highlights the profound impacts of alcohol exposure.

Individuals with FASD display impairments across all domains of executive functioning compared to non-exposed individuals (Khoury et al. 2015; Kodali et al. 2017; Moore et al. 2021). Executive functioning encompasses a collection of processes including working memory, self-regulation, problem solving, anticipation, and inhibition that are involved in goal-directed actions (Anderson 2002). Executive functioning domains that are impacted by PAE include working memory, inhibition, and set shifting (Khoury et al. 2015). Children with FASD display lower performance on response inhibition tasks, and neuroimaging research supports this finding with evidence of different patterns of activation in brain regions involved in inhibition (Kodali et al. 2017). Furthermore, working memory is impacted by PAE as evidenced by worsened performance on tests of working memory with larger differences seen as age increases (Moore et al. 2021). In a review of impairments of executive functioning in children with PAE, planning, fluency, and set-shifting were the most impacted (Kingdon et al. 2016). Children with FASD exhibit impairments in both verbal and nonverbal fluency as evidenced by worsened performance on set-shifting fluency tasks (Schonfeld et al. 2001).

3.2 Behavior

Challenging behaviors that have been associated with FASD include aggressive behavior, trouble with the law, academic difficulties, inappropriate sexual behavior, and substance abuse (Joseph et al. 2022; Lynch et al. 2017; Streissguth et al. 1996, 2004; Tsang et al. 2016). Children and adolescents diagnosed with FASD have increased vulnerability for aggressive behavior and difficulties with emotion regulation (Joseph et al. 2022), and increases are noted in disruptive behaviors such as hyperactivity, inattention, lying, cheating, lack of guilt, and disobedience (Nash et al. 2006). Oppositional behaviors and those associated with conduct disorder are more common in individuals with FASD (Doyle et al. 2019), and FASD has been associated with elevated levels of conduct disorder symptoms in adolescents even when controlling for additional environmental factors (Disney et al. 2008). The results of one study demonstrated that FASD is related to problem behaviors in early adulthood, such as mental health challenges, substance use, and difficulties with the legal system. Notably, individuals with PAE who did not display dysmorphic facial characteristics of FASD exhibited the highest rates of problem behaviors as adults (Lynch et al. 2017). Moreover, many children and adolescents diagnosed with FASD report frequent sleep problems including trouble falling asleep, staying asleep, and waking up early (Hayes et al. 2020). Parents of children and adolescents with FASD who experience more sleep problems report elevated challenging behaviors and poorer child and family outcomes (Hayes et al. 2020).

4 Impact on Psychopathology

Prenatal exposure to environmental toxins, such as alcohol, increases the risk for neurodevelopmental conditions, such as autism and ADHD (Sonuga-Barke et al. 2022). Children with PAE are at higher risk for psychiatric disorders and are more likely to experience symptoms of disorders such as ADHD, depressive disorders, oppositional defiant disorder (ODD), conduct disorder (CD), and specific phobia (Fryer et al. 2007), and similar comorbidities have been seen in adults with FASD (Famy et al. 1998). Other co-occurring conditions include language disorders, specific learning disorders, epilepsy, and sleep disturbances (Mattson et al. 2019; Wozniak et al. 2019). The most common psychiatric diagnoses associated with FASD are ADHD, ODD, CD, anxiety disorders, and depression (Mattson et al. 2019), and these associations are described below.

4.1 *Attention-Deficit/Hyperactivity Disorder*

ADHD is defined by its two overarching symptoms, inattention and hyperactivity/impulsivity (American Psychiatric Association 2013). PAE increases the risk of developing ADHD (Pagnin et al. 2019), and similar brain regions are affected in FASD and ADHD, such as the basal ganglia (Sonuga-Barke et al. 2022). While there is significant clinical overlap between FASD and ADHD, differences are also notable (Mattson et al. 2019). A study by Glass et al. (2013) compared the neuropsychological functioning of children with and without PAE and with and without ADHD. Overall, children with PAE with or without ADHD were more impaired than children without PAE regardless of ADHD diagnosis. A similar study of behavioral outcomes had different results; the combination of PAE and ADHD resulted in worse outcomes than either condition alone (Ware et al. 2013). Thus, it appears that PAE and ADHD have differential effects on neuropsychological and behavioral functioning, and both conditions should be evaluated.

4.2 *Oppositional Defiant Disorder and Conduct Disorder*

ODD is defined by angry and irritable mood, argumentative and defiant behavior, and vindictiveness that presents itself in childhood (American Psychiatric Association 2013). CD is presented by a pattern of behavior that violates the basic rights of others, taking the form of aggression to people and animals, destruction of property, deceitfulness or theft, and serious violations of rules (American Psychiatric Association 2013). Both ODD and CD are associated with negative life outcomes, such as incarceration (Doyle et al. 2019) and both disorders commonly co-occur with FASD. The risk of developing conduct disorder symptoms is increased by PAE, even when controlling for parental alcoholism, parental drug abuse/dependence, and externalizing disorders (Disney et al. 2008). A meta-analysis by Popova et al. (2016) showed that conduct disorder was one of the most prevalent co-occurring disorders in individuals with FASD.

4.3 Anxiety and Depressive Disorders

Anxiety and depressive disorders are characterized by persistent worrying and depressed mood (American Psychiatric Association 2013). Children with FASD have an increased risk of anxiety and depressive disorders, and this increased risk is seen in adults and children (Famy et al. 1998; Fryer et al. 2007; Hellemans et al. 2010; Mattson et al. 2019). The increased prevalence of anxiety and depressive disorders associated with PAE may be due to a dysregulated HPA axis, as HPA axis dysregulation has been shown in both animals and humans prenatally exposed to alcohol (Hellemans et al. 2010).

5 Impact on Brain Development

The range of developmental deficits of PAE can be attributed to altered neural development. In animal studies, the severity of brain changes and resulting neurobehavioral difficulties are affected by the amount of PAE (Petrelli et al. 2018). In humans, there are likely a multitude of factors that interact with PAE to cause brain changes. Over the last 30 years, differences in brain structure and function in people with FASD have been studied using brain imaging (Archibald et al. 2001; Donald et al. 2015; Lebel et al. 2019; Mattson et al. 2001; Moore and Xia 2022).

5.1 Brain Volume

Multiple studies have found that, compared to children with no PAE, children with PAE have smaller total brain volume (Andre et al. 2020; Donald et al. 2015; Lebel et al. 2011; Moore and Xia 2022; Norman et al. 2009; Yang et al. 2012). This reduction is primarily due to reductions in the frontal, temporal, and parietal lobe (Norman et al. 2009). Despite this difference, children with PAE have similar brain volume trajectories to controls (Moore and Xia 2022). This difference in brain volume may be attributed to specific areas such as the corpus callosum, basal ganglia, and cerebellum.

5.2 Cortical Thickness

Cortical thinning is essential in development to refine neural networks (Treit et al. 2014). Children with FASD may have selective cortical thinning and thickening (Wozniak et al. 2019), and white matter volume is more affected than gray matter volume in FASD (Archibald et al. 2001). Specifically, studies have shown that there is less white matter in the perisylvian cortices of the temporal and parietal regions of children with FASD (Norman et al. 2009; Zhou et al. 2018). Children with FASD have been shown to have greater cortical thickness in the lateral temporal, parietal, frontal, and occipital cortices (Norman et al. 2009), and this difference was also shared by children with PAE that did not have facial dysmorphology (Sowell et al. 2008). A study by Yang et al. (2012) found that cortical thickness in children with FASD is related to palpebral fissure length; compared to controls, children with FASD had increased cortical thickness that was most prominent in the inferior frontal gyrus. In a longitudinal MRI study of children between the ages of 5 and 15 with and without FASD, Treit et al. (2014) found that controls showed more cortical thinning over a 2- to 4-year period compared to children with PAE. However, some studies show no group differences between age and cortical thickness (Moore and Xia 2022). Increased cortical thickness in children with FASD may lead to deficits on tests of verbal abilities such as verbal recall and visuospatial functioning (Sowell et al. 2008).

5.3 Specific Affected Regions

In addition to overall reductions in brain volume, PAE affects some parts of the brain to a greater degree than others. In particular, the corpus callosum, basal ganglia, and cerebellum appear to be affected (Bookstein et al. 2006; Mattson et al. 1992; Riley et al. 1995).

Corpus Callosum PAE particularly affects the corpus callosum, which is a bundle of nerve fibers that join the right and left hemispheres to allow for the transfer of information between hemispheres (Riley et al. 1995; Wozniak and Muetzel 2011). This structure is associated with neuropsychological functioning, integrating motor and sensory information and cognitive processing (Inkelis et al. 2020). Overall, the volume of the corpus callosum is smaller in children with FASD, and the shape of the corpus callosum also tends to vary, especially the mid and posterior areas which are the most affected due to PAE (Riley et al. 1995). If severe, PAE may result in agenesis of the corpus callosum (Mattson et al. 2001; Norman et al. 2009). The reduction in size of the corpus callosum in children with FASD has been linked to impairment in motor functioning, attention, verbal learning, and executive function (Biffen et al. 2017; Norman et al. 2009).

Basal Ganglia The basal ganglia include the caudate, putamen, and pallidum and are connected to the frontal lobes (Fazl and Fleisher 2018). This group of gray matter structures are associated with motor and cognitive functioning (Fazl and Fleisher 2018). The earliest MRI studies of children with FASD (Mattson et al. 1992, 1994, 1996) documented reductions in basal ganglia volume (Mattson et al. 1992), and more recent studies support this finding (Donald et al. 2015). The reduced size of the caudate nucleus may account for some of the impairments in cognitive functioning seen in FASD, specifically in boys (Woods et al. 2018). Because the caudate also has connections to the frontal lobes, decreased volume of the caudate may have downstream effects on the function of the frontal-subcortical networks, resulting in executive functioning impairment in children with FASD (Khoury et al. 2015; Mattson et al. 1996).

Cerebellum The cerebellum is known for motor function and cognitive skills and plays a role in emotional processing (Schmahmann et al. 2019). The cerebellum is vulnerable to PAE (Bookstein et al. 2006), and children with FASD have shown smaller volume in the cerebellum compared to controls (Archibald et al. 2001), specifically in the anterior vermis region (Sowell et al. 1996). This region of the cerebellum develops early and is more affected than posterior regions of the cerebellum that develop later (Sowell et al. 1996). The effects of cerebellar abnormalities have been linked to deficits in attention and balance (Schmahmann et al. 2019), and reduction in volume of this region relates to impaired balance and coordination and attention deficits seen in children with FASD (Mattson et al. 2006; Roebuck et al. 1998). Deficits due to this region have also been seen in a conditioned eye-blink response task used to assess conditioning processes (Spottiswoode et al. 2011).

Other Subcortical Structures Neuroimaging studies have also shown an effect of PAE on the cingulate gyrus and hippocampus (Bjorkquist et al. 2010; Roediger et al. 2021). The cingulate gyrus is associated with attention, cognitive control, executive functioning, and emotion (Bubb et al. 2018). In children with PAE, worse performance on working memory and attention measures was associated with reduced cingulate volume (Bjorkquist et al. 2010). The hippocampus has also been shown to be affected by PAE (Roediger et al. 2021). Some studies have shown reduced hippocampal volume when compared to children without PAE, while other studies show no difference between the two groups (Archibald et al. 2001; Norman et al. 2009; Roediger et al. 2021). Reduction in hippocampal volume could be associated with short- and long-delay memory deficits seen in children with FASD (Roediger et al. 2021).

6 Lifespan Consideration

The physical and neuropsychological impairments resulting from prenatal exposure to alcohol are long-lasting and persist throughout the lifespan (Coles et al. 2022; Lynch et al. 2017; Moore and Riley 2015; Streissguth et al. 1996). While most research has focused on children and adolescents with FASD, limited results suggest that the physical, cognitive, and behavioral impairments remain in adulthood (Moore and Riley 2015). These life-long impairments are supported by neuroimaging research which shows that individuals with FASD have atypical cortical and white matter microstructural developmental trajectories that persist from childhood through adolescence (Moore and Xia 2022). Additionally, negative mental health and behavioral outcomes are commonly reported in adults with FASD (Lynch et al.

2017; Streissguth et al. 1996). Adults with FASD experience significant impairments in daily living including challenges with independent living, housing, and employment (McLachlan et al. 2020). Prenatal exposure to alcohol is associated with elevated rates of mental health challenges such as anxiety and depression in adulthood (Coles et al. 2022). In one sample, mental health challenges were reported by 94% of adults with FASD and 23% reported suicide attempts (Streissguth et al. 1996). The most commonly reported diagnoses according to the DSM (American Psychiatric Association 2013) in a sample of adults with FASD were alcohol or drug dependence, depression, psychotic disorders, and avoidant, antisocial, and dependent personality disorders (Famy et al. 1998). Additionally, the lifelong cognitive and behavioral challenges of individuals with FASD may contribute to increased involvement in the criminal justice system as witnesses, complainants, and defendants (Fast and Conry 2009). Adults with FASD display higher rates of conduct and oppositional problem behaviors than non-exposed controls (Lynch et al. 2017). Moreover, adults with FASD experience high rates of confinement (Streissguth et al. 1996). In one sample among adults with FASD, 40% had been incarcerated, 30% had been committed to psychiatric treatment, and 20% had received in-patient substance use treatment (Streissguth et al. 1996).

7 Treatment and Intervention

Although there have been recent efforts to explore postnatal interventions for individuals with FASD, there is relatively little research on the availability and effectiveness of interventions in this population (Flannigan et al. 2020; Pei et al. 2017; Reid et al. 2015). The most common evidence-based interventions for individuals with FASD include early interventions for infant development and targeted treatments to address self-regulation, attentional control, specific skills, and social skills (Petrenko et al. 2017; Reid et al. 2015). Family-focused interventions such as parent education, support and advocacy services for parents, and mentorship services for caregivers have also been explored (Petrenko et al. 2017; Reid et al. 2015). Early intervention services may be especially critical in targeting cognitive and behavioral impairments in children with neurodevelopmental disorders (Reid et al. 2015). However, diagnostic challenges and high rates of undiagnosed or misdiagnosed children with PAE remain significant barriers to early intervention in this population (Olson et al. 2007; Reid et al. 2015). Pharmacological treatments are also used to target ADHD symptoms, emotion dysregulation, and sleep disturbances in children with FASD (Ritfeld et al. 2022). However, Ritfeld et al. (2022) found conflicting evidence on the effectiveness of pharmacological treatments in children with FASD with limited available research. Additionally, micronutrient supplements consumed pre- and postnatally may improve outcomes in alcohol-exposed individuals (Coles et al. 2015; Wozniak et al. 2015, 2020). In a preliminary study conducted by Coles et al. (2015), micronutrient supplements consumed prenatally had a positive effect on infant developmental outcomes. There is evidence to suggest that the micronutrient supplement choline consumed postnatally may improve developmental outcomes in children with FASD (Coles et al. 2015; Wozniak et al. 2015, 2020). Although there have been recent efforts to address the research gap in evidence-based interventions for individuals with FASD throughout the lifespan, more research is needed to discern the effectiveness of these different treatment approaches.

8 Summary

FASD includes a range of disorders affected by PAE (Mattson et al. 2019; Moore and Xia 2022). Studies have shown that children with PAE exhibit smaller total brain volume than children without PAE and increased cortical thickness (Donald et al. 2015; Lebel et al. 2011; Moore and Xia 2022; Norman et al. 2009; Sowell et al. 2008; Treit et al. 2014; Yang et al. 2012) with selective reductions to the corpus callosum, basal ganglia, and cerebellum (Archibald et al. 2001; Donald et al. 2015; Mattson et al. 1992; Riley et al. 1995; Sowell et al. 1996). It has been well-established that PAE can cause profound cognitive and behavioral impairments that persist throughout the lifespan (Mattson et al. 2019; Moore and Riley 2015; Streissguth et al. 1996). Children with FASD may experience co-occurring disorders such as ADHD, oppositional defiant disorder, conduct disorder, anxiety disorders, and depressive disorders (Mattson et al. 2019). There are different treatment approaches for individuals with FASD including behavioral interventions, family-focused interventions, and pharmacological treatments (Reid et al. 2015; Ritfeld et al. 2022). In light of the broad array of potential negative outcomes, the impact on brain development, the high rate of missed diagnoses, and the relative paucity of treatment options, more research is needed to advance diagnostic techniques and postnatal interventions to improve outcomes in individuals that have been prenatally exposed to alcohol.

Acknowledgments Preparation of this chapter was supported by grants U01 AA014834 and T32 AA013525 from the National Institute on Alcohol Abuse and Alcoholism.

References

American Psychiatric Association (2013) In: Diagnostic and statistical manual of mental disorders, fifth edition (DSM-5). 5th edn

Anderson P (2002) Assessment and development of executive function (EF) during childhood. Child Neuropsychol 8(2):71–82. https://doi.org/10.1076/chin.8.2.71.8724

Andre QR, McMorris CA, Kar P, Ritter C, Gibbard WB, Tortorelli C, Lebel C (2020) Different brain profiles in children with prenatal alcohol exposure with or without early adverse exposures. Hum Brain Mapp 41(15):4375–4385. https://doi.org/10.1002/hbm.25130

Archibald SL, Fennema-Notestine C, Gamst A, Riley EP, Mattson SN, Jernigan TL (2001) Brain dysmorphology in individuals with severe prenatal alcohol exposure. (0012–1622 (Print))

Biffen SC, Warton CMR, Lindinger NM, Randall SR, Lewis CE, Molteno CD, Jacobson JL, Jacobson SW, Meintjes EM (2017) Reductions in corpus callosum volume partially mediate effects of prenatal alcohol exposure on IQ. Front Neuroanat 11:132. https://doi.org/10.3389/fnana.2017.00132

Bjorkquist OA, Fryer SL, Reiss AL, Mattson SN, Riley EP (2010) Cingulate gyrus morphology in children and adolescents with fetal alcohol spectrum disorders. Psychiatry Res Neuroimaging 181(2):101–107. https://doi.org/10.1016/j.pscychresns.2009.10.004

Bookstein FL, Streissguth AP, Connor PD, Sampson PD (2006) Damage to the human cerebellum from prenatal alcohol exposure: the anatomy of a simple biometrical explanation. Anat Rec 289(5):195–209. https://doi.org/10.1002/ar.b.20114

Bubb EJ, Metzler-Baddeley C, Aggleton JP (2018) The cingulum bundle: anatomy, function, and dysfunction. Neurosci Biobehav Rev 92:104–127. https://doi.org/10.1016/j.neubiorev.2018.05.008

Carr JL, Agnihotri S, Keightley M (2010) Sensory processing and adaptive behavior deficits of children across the fetal alcohol spectrum disorder continuum. Alcohol Clin Exp Res 34(6): 1022–1032. https://doi.org/10.1111/j.1530-0277.2010.01177.x

Chasnoff IJ, Wells AM, King L (2015) Misdiagnosis and missed diagnoses in foster and adopted children with prenatal alcohol exposure. Pediatrics 135(2):264–270. https://doi.org/10.1542/peds.2014-2171

Coles C (1994) Critical periods for prenatal alcohol exposure: evidence from animal and human studies. Alcohol Health Res World 18(1):22–29

Coles CD, Kable JA, Keen CL, Jones KL, Wertelecki W, Granovska IV, Pashtepa AO, Chambers CD (2015) Dose and timing of prenatal alcohol exposure and maternal nutritional supplements: developmental effects on 6-month-old infants. Matern Child Health J 19(12):2605–2614. https://doi.org/10.1007/s10995-015-1779-x

Coles CD, Grant TM, Kable JA, Stoner SA, Perez A (2022) Prenatal alcohol exposure and mental health at midlife: a preliminary report on two longitudinal cohorts. Alcohol Clin Exp Res 46(2): 232–242. https://doi.org/10.1111/acer.14761

Connor PD, Sampson PD, Streissguth AP, Bookstein FL, Barr HM (2006) Effects of prenatal alcohol exposure on fine motor coordination and balance: a study of two adult samples. Neuropsychologia 44(5):744–751. https://doi.org/10.1016/j.neuropsychologia.2005.07.016

Crocker N, Vaurio L, Riley EP, Mattson SN (2011) Comparison of verbal learning and memory in children with heavy prenatal alcohol exposure or attention-deficit/hyperactivity disorder. Alcohol Clin Exp Res 35(6):1114–1121. https://doi.org/10.1111/j.1530-0277.2011.01444.x

Disney ER, Iacono W, McGue M, Tully E, Legrand L (2008) Strengthening the case: prenatal alcohol exposure is associated with increased risk for conduct disorder. Pediatrics 122(6): e1225–e1230. https://doi.org/10.1542/peds.2008-1380

Donald KA, Roos A, Fouche J-P, Koen N, Howells FM, Woods RP, Zar HJ, Narr KL, Stein DJ (2015) A study of the effects of prenatal alcohol exposure on white matter microstructural integrity at birth. Acta Neuropsychiatrica 27(4):197–205. https://doi.org/10.1017/neu.2015.35

Doney R, Lucas BR, Jones T, Howat P, Sauer K, Elliot EJ (2014) Fine motor skills in children with prenatal alcohol exposure or fetal alcohol spectrum disorder. J Dev Behav Pediatr 35(9): 598–609. https://doi.org/10.1097/DBP.0000000000000107

Doyle LR, Mattson SN (2015) Neurobehavioral disorder associated with prenatal alcohol exposure (ND-PAE): review of evidence and guidelines for assessment. (2196-2987 (print))

Doyle LR, Glass L, Wozniak JR, Kable JA, Riley EP, Coles CD, Sowell ER, Jones KL, Mattson SN (2019) Relation between oppositional/conduct behaviors and executive function among youth with histories of heavy prenatal alcohol exposure. Alcohol Clin Exp Res 43(6):1135–1144. https://doi.org/10.1111/acer.14036

Famy C, Streissguth AP, Unis AS (1998) Mental illness in adults with fetal alcohol syndrome or fetal alcohol effects. Am J Psychiatry 155(4):552–554. https://doi.org/10.1176/ajp.155.4.552

Fast DK, Conry J (2009) Fetal alcohol spectrum disorders and the criminal justice system. Dev Disabil Res Rev 15(3):250–257. https://doi.org/10.1002/ddrr.66

Fazl A, Fleisher J (2018) Anatomy, physiology, and clinical syndromes of the basal ganglia: a brief review. Semin Pediatr Neurol 25:2–9. https://doi.org/10.1016/j.spen.2017.12.005

Flannigan K, Coons-Harding KD, Anderson T, Wolfson L, Campbell A, Mela M, Pei J (2020) A systematic review of interventions to improve mental health and substance use outcomes for individuals with prenatal alcohol exposure and fetal alcohol spectrum disorder. Alcohol Clin Exp Res 44(12):2401–2430. https://doi.org/10.1111/acer.14490

Fryer SL, McGee CL, Matt GE, Riley EP, Mattson SN (2007) Evaluation of psychopathological conditions in children with heavy prenatal alcohol exposure. Pediatrics 119(3):e733–e741. https://doi.org/10.1542/peds.2006-1606

Glass L, Ware AL, Crocker N, Deweese BN, Coles CD, Kable JA, May PA, Kalberg WO, Sowell ER, Jones KL, Riley EP, Mattson SN (2013) Neuropsychological deficits associated with heavy prenatal alcohol exposure are not exacerbated by ADHD. Neuropsychology 27(6):713–724. https://doi.org/10.1037/a0033994

Hayes N, Moritz KM, Reid N (2020) Parent-reported sleep problems in school-aged children with fetal alcohol spectrum disorder: association with child behaviour, caregiver, and family functioning. Sleep Med 74:307–314. https://doi.org/10.1016/j.sleep.2020.07.022

Hellemans KG, Sliwowska JH, Verma P, Weinberg J (2010) Prenatal alcohol exposure: fetal programming and later life vulnerability to stress, depression and anxiety disorders. Neurosci Biobehav Rev 34(6):791–807. https://doi.org/10.1016/j.neubiorev.2009.06.004

Hoyme HE, Kalberg WO, Elliott AJ, Blankenship J, Buckley D, Marais AS, Manning MA, Robinson LK, Adam MP, Abdul-Rahman O, Jewett T, Coles CD, Chambers C, Jones KL, Adnams CM, Shah PE, Riley EP, Charness ME, Warren KR, May PA (2016) Updated clinical guidelines for diagnosing fetal alcohol spectrum disorders. Pediatrics 138(2). https://doi.org/10.1542/peds.2015-4256

Inkelis SM, Moore EM, Bischoff-Grethe A, Riley EP (2020) Neurodevelopment in adolescents and adults with fetal alcohol spectrum disorders (FASD): a magnetic resonance region of interest analysis. Brain Res 1732:146654. https://doi.org/10.1016/j.brainres.2020.146654

Jones KL, Smith DW, Ulleland CN, Streissguth AP (1973) Pattern of malformation in offspring of chronic alcoholic mothers. Lancet 301(7815):1267–1271. https://doi.org/10.1016/S0140-6736(73)91291-9

Joseph JJ, Mela M, Pei J (2022) Aggressive behaviour and violence in children and adolescents with FASD: a synthesizing review. Clin Psychol Rev 94:1–12. https://doi.org/10.1016/j.cpr.2022.102155

Kable JA, O'Connor MJ, Olson HC, Paley B, Mattson SN, Anderson SM, Riley EP (2016) Neurobehavioral disorder associated with prenatal alcohol exposure (ND-PAE): proposed DSM-5 diagnosis. Child Psychiatry Hum Dev 47(2):335–346. https://doi.org/10.1007/s10578-015-0566-7

Khoury JE, Milligan K, Girard TA (2015) Executive functioning in children and adolescents prenatally exposed to alcohol: a meta-analytic review. Neuropsychol Rev 25(2):149–170. https://doi.org/10.1007/s11065-015-9289-6

Kingdon D, Cardoso C, McGrath JJ (2016) Research review: executive function deficits in fetal alcohol spectrum disorders and attention-deficit/hyperactivity disorder – a meta-analysis. J Child Psychol Psychiatry 57(2):116–131. https://doi.org/10.1111/jcpp.12451

Kodali VN, Jacobson JL, Lindinger NM, Dodge NC, Molteno CD, Meintjes EM, Jacobson SW (2017) Differential recruitment of brain regions during response inhibition in children prenatally exposed to alcohol. Alcohol Clin Exp Res 41(2):334–344. https://doi.org/10.1111/acer.13307

Lebel C, Roussotte F, Sowell ER (2011) Imaging the impact of prenatal alcohol exposure on the structure of the developing human brain. Neuropsychol Rev 21(2):102–118. https://doi.org/10.1007/s11065-011-9163-0

Lebel CA, McMorris CA, Kar P, Ritter C, Andre Q, Tortorelli C, Gibbard WB (2019) Characterizing adverse prenatal and postnatal experiences in children. Birth Defects Res 111(12): 848–858. https://doi.org/10.1002/bdr2.1464

Lee K, Mattson S, Riley E (2004) Classifying children with heavy prenatal alcohol exposure using measures of attention. J Int Neuropsychol Soc 10(2):271–277. https://doi.org/10.1017/S1355617704102142

Lewis CE, Thomas KG, Dodge NC, Molteno CD, Meintjes EM, Jacobson JL, Jacobson SW (2015) Verbal learning and memory impairment in children with fetal alcohol spectrum disorders. Alcohol Clin Exp Res 39(4):724–732. https://doi.org/10.1111/acer.12671

Lynch ME, Kable JA, Coles CD (2017) Effects of prenatal alcohol exposure in a prospective sample of young adults: mental health, substance use, and difficulties with the legal system. Neurotoxicol Teratol 64:50–62. https://doi.org/10.1016/j.ntt.2017.10.001

Maenner, M. J., Shaw KA, Bakian AV, Bilder DA, Durkin MS, Esler A, Furnier SM, Hallas L, Hall-Lande J, Hudson A, Hughes MM, Patrick M, Pierce K, Poynter JN, Salinas A, Shenouda J, Vehorn A, Warren Z, Constantino JN, DiRienzo M, Fitzgerald RT, Grzybowski A, Spivey MH, Pettygrove S, Zahorodny W, Ali A, Andrews JG, Baroud T, Gutierrez J, Hewitt A, Lee LC, Lopez M, Mancilla KC, McArthur D, Schwenk YD, Washington A, Williams S, Cogswell ME (2021). Prevalence and characteristics of autism spectrum disorder among children aged 8 years – autism and developmental disabilities monitoring network, 11 sites, United States, 2018. (1545-8636 (electronic))

Mattson SN, Riley EP (1998) A review of the neurobehavioral deficits in children with fetal alcohol syndrome or prenatal exposure to alcohol. Alcohol Clin Exp Res 22(2):279–294. https://doi.org/10.1111/j.1530-0277.1998.tb03651.x

Mattson SN, Riley EP (2011) The quest for a neurobehavioral profile of heavy prenatal alcohol exposure. Alcohol Res Health 34(1):51–55

Mattson SN, Riley EP, Jernigan TL, Ehlers CL, Delis DC, Jones KL, Stern C, Johnson KA, Hesselink JR, Bellugi U (1992) Fetal alcohol syndrome: a case report of neuropsychological, MRI and EEG assessment of two children. Alcohol Clin Exp Res 16(5):1001–1003. https://doi.org/10.1111/j.1530-0277.1992.tb01909.x

Mattson SN, Riley EP, Jernigan TL, Garcia A, Kaneko WM, Ehlers CL, Jones KL (1994) A decrease in the size of the basal ganglia following prenatal alcohol exposure: a preliminary report. Neurotoxicol Teratol 16(3):283–289. https://doi.org/10.1016/0892-0362(94)90050-7

Mattson SN, Riley EP, Sowell ER, Jernigan TL, Sobel DF, Jones KL (1996) A decrease in the size of the basal ganglia in children with fetal alcohol syndrome. Alcohol Clin Exp Res 20(6): 1088–1093. https://doi.org/10.1111/j.1530-0277.1996.tb01951.x

Mattson SN, Schoenfeld AM, Riley EP (2001) Teratogenic effects of alcohol on brain and behavior. Alcohol Res Health 25(3):185–191. https://www.ncbi.nlm.nih.gov/pubmed/11810956

Mattson SN, Calarco KE, Lang AR (2006) Focused and shifting attention in children with heavy prenatal alcohol exposure. Neuropsychology 20(3):361–369. https://doi.org/10.1037/0894-4105.20.3.361

Mattson SN, Bernes GA, Doyle LR (2019) Fetal alcohol spectrum disorders: a review of the neurobehavioral deficits associated with prenatal alcohol exposure. Alcohol Clin Exp Res 43(6):1046–1062. https://doi.org/10.1111/acer.14040

May PA, Chambers CD, Kalberg WO, Zellner J, Feldman H, Buckley D, Kopald D, Hasken JM, Xu R, Honerkamp-Smith G, Taras H, Manning MA, Robinson LK, Adam MP, Abdul-Rahman-O, Vaux K, Jewett T, Elliott AJ, Kable JA et al (2018) Prevalence of fetal alcohol spectrum disorders in 4 US communities. JAMA 319(5):474–482. https://doi.org/10.1001/jama.2017.21896

McLachlan K, Flannigan K, Temple V, Unsworth K, Cook JL (2020) Difficulties in daily living experienced by adolescents, transition-aged youth, and adults with fetal alcohol spectrum disorder. Alcohol Clin Exp Res 44(8):1609–1624. https://doi.org/10.1111/acer.14385

Moore EM, Riley EP (2015) What happens when children with fetal alcohol spectrum disorders become adults? Curr Dev Disord Rep 2(3):219–227. https://doi.org/10.1007/s40474-015-0053-7

Moore EM, Xia Y (2022) Neurodevelopmental trajectories following prenatal alcohol exposure. Front Hum Neurosci 15. https://doi.org/10.3389/fnhum.2021.695855

Moore EM, Glass L, Infante MA, Coles CD, Kable JA, Jones KL, Riley EP, Mattson SN (2021) Cross-sectional analysis of spatial working memory development in children with histories of heavy prenatal alcohol exposure. Alcohol Clin Exp Res 45(1):215–223. https://doi.org/10.1111/acer.14506

Nanson JL, Hiscock M (1990) Attention deficits in children exposed to alcohol prenatally. Alcohol Clin Exp Res 14(5):656–661

Nash K, Rovet J, Greenbaum R, Fantus E, Nulman I, Koren G (2006) Identifying the behavioural phenotype in fetal alcohol spectrum disorder: sensitivity, specificity and screening potential. Arch Womens Ment Health 9(4):181–186. https://doi.org/10.1007/s00737-006-0130-3

Norman AL, Crocker N, Mattson SN, Riley EP (2009) Neuroimaging and fetal alcohol spectrum disorders. Dev Disabil Res Rev 15(3):209–217. https://doi.org/10.1002/ddrr.72

Olson HC, Jirikowic T, Kartin D, Astley S (2007) Responding to the challenge of early intervention for fetal alcohol spectrum disorders. Infants Young Child 20(2):172–189. https://doi.org/10.1097/01.IYC.0000264484.73688.4a

Pagnin D, Zamboni Grecco ML, Furtado EF (2019) Prenatal alcohol use as a risk for attention-deficit/hyperactivity disorder. Eur Arch Psychiatry Clin Neurosci 269(6):681–687. https://doi.org/10.1007/s00406-018-0946-7

Pei J, Baugh L, Andrew G, Rasmussen C (2017) Intervention recommendations and subsequent access to services following clinical assessment for fetal alcohol spectrum disorders. Res Dev Disabil 60:176–186. https://doi.org/10.1016/j.ridd.2016.11.007

Petrelli B, Weinberg J, Hicks GG (2018) Effects of prenatal alcohol exposure (PAE): insights into FASD using mouse models of PAE. Biochem Cell Biol 96(2):131–147. https://doi.org/10.1139/bcb-2017-0280

Petrenko CLM, Pandolfino ME, Robinson LK (2017) Findings from the families on track intervention pilot trial for children with fetal alcohol spectrum disorders and their families. Alcohol Clin Exp Res 41(7):1340–1351. https://doi.org/10.1111/acer.13408

Popova S, Lange S, Shield K, Mihic A, Chudley AE, Mukherjee RAS, Bekmuradov D, Rehm J (2016) Comorbidity of fetal alcohol spectrum disorder: a systematic review and meta-analysis. Lancet 387(10022):978–987. https://doi.org/10.1016/S0140-6736(15)01345-8

Rasmussen C, Tamana S, Baugh L, Andrew G, Tough S, Zwaigenbaum L (2013) Neuropsychological impairments on the NEPSY-II among children with FASD. Child Neuropsychol 19(4):337–349. https://doi.org/10.1080/09297049.2012.658768

Reid N, Dawe S, Shelton D, Harnett P, Warner J, Armstrong E, LeGros K, O'Callaghan F (2015) Systematic review of fetal alcohol spectrum disorder interventions across the life span. Alcohol Clin Exp Res 39(12):2283–2295. https://doi.org/10.1111/acer.12903

Riley EP, Mattson SN, Sowell ER, Jernigan TL, Sobel DF, Jones KL (1995) Abnormalities of the corpus callosum in children prenatally exposed to alcohol. Alcohol Clin Exp Res 19(5):1198–1202. https://doi.org/10.1111/j.1530-0277.1995.tb01600.x

Ritfeld GJ, Kable JA, Holton JE, Coles CD (2022) Psychopharmacological treatments in children with fetal alcohol spectrum disorders: a review. Child Psychiatry Hum Dev 53(2):268–277. https://doi.org/10.1007/s10578-021-01124-7

Roebuck TM, Simmons RW, Mattson SN, Riley EP (1998) Prenatal exposure to alcohol affects the ability to maintain postural balance. Alcohol Clin Exp Res 22(1):252–258

Roediger DJ, Krueger AM, de Water E, Mueller BA, Boys CA, Hendrickson TJ, Schumacher MJ, Mattson SN, Jones KL, Lim KO, Wozniak JR (2021) Hippocampal subfield abnormalities and memory functioning in children with fetal alcohol spectrum disorders. Neurotoxicol Teratol 83. https://doi.org/10.1016/j.ntt.2020.106944

Schmahmann JD, Guell X, Stoodley CJ, Halko MA (2019) The theory and neuroscience of cerebellar cognition. Annu Rev Neurosci 42:337–364. https://doi.org/10.1146/annurev-neuro-070918-050258

Schonfeld A, Mattson S, Lang A, Delis D, Riley E (2001) Verbal and nonverbal fluency in children with heavy prenatal alcohol exposure. J Stud Alcohol 62(2):239–246. https://doi.org/10.15288/jsa.2001.62.239

Sonuga-Barke EJS, Becker SP, Bölte S, Castellanos FX, Franke B, Newcorn JH, Nigg JT, Rohde LA, Simonoff E (2022) Annual research review: perspectives on progress in ADHD science – from characterization to cause. J Child Psychol Psychiatry. https://doi.org/10.1111/jcpp.13696

Sowell ER, Jernigan TL, Mattson SN, Riley EP, Sobel DF, Jones KL (1996) Abnormal development of the cerebellar vermis in children prenatally exposed to alcohol: size reduction in lobules I-V. Alcohol Clin Exp Res 20(1):31–34. https://doi.org/10.1111/j.1530-0277.1996.tb01039.x

Sowell ER, Mattson SN, Kan E, Thompson PM, Riley EP, Toga AW (2008) Abnormal cortical thickness and brain-behavior correlation patterns in individuals with heavy prenatal alcohol exposure. Cereb Cortex 18(1):136–144. https://doi.org/10.1093/cercor/bhm039

Spottiswoode BS, Meintjes EM, Anderson AW, Molteno CD, Stanton ME, Dodge NC, Gore JC, Peterson BS, Jacobson JL, Jacobson SW (2011) Diffusion tensor imaging of the cerebellum and eyeblink conditioning in fetal alcohol spectrum disorder. Alcohol Clin Exp Res 35(12): 2174–2183. https://doi.org/10.1111/j.1530-0277.2011.01566.x

Streissguth AP, Barr HM, Kogan J, Bookstein FL (1996) Final report: understanding the occurrence of secondary disabilities in clients with fetal alcohol syndrome (FAS) and fetal alcohol effects (FAE). Final Report. University of Washington Publication Services

Streissguth AP, Bookstein FL, Barr HM, Sampson PD, O'Malley K, Young JK (2004) Risk factors for adverse life outcomes in fetal alcohol syndrome and fetal alcohol effects. J Dev Behav Pediatr 25(4):228–238. http://www.ncbi.nlm.nih.gov/entrez/query.fcgi?cmd=Retrieve&db=PubMed&dopt=Citation&list_uids=15308923

Tan CH, Denny CH, Cheal NE, Sniezek JE, Kanny D (2015) Alcohol use and binge drinking among women of childbearing age – United States, 2011-2013. Morb Mortal Wkly Rep 64(37): 1042–1046. https://doi.org/10.15585/mmwr.mm6437a3

Treit S, Zhou D, Lebel C, Rasmussen C, Andrew G, Beaulieu C (2014) Longitudinal MRI reveals impaired cortical thinning in children and adolescents prenatally exposed to alcohol. Hum Brain Mapp 35(9):4892–4903. https://doi.org/10.1002/hbm.22520

Tsang TW, Lucas BR, Olson HC, Pinto RZ, Elliott EJ (2016) Prenatal alcohol exposure, FASD, and child behavior: a meta-analysis. Pediatrics 137(3):1–20. http://libproxy.sdsu.edu/login?url=https://search.ebscohost.com/login.aspx?direct=true&db=psyh&AN=2016-23950-013&site=ehost-live&scope=site

Ware AL, O'Brien JW, Crocker N, Deweese BN, Roesch SC, Coles CD, Kable JA, May PA, Kalberg WO, Sowell ER, Jones KL, Riley EP, Mattson SN (2013) The effects of prenatal alcohol exposure and attention-deficit/hyperactivity disorder on psychopathology and behavior. Alcohol Clin Exp Res 37(3):507–516. https://doi.org/10.1111/j.1530-0277.2012.01953.x

Williams JF, Smith VC, Committee on Substance Abuse (2015) Fetal alcohol spectrum disorders. Pediatrics 136(5):e20153113. https://doi.org/10.1542/peds.2015-3113

Woods KJ, Thomas KGF, Molteno CD, Jacobson JL, Jacobson SW, Meintjes EM (2018) Prenatal alcohol exposure affects brain function during place learning in a virtual environment differently in boys and girls. Brain Behav 8(11):e01103. https://doi.org/10.1002/brb3.1103

Wozniak JR, Muetzel RL (2011) What does diffusion tensor imaging reveal about the brain and cognition in fetal alcohol spectrum disorders? Neuropsychol Rev 21(2):133–147. https://doi.org/10.1007/s11065-011-9162-1

Wozniak JR, Fuglestad AJ, Eckerle JK, Fink BA, Hoecker HL, Boys CJ, Radke JP, Kroupina MG, Miller NC, Brearley AM, Zeisel SH, Georgieff MK (2015) Choline supplementation in children with fetal alcohol spectrum disorders: a randomized, double-blind, placebo-controlled trial. Am J Clin Nutr 102(5):1113–1125. https://doi.org/10.3945/ajcn.114.099168

Wozniak JR, Riley EP, Charness ME (2019) Clinical presentation, diagnosis, and management of fetal alcohol spectrum disorder. (1474-4465 (electronic))

Wozniak JR, Fink BA, Fuglestad AJ, Eckerle JK, Boys CJ, Sandness KE, Radke JP, Miller NC, Lindgren C, Brearley AM, Zeisel SH, Georgieff MK (2020) Four-year follow-up of a randomized controlled trial of choline for neurodevelopment in fetal alcohol spectrum disorder. J Neurodev Disord 12(1):9. https://doi.org/10.1186/s11689-020-09312-7

Yang Y, Roussotte F, Kan E, Sulik KK, Mattson SN, Riley EP, Jones KL, Adnams CM, May PA, O'Connor MJ, Narr KL, Sowell ER (2012) Abnormal cortical thickness alterations in fetal alcohol spectrum disorders and their relationships with facial dysmorphology. Cereb Cortex 22(5):1170–1179. https://doi.org/10.1093/cercor/bhr193

Zhou D, Rasmussen C, Pei J, Andrew G, Reynolds JN, Beaulieu C (2018) Preserved cortical asymmetry despite thinner cortex in children and adolescents with prenatal alcohol exposure and associated conditions. Hum Brain Mapp 39(1):72–88. https://doi.org/10.1002/hbm.23818

Functional Consequences of Paternal Preconceptual Alcohol Consumption

Sahir Hussain, Darren Day, and Bart Ellenbroek

Contents

Abstract Alcohol is the most harmful drug of abuse, making alcoholism a major economic and public health crisis. Unsurprisingly, this has led to the majority of the neurobiological research on alcohol focusing on its direct effects on an individual, including those affected by foetal alcohol spectrum disorders (FASDs). However, research has shown that heavy paternal drinking predicts earlier and heavier adolescent drinking in the offspring, accompanied by other behavioural and molecular changes. While alcohol use disorder (AUD) is highly heritable, research on genetic variants alone does not sufficiently account for AUD risk and the FASDs-like symptoms seen in offspring of alcoholic fathers. Recently, there has been an increase in appreciation of the importance of epigenetic mechanisms of inheritance, which transfer changes due to parental experiences through the germline. This chapter aims to present an overview of the current knowledge on the inter- and transgenerational impacts of preconceptual paternal alcohol consumption (PPAC), the outcomes seen across generations and the mechanisms by which these changes may be passed down generations.

S. Hussain, D. Day, and B. Ellenbroek (✉)
Te Herenga Waka-Victoria University of Wellington, Wellington, New Zealand
e-mail: Bart.ellenbroek@vuw.ac.nz

Curr Topics Behav Neurosci (2025) 75: 117–147
https://doi.org/10.1007/7854_2025_601
Published Online: 23 August 2025

1 Introduction

Alcohol is the most widely used substance of abuse. The World Health Organization (WHO) reports approximately three million deaths every year resulting from harmful use of alcohol (WHO 2022). This represents 5.3% of all deaths per year, and 5.1% of the global burden of disease and injury is attributable to alcohol, as measured in disability-adjusted life years (WHO 2022). A study in Scandinavian countries showed individuals with alcohol use disorder (AUD) have 24–28 years shorter life expectancy (Westman et al. 2015). Therefore, AUD (also referred to as alcoholism or alcohol abuse) represents a major economic and public health crisis. There is widespread agreement among experts in the field that alcohol is the most harmful addictive substance, particularly if one combines harm to the individual with harm to society (Crossin et al. 2023; van Amsterdam et al. 2015). As a result, alcoholism places a significant burden on health and many national economies. For example, a recent New Zealand (Ministry of Health 2022) report stated 1 in 5 New Zealanders show hazardous drinking habits, with an even higher number in the age group of 18–34 years (up to 32–38%). Similar behavioural statistics are seen in other Western nations like Australia (Australian Bureau of Statistics 2021) and the United Kingdom (ONS 2022). This excessive alcohol use is directly related to reduced productivity and increased medical claims, road accidents, violence, and mortality. Reducing the damage caused by alcoholism will therefore not only benefit the patient and their family but can also substantially reduce the economic and judicial costs associated with hazardous drinking.

Data indicates national, cultural and gender differences in drinking behaviours with Europe and the regions of the Americas having the highest levels of drinking, while the Eastern Mediterranean nations consume the lowest amounts of alcohol per capita (WHO 2023). Differences in trends between regions across the world are evident too, with drinking showing a decline in Europe while we see rises in South-East Asia and Western Pacific regions. There are also cultural differences within countries; for New Zealand, adjusted for age and gender, Māori are the most likely to show hazardous drinking patterns, while Asians show the lowest (Ministry of Health 2022). In the Netherlands, the prevalence of drinking was substantially lower in all ethnic minority groups compared to the Dutch, and minorities showed a lower risk of alcohol abuse (van Amsterdam et al. 2020). Lastly, men are overwhelmingly the highest consumers of alcohol, with them drinking nearly four times more pure alcohol per capita than women across the world in 2019 (WHO 2023).

2 Gene X Environment in Alcohol Use

The development and severity of psychiatric disorders are characterized as being a product of both nature (family genetics) and nurture (the influence of one's surrounding environment across a lifetime), with these being termed gene x environment interactions (Ellenbroek and Youn 2016). Genetic factors play a large role in alcohol drinking behaviour; one major example is alcohol dehydrogenase (ADH) and aldehyde dehydrogenase (ALDH) polymorphisms, which are associated with different rates of alcohol metabolism. In particular, the *ALDH2*2* variant, which is common in Asian populations, produces an inactive ALDH enzyme, which leads to slower metabolism of alcohol and leads to the toxic accumulation of acetaldehyde in the blood and tissues, causing the alcohol-flush reaction and significantly altering drinking behaviour (Thomasson et al. 1993). Twin and adoption studies demonstrate a clear genetic component for the development of alcohol abuse; however, the estimated heritability ranges from 40% to 70% (Hottenga et al. 2018; Verhulst et al. 2015; Young-Wolff et al. 2011), which indicates an environmental influence in the development of these behaviours. Factors like availability of alcohol, childhood neglect, parental dinking and socioeconomic status, among others, have been associated (Dasgupta 2016). Importantly, the consequences to environmental factors seem to be influenced by genes associated with substance abuse. Examples of environmental factors include carriers of the s-allele of the 5-HTTLPR (serotonin-transporter-linked promoter region), which is associated with lower transcriptional efficiency, showing increased alcohol use, especially after childhood maltreatment (Kaufman et al. 2007; van der Zwaluw et al. 2010) and the protective effect of a rs1229984 allele is attenuated by peer drinking (Olfson et al. 2014). These indicate that the environment can moderate gene expression, resulting in complex phenotypes. These are epigenetic changes and refer to the modification of gene expression by environmental factors without a change to the genetic code itself.

Until recently, virtually all research in the field of alcoholism focused on the affected individual, trying to unravel the neurobiology of addictive behaviour or identify genetic risk factors involved in the development of alcoholism. This includes research on foetal alcohol spectrum disorders (FASDs), as this investigates the effects of alcohol exposure on the unborn foetus because of maternal alcohol intake. At least 50% of children with FASD symptoms, however, also have an alcoholic father (Abel 2004), and, as early as 1968, researchers have reported FASD symptoms in children of alcoholic fathers whose mothers did not drink alcohol, which suggests paternal drinking also plays a role in FASD symptoms (Calhoun and Warren 2007; Lemoine et al. 2003). While overall research in this area is still limited, there are a growing number of studies using human and animal models to investigate the consequences of paternal alcohol exposure on subsequent generations. This chapter aims to summarize the work on the effects of paternal alcohol exposure, which are transmitted through the germline due to preconceptual exposure to ethanol. Governments often post strict guidelines for women around drinking just before and during pregnancy (Armstrong 2017); in contrast, not much focus is put on men during the period of conception. As drinking in men is more prevalent, it is important to assess the impact of male drinking on future generations. Additionally, while maternal preconception drinking can have effects due to changes in in-utero conditions and maternal care (Rompala and Homanics 2019), any changes seen due to paternal drinking would have to be directly through the germline or through social factors. As the social effects of paternal drinking have been well documented (Scholte et al. 2008; Thor et al. 2022), the rest of the chapter will primarily focus on intergenerational inheritance of phenotypes due to Paternal Preconceptual Alcohol Consumption (PPAC) and the potential mechanisms by which this occurs.

3 Effects of Preconceptual Alcohol Exposure on the First Generation

There is a growing amount of research showing intergenerational effects of PPAC. While some of the research has been performed in humans, most rely on rodent models, as it allows for a more stringent control over experimental variables, such as the amount and the timing of the exposure. A complicating factor in assessing the exact impact of PPAC in rodents is the large variety of experimental designs that differ in mode, dose and duration of ethanol administration, species, strains and other experimental variations. Nonetheless, some general conclusions can be drawn from the current literature.

3.1 Birth and Early Development

The effects of paternal ethanol exposure can be seen as early as during pregnancy, with heavy PPAC being shown to increase the risk of early pregnancy loss and stillbirths in humans (Henriksen et al. 2004). There are also reports on the influence on maternal fertility, although the results are mixed, with some studies reporting an increase in placental weights and gestation length in dams of litters sired by rodents chronically exposed to ethanol (Abel 1995; Chang et al. 2019; Klassen and Persaud 1976). Meek et al. (2007), on the other hand, reported the opposite with decreased gestation lengths in dams mated with acute ethanol-exposed sires. The study also showed acute alcohol exposure in male mice before mating had lower numbers of pups with reduced birth weights and increased mortality after birth (Meek et al. 2007). Other studies looking at paternal ethanol exposure confirmed these effects on litter characteristics with reported alterations in litter sizes and pup sex ratios (Abel 1989, 1995; Stockard and Papanicolaou 1916a, b), fewer viable litters (Rathod et al. 2020), an increased number of runts (underweight births) and low weights in alcohol-sired litters (Asimes 2018; Bielawski and Abel 1997; Bielawski et al. 2002; Ceccanti et al. 2016; Chang et al. 2017; Ledig et al. 1998). However, some experiments have found no effect of paternal drinking on the number of pups, sex ratio, and mortality rate of litters (Abel and Lee 1988; Abel and Tan 1988; Liang et al. 2014; Nieto et al. 2022; Randall et al. 1982; Rompala et al. 2016), which may be due to the differences in experimental designs. The studies mentioned above vary markedly in aspects such as ethanol exposure duration (acute vs chronic), ethanol dosage (high vs low), method of exposure (vapour vs forced diet vs free choice), time between end of drinking and conception (immediately after exposure vs allowing for time between last exposure and mating) and age of the sires at ethanol exposure and mating. Notably, the inconsistency also extends to human studies, where some groups have reported no effect of PPAC on birth weights or early motor development in infants (Hutchinson et al. 2019; McCormack et al. 2018; Passaro et al. 1998).

Rodent studies, where the males were removed after mating, have also reported differences in organ weights based on paternal ethanol exposure (Abel 1993; Chang et al. 2017; Lee et al. 2013), while studies in humans indicate potential differences in anogenital distance (Xia et al. 2018) and an increased risk of congenital heart disease (Zhang et al. 2020). Few studies of PPAC look at early developmental milestones, and those that do show mixed results, with one showing a delay in eye opening and achievement of surface righting abilities in ethanol-sired pups (Meek et al. 2007), while another study reports ethanol-sired pups developing the righting reflex earlier than controls (Jamerson et al. 2004). Additionally, recent work from our lab has also found slower development of eye-opening and coat growth in offspring of ethanol-exposed fathers (Hussain et al. 2025). However, other studies have found no difference in the achievement of developmental landmarks between offspring of ethanol-exposed rats and offspring (Wozniak et al. 1991). Taken together, these highlight the lack of consensus with regard to findings in current studies of PPAC.

3.2 Behavioural Impacts on Offspring Phenotypes

While birthweights and litter characteristics show the impact of PPAC, most research done on paternal exposure has looked at cognitive, behavioural, and epigenetic changes seen in the offspring. Below we go through the potential effects of PPAC on the development of a broad range of impairments in the next generation.

3.2.1 Cognition and Perception

There is evidence demonstrating children of alcoholics show delayed cognitive performance and deficits in memory, language processing and motor skills compared to children of non-alcoholics, with evidence of these deficits being greater in sons (Aronson and Gilbert 1963; Ervin et al. 1984). Additionally, preschool-going children of alcoholic fathers show poorer visuospatial processing of information compared to controls (Schandler et al. 1988). Another study showed that sons of active alcoholics perform significantly worse on visuospatial, memory, and attentional tasks as well as general intellectual functioning than sons of controls as well as sons of social drinking fathers (Ozkaragoz et al. 1997). Unfortunately, the study did not investigate daughters.

Rodent models support these findings, with ethanol-sired offspring showing reduced spatiotemporal learning (Wozniak et al. 1991) and memory impairments associated with changes in the hippocampus and prefrontal cortex (PFC) (Ledig et al. 1993). Learning and Memory have been assessed using multiple different experimental designs, with poorer performances being seen by the offspring of ethanol-sired rodents in the passive avoidance test (Abel and Lee 1988), radial arm maze (Wozniak et al. 1991), two-way shock avoidance (Abel and Tan 1988) and Morris water maze (Liang et al. 2014). PPAC mice have trouble completing T-maze performance tests, indicative of reduced cognitive functioning (Hollander et al. 2019). While in all these studies the sires underwent chronic ethanol exposure, the time between final exposure and mating ranged from 10 days to 1 month. These differences in the time from exposure to mating are likely to impact spermatogenesis and epigenetic changes differently as one spermatogenesis cycle in rodents takes 8 weeks (van Haaster and de Rooij 1993). Recent work by Nieto et al. (2022) indicates that such behavioural differences may be present in offspring produced even after the completion of a spermatogenesis cycle, illustrating the long-lasting nature of the alcohol-induced alterations in the fathers, which ultimately affect behaviour in the next generation. Other studies report increased immobility of offspring of ethanol-exposed fathers in the forced swim test and reduced grooming behaviour despite no differences being seen in the fathers themselves, suggesting offspring sired by alcohol-consuming fathers have a blunted response to conditions that spontaneously elicit behaviour (Abel 1991a, b). Lastly, there is evidence to suggest chronic ethanol exposure in mice has dose-dependent effects on offspring's hearing loss (Liang et al. 2015).

3.2.2 Locomotor Activity and Motor Coordination

PPAC has also been shown to affect motor skills of the offspring. Abel (1993) showed an increase in open-field locomotor activity in PPAC F1 offspring at postnatal day (PND) 21, PND 42 and PND 90 (Abel 1993), while similar effects of PPAC were seen in 1-year-old rats (Ledig et al. 1998). There is some indication that these effects may be dose dependent as Abel showed an increased activity in offspring of sires subject to high ethanol intake but decreased activity in offspring of rats with a lower alcohol intake (Abel 1989). There are also studies that have found no differences between the paternal drinking conditions and locomotor (Beeler et al. 2019; Nieto et al. 2022; Rompala et al. 2017). This heterogeneity of outcomes is also seen in tests of motor coordination, with poorer performances of PPAC offspring being observed on balance and sensorimotor tasks in some studies (Bottom et al. 2022; Conner et al. 2020) but not in others (Nieto et al. 2022).

Although these studies show differences in baseline locomotor activity and coordination, there is convincing evidence that PPAC offspring also respond differently to the effects of alcohol on motor behaviour. Our lab has shown a reduced sensitivity to the motor-impairing effects of ethanol at 1 g/kg and 2 g/kg doses, with PPAC offspring showing lower drops in performance on the accelerating rotarod task compared to controls (Hussain et al. 2022). Other studies have found sex-dependent effects when under the influence of ethanol, with alcohol-sired males showing attenuated effects (Finegersh and Homanics 2014; Nieto et al. 2022). In females, PPAC has either shown no effects (Finegersh and Homanics 2014) or an increase in sensitivity (Nieto et al. 2022) to the motor disruptive effects of ethanol. These differences may in part be due to differences in animal strain as Abel found that the reduction in activity due to PPAC was more prominent in Long Evans than Sprague-Dawley rats (Abel 1989). Taken together, while there remains some ambiguity of the effects of PPAC on motor skills at baseline levels, there are clear differences in response to ethanol sensitivity, which may even be sex dependent.

3.2.3 Alcohol and Reward

One of the major areas of focus in drug abuse research is the changes it brings about in reward systems. In relation to the topic of the present chapter, research has shown that exposure to ethanol by either parent pre-conception can alter the neural circuitry associated with reward processing. In a human study, offspring of parents with an ethanol abuse history showed a heightened sensitivity to the rewarding effects of alcohol and preference for alcohol-related cues (Orford and Velleman 1990). Rodent studies have also found that PPAC induces a higher susceptibility to the rewarding effects of ethanol, affecting *BDNF* (Brain Derived Neurotropic Factor) and nerve growth factor (*NGF*) in the frontal cortex, striatum, hippocampus, and olfactory lobes (Ceccanti et al. 2016). *BDNF* and *NGF* are neurotrophic factors involved in, among others, cell growth, differentiation, and maturation, and cell migration, metabolism, and apoptotic cell death. They play a significant role in regulating structural and behavioural plasticity associated with drug addiction (Russo et al. 2009). Ceccanti et al. found alterations in *NGF* and *BDNF* in reward-related brain regions in offspring of mice exposed to alcohol (Ceccanti et al. 2016). These findings, particularly *BDNF* dysregulation in reward pathways, have also been reported by other studies (Finegersh and Homanics 2014; Rompala et al. 2017).

In terms of behaviour, one study reported a bidirectional effect on a conditioned place preference (CPP) task in male offspring depending on the paternal alcohol exposure dosage, where a lower dose (0.5 g/kg) increased preference, but a high dose (1.5 g/kg) decreased preference (Ceccanti et al. 2016). An increasing number of studies over the past decade have looked at alcohol consumption in the offspring of PPAC. One such study found that adult male mice exposed intermittently to alcohol vapour over 5 weeks sired male offspring with decreased alcohol drinking preference and consumption (Finegersh and Homanics 2014). These key findings were replicated with the male offspring of chronic ethanol-exposed mice via both vapour (Rompala et al. 2017) and free-choice ethanol drinking (Beeler et al. 2019), showing reduced ethanol preference and consumption. However, another study exposing mice to ethanol vapours found lower ethanol consumption in females but not male offspring of the ethanol-sired mice (Rathod et al. 2020). Alternatively, one study found that paternal binge-alcohol exposure over 8 days increased alcohol intake by intraoral infusion in postnatal offspring (Hollander et al. 2019). While these studies looked at chronic ethanol exposure within 48 h of conception, studies with longer delays between ethanol exposure and mating have shown different effects of PPAC on offspring drinking. When Wistar rats were chronically exposed to ethanol in vapour chambers for 6 weeks and then mated with naïve females 8 weeks after treatment, it was seen that ethanol-sired offspring consumed less alcohol in a self-administration task without any differences in their response to food (Nieto et al. 2022).

In a previous study, we exposed serotonin transporter (SERT) knockout male rats to alcohol using an intermittent voluntary two-bottle choice paradigm, with mating taking place 1 month after the last ethanol exposure (Hussain et al. 2022). We found overall reductions in ethanol drinking in ethanol-sired offspring, with increases in SERT activity leading to greater reductions in ethanol intake in both sexes, although there were no clear effects of genotype on drinking behaviours in the fathers. As serotonin is known to play a role in reward circuitry (Fischer and Ullsperger 2017), these results link well with other reports on the influence of PPAC on offspring reward. In one such study, analysis of gene expression showed that the serotonin reuptake transporter gene Solute Carrier Family 6 Member 4 (*Slc6a4*) and the low-affinity pan-neurotrophins receptor (*p75ntr*) genes were elevated and the 5-Hydroxytryptamine Receptor 2C (*Htr2c*) was reduced in the brainstem of offspring of chronic alcohol-exposed sires (Ferraguti et al. 2020). Taken together with the behavioural findings, it appears PPAC may interfere with ethanol sensitivity in future generations by epigenetic regulations of pathways involved in mood, emotion and neurotrophins and provide some protection against the initial reinforcing effects of alcohol.

3.3 Psychological Alterations

3.3.1 Depression and Anxiety

There are a host of studies that report anxiety- and depression-like behaviours in offspring of parents with alcohol exposure. Unfortunately, most of these studies looked at the maternal effects on offspring behaviour, and it has been noted that preference in intergenerational studies is often given to conduct disorders (Easey and Sharp 2021). Nevertheless, one systematic review of human studies found significant links between fathers with AUD and depressive symptoms in their offspring (particularly males) later in life (Mahesh et al. 2021). Rodent studies parse this association further, with mice of ethanol-sired rats showing increased anxiety and an increase in depressive-like behaviours, particularly in females (Kim et al. 2014; Liang et al. 2014). Another study found no baseline differences in anxiety-like behaviours but increases in the anxiolytic effects of ethanol in the male offspring of mice with chronic ethanol exposure (Finegersh and Homanics 2014). Conversely, two studies, one in mice and the other in rats, looking at anxiety behaviours in the open field test showed that anxiety decreased in the male progeny of ethanol-sired rats but not in the female (Beeler et al. 2019; Koabel et al. 2021). While not directly PPAC, a study did find offspring of male mice who were prenatally exposed to ethanol show increased depressive-like behaviours on the forced swim task, indicating nongenomic inheritance of traits through the male germline (Bottom et al. 2022). These findings show that parental ethanol exposure has complex sex-dependent effects on offspring anxiety- and depression-like behaviours and these changes are dependent on multiple factors.

3.3.2 Attention Deficits and Hyperactivity Disorders

Conduct issues have also been associated with PPAC, a large population study found both paternal and maternal alcohol exposure being associated with attention deficit hyperactivity disorder (ADHD) especially when combined with tobacco use (Han et al. 2015). In animal models, one acute dose of alcohol before mating was found to induce behavioural abnormalities such as significantly lower risk assessment behaviour and increased aggression along with fewer fearful behaviours as adults when compared to a control group (Meek et al. 2007). Even in chronic PPAC 2 months before gestation, offspring showed increased impulsivity in the novelty-seeking tests, and an increased time spent in the light compartment during the light-dark test (Ledig et al. 1998). The paternal effect may be mediated by catecholaminergic activity with increased hyperactivity in PPAC offspring when compared to controls being altered in response to amphetamines (Abel 1993). Extending these findings, another study found paternal ethanol exposure induced ADHD-like hyperactive, inattentive, and impulsive behaviours in offspring and found that both protein and messenger ribonucleic acid (mRNA) expression of dopamine transporter (*Slc6a3*) were significantly decreased in the cerebral cortex and striatum (Kim et al. 2014). This implies that in addition to the behavioural differences, clear alterations in gene expression and development of cortical structures are induced by PPAC, which likely impact development of offspring. Therefore, it is important to understand which mechanisms may underlie these PPAC-induced changes in brain and behaviour.

4 Mechanisms of Preconception Environment Heritability

The concept of parental and ancestral environment playing a role in phenotypes of subsequent generations has been mentioned throughout history. There is some evidence of Greek and Roman philosophers like Alus Gellius, Plato and Plutarch showing awareness of the impacts of alcohol on procreation, although the specificity of these quotes has often been misinterpreted (Abel 1999). In the 1800s, biologist Jean-Baptiste Lamark mentioned ancestral environment as a driver of phenotypes, but these views had previously been rejected for the singular genetic basis of inheritance (Skinner 2016). However, over the years there is an increasing amount of evidence that supports the idea of nongenomic inheritance. A particular point of importance for the influence of the parental environment seems to be around conception, with epigenetic changes being crucial at fertilization to drive early embryonic development (Greeson et al. 2023; Rompala and Homanics 2019).

Some of the best human evidence of intergenerational effects comes from research done on food scarcity. The Dutch famine of 1944–45 showed an increase in cardiovascular, metabolic and mental health conditions in offspring of mothers who suffered from malnutrition during this time (Painter et al. 2006; Roseboom et al. 2011). But interestingly, the offspring of fathers exposed to this prenatal stress were also affected, leading to higher weights and body mass index (BMI) scores compared to offspring of non-exposed fathers (Veenendaal et al. 2013). While the famine would have affected both parents, the studies suggest different influences from each parent. Other analyses on this cohort have noted persistent epigenetic changes caused by this famine exposure, particularly in the insulin-like growth factor 2 gene (IGF2) (Heijmans et al. 2008).

As reporting on paternal trends has increased over the past 10 years, numerous preclinical studies have emerged that support the evidence of environmental heritability in humans. For example, paternal experiences like diet, body mass index and smoking have been shown to affect deoxyribonucleic acid (DNA) methylation in the offspring (Billah et al. 2022; Merrill et al. 2021; Yin et al. 2022). In addition, reviews on paternal and maternal trauma and post-traumatic stress disorder show strong transgenerational effects (Burchert et al. 2017; Harkness 1993; Youssef et al. 2018). While generally controlled for, human studies come with a multitude of confounding factors, which include socioeconomic conditions, influence of parents during child development, infections, and genetic variations within sample populations, to name a few.

As mentioned earlier, rodent sires are generally not involved in any of the stages of rearing, which allows for better separation of social and biological factors. The range of the behavioural and neurochemical effects seen after PPAC in rodents described here and reviewed by others (Rompala and Homanics 2019; Terracina et al. 2022) clearly points to a biological role in the manifestation of these phenotypes.

4.1 Epigenetic Mechanisms

Epigenetic changes can be described as cellular and physiological trait variations that are not caused by changes in the DNA (reviewed in. Ellenbroek and Youn 2016). This is unlike the qualitative change seen due to genetic mutations, which cause changes in the nucleotide sequence of messenger ribonucleic acid (mRNA) and the amino-acid sequence of proteins produced. Additionally, most epigenetic changes are reversible and can last from anywhere between a few seconds to years.

However, some epigenetic changes are known to last a lifetime and even influence future generations. There are multiple mechanisms through which this can occur, such as histone modification, DNA methylation and noncoding RNA (ribonucleic acid) interference. While the mechanism of each of these is different, all of them work to alter gene expression without altering the underlying DNA code. These epigenetic mechanisms are essential for controlling normal development and provide means for specific cells, and the organism, to react to environmental cues (Oppermann 2013), and they also help explain the variances and plasticity seen in phenotypes of a disease (Feinberg 2007).

4.1.1 DNA Methylation

DNA methylation is the most often investigated form of epigenetic change and is characterized by the transfer of a methyl group from S-adenyl methionine (SAM) to the fifth carbon of a cytosine residue to form 5-methylcytosine (5mC) (Moore et al. 2013). This process is catalysed by a family of DNA methyltransferases (DNMTs) and occurs most often in CpG nucleotides that are distributed unevenly across the genome, making its patterns cell and tissue specific allowing for complex patterns of gene expression. Subsequently the 5-methylcytosine can be oxidated to 5-hydroxymethylcytosine (5hmC) (Fig. 1) (Ellenbroek and Youn 2016). The methylation of CpGs in the promoter region of a gene is typically associated with gene

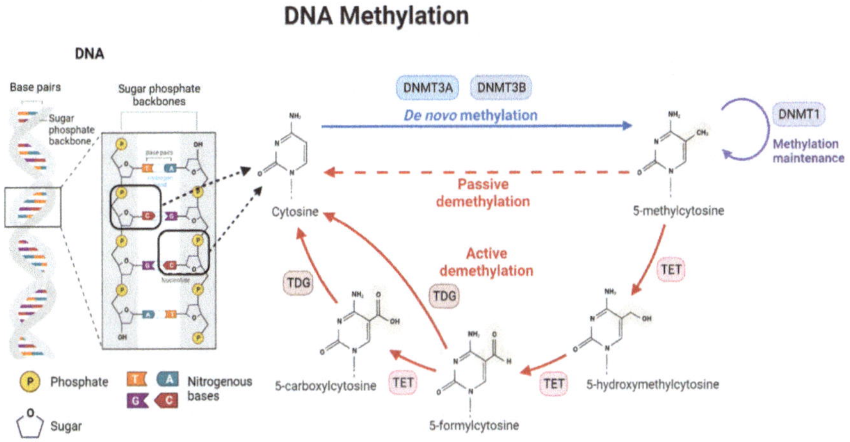

Fig. 1 DNA methylation and demethylation. DNA methylation occurs on the fifth carbon of a cytosine base catalysed by DNA methyltransferases (DNMTs). Passive DNA demethylation can occur when cells divide. A replication-independent active demethylation occurs when ten-eleven translocation (TET) methyl cytosine dioxygenases enzymes mediate DNA demethylation by oxidizing 5-methylcytosine (5mC) in DNA to 5-hydroxymethylcytosine (5hmC), 5-formylcytosine (5fC), and 5-carboxylcytosine (5caC). This active DNA demethylation involves excision of 5fC and 5caC by the DNA repair enzyme thymine DNA glycosylase (TDG). (Created in BioRender. Ellenbroek (2025) https://BioRender.com/7dyukho)

repression, while hypomethylation in the promoter region and hypermethylation in the gene body are often associated with enhanced gene expression (Lanata et al. 2018). It is important to remember that altered methylation can cause region-specific up- and down-regulations of activity; for example, lower genomic DNA methylation alterations in alcohol drinkers are associated with the upregulation of N-methyl-D-aspartate 2b receptor subtype (NR2B) which is associated with high levels of lifetime drinking and daily alcohol intake (Biermann et al. 2009) and elevated DNA hypermethylation in the promoter regions of the alpha synuclein gene leads to a downregulation of alpha synuclein expression in alcohol drinkers, followed by disturbed dopaminergic neurotransmission (Bönsch et al. 2006).

4.1.2 Histone Modification and Non-coding RNA

In addition to DNA methylation, histone modifications can mediate the effects of environmental stressors on gene expression. As eukaryotic DNA is tightly wrapped around histones, they play an integral role in maintaining the structure of DNA. The modification of histones can loosen the bonds with DNA, allowing for an increase in gene expression, while the opposite can occur through modification that tightens these bonds. While most modifications occur on histone tails, they can also occur on the globular portion, and different residues on each of the histones can be modified (Fig. 2). There are many identified histone modifications (Tweedie-Cullen et al. 2012); however, the three most common ones are histone acetylation and phosphorylation, which are associated with transcriptional activation, and histone methylation, which can be associated with both transcriptional repression and activation (Reviewed in. Ellenbroek and Youn 2016).

Lastly, a host of different non-coding RNAs (ncRNAs), including long non-coding RNAs (lncRNAs), piwi-interacting RNAs (piRNA), small interfering RNAs (siRNA) and micro RNAs (miRNAs), are all involved in epigenetic activity through modifying chromatin structure and silencing gene transcription as well as mediating the recruitment of histones and DNMTs (Wei et al. 2017). They even appear to form memories of these events through self-reinforcing epigenetic loops (Holoch and Moazed 2015). The most abundant internal mRNA/lncRNA modification is made of N6-methyl adenosine (m6A), while some of the other common modifications are 5mC, N1-methyladenosine (m1A) and pseudouridines (Ψ) (Liu and Pan 2015). M6A modifications in the m6A methyltransferase *METTL14*, which is important for cell development, are associated with alcohol dependence and alcoholism (Niu et al. 2013; Wang et al. 2016). Similarly, variants within introns of m6A demethylase gene *FTO* have been associated with risk of type 2 diabetes and childhood obesity (Bell et al. 2010).

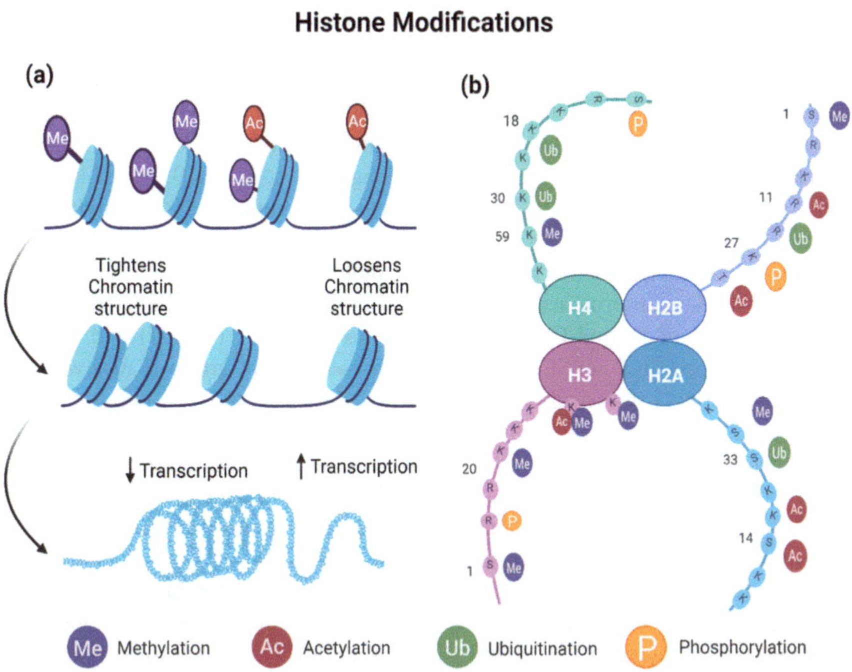

Fig. 2 Depiction of histone modifications (**a**) Histone modifications like methylation are believed to tighten the chromatin structure by increasing binding between histones and thus reduce the amount of transcription from genes in that region, while histone acetylation reduces binding between histones and increases transcription. (**b**) All 4 histones can be modified at different residues. (Created in BioRender. Ellenbroek (2025) https://BioRender.com/o1v4rby)

4.2 *Epigenetic Changes Due to Alcohol Exposure*

With regard to alcohol, research suggests ethanol consumption plays a role in DNA methylation and histone modifications, thus influencing multiple epigenetic changes throughout the body (Ciafrè et al. 2019). Some of these changes have been seen in the germline (Chastain and Sarkar 2017), and epigenetic changes have been identified in the subsequent generations (Finegersh et al. 2015). Below we detail the effects of ethanol on epigenetic mechanisms in the fathers and introduce the complexities of the patterns of alterations seen in the offspring.

4.2.1 Epigenetic Changes in the Fathers

Human Studies

Most studies in humans looking at the effects of ethanol focus on DNA methylation by looking at blood samples. Genome-wide methylation analysis studies of mononuclear blood cells have found significant differences between AUD patients and controls in total DNA methylation levels (Philibert et al. 2018), with other reports seeing a 10% increase in genomic DNA methylation in blood cells in patients with AUD (Bönsch et al. 2004). Studies that examine specific genes in individuals have seen an increase in DNA methylation and a reduction in gene expression of Period Circadian Regulator 2 (*PER2*) and Proopiomelanocortin (*POMC*) in heavy drinkers, which were predictors of alcohol craving (Gangisetty et al. 2019; Muschler et al. 2010). Assessment of DNMTs shows an increase of genomic DNA methylation in patients with alcoholism being significantly associated with their lowered DNMT-3b mRNA expression (Bönsch et al. 2006), and another study of DNMTs and methyl-CpG-binding domain proteins in the blood of individuals with AUD also shows a downregulation of methyl-CpG-binding domain protein 1 (*MBD1*) and methyl-CpG-binding protein 2 (*MECP2*) in the alcohol-dependent group (Krasteva et al. 2020). Interestingly, these differences in methylation seem to be related to patterns of consumption, with one study finding significantly lower DNA methylation levels in the leukocytes of heavy drinkers compared to non-drinkers and moderate drinkers when looked at cross-sectionally (Maeda et al. 2022) and another reporting differential blood DNA methylation in certain CpG islands in former drinkers compared to current drinkers in a longitudinal study (Dugué et al. 2021).

Epigenome-wide analysis of alcohol dependence from human brain tissues identified differentially methylated regions in the dorsolateral PFC (dlPFC) and NAc with reduced methylation in drinkers (Meng et al. 2021). A study of post-mortem brain tissue of 104 individuals showed a number of differentially methylated regions in glucocorticoid signalling and inflammation-related genes in the amygdala and PFC associated with AUD (Lohoff et al. 2021). However, others found no differences in DNA methylation patterns between alcoholics and matched controls in the PFC, highlighting the tissue-specific nature of methylation changes. Apart from methylation, epigenetic regulation in the brain has also been reported through alterations in lncRNAs and histone modifications (Manzardo et al. 2012). The lncRNA brain-derived neurotrophic factor-antisense (*BDNF-AS*) acts as an epigenetic regulator in the human amygdala only in early onset AUD but shows no increase in levels in late onset AUD, suggesting the regulation is age-dependent (Bohnsack et al. 2019). Lastly, histone modifications are also reported in the frontal cortex of individuals with AUD (Manzardo et al. 2012; Ponomarev et al. 2012).

While these studies show an effect of alcohol on epigenetic change, the only way these changes would pass down the germline is through the sperm cells. To date, there have been multiple studies in humans on the effects of alcohol on overall sperm health and structure, with reports associating alcohol drinking with reduced sperm count, sperm motility and fragmentation (Joo et al. 2012; Ricci et al. 2017). When looking at sperm DNA, alcohol is linked with greater oxidative stress, which in turn is associated with greater DNA damage and increased global DNA hypomethylation in human sperm (Shen et al. 1999), Non-drinkers show higher sperm DNA stainability compared to drinkers, with lower scores indicating the presence of immature sperm nuclei with abnormal proteins and/or altered protamine/histone ratios (Blaseg et al. 2022). However, to the best of our knowledge, only one study directly investigated sperm DNA methylation in humans. The authors found a reduction in

DNA methylation of the regulatory gene *H19/igf2* (insulin-like growth factor 2) intergenic differentially methylated region (*IG-DMR*), where individuals who heavily consumed alcohol showed significant demethylation compared to those who did not drink alcohol (Ouko et al. 2009). The study showed a correlation between chronic alcohol use and demethylation of regions of sperm DNA that are typically hypermethylated, indicating epigenetic changes in specific imprinted genes that could be transmitted to the next generation. In addition, while no studies on humans have directly tested the effects of ethanol on ncRNAs in sperm, ethanol consumption is known to alter miRNA expression in the gut (Tang et al. 2008).

It is important to mention that the structure of sperm cell DNA is different from somatic cell DNA, as it is much more tightly bound, with most histones being replaced with protamines (Balhorn 2007). This was thought to mean that epigenetic changes through histone modifications were not a way of paternal epigenetic inheritance; however, recent work shows a fraction of the genome escapes this remodelling and keeps its histones, which makes changes in chromatin a somewhat more viable route for epigenetic inheritance (Sharma 2019). Together, this means all major mechanisms of epigenetic changes can occur in epigenetic inheritance; however, based on current knowledge, the most likely and best-studied routes for the transmission of any epigenetic changes are through DNA methylation of select genes or small RNA modifications.

Animal Studies

The amount of research looking at the effects of ethanol on epigenetic modifications in animal models outweighs the work done in humans. Much like the reports on blood samples and brain region changes mentioned above, rodent studies also found changes in gene expression in the brain. Adolescent alcohol exposure was associated with dysregulation in DNA methylation mechanisms in the amygdala through altering DNMT activity during ethanol exposure as well as in withdrawal (Sakharkar et al. 2019). The study also mentions the decrease in DNMT enzyme activity is accompanied by an increase in DNA-damage-inducible (*Gadd45g*) mRNA expression, which facilitates active DNA demethylation (Sakharkar et al. 2019). Rats exposed to ethanol also show increased hypermethylation of the PFC (Barbier et al. 2015; Otero et al. 2012). Additionally, rodent studies on ethanol exposure have identified modifications of multiple histones; examples of these are the increased levels of acetyl-histone histone 4-lysine 5 (*acH4K5*) in the medial prefrontal cortex (mPFC) after long-term exposure (Montesinos et al. 2016) and acetylation of histone 3-lysine 9 (*H3K9ac*) in hypothalamic microglia after early life exposure to ethanol (Chastain et al., 2019). Lastly, the effects of alcohol on miRNAs are also seen with prolonged alcohol exposure affecting specific miRNAs in the amygdala, hippocampus and cerebellum (Barney et al. 2022; Kyzar et al. 2019; Rossetto et al. 2019).

Of greater interest to understand the intergenerational impact of alcohol drinking is the research linking ethanol to epigenetic modifications in the sperm. In mice 2 weeks of ethanol injections increased oxidative stress in the testes and decreased sperm count (Jana et al. 2010). As ethanol induces oxidative stress, it may have a particularly robust effect on DNA methylation during spermatogenesis due to a negative regulation of SAM (Kruman and Fowler 2014), the primary methyl donor for DNMTs. Studies have found reduced DNMT1 RNA levels in the sperm of chronic alcohol treated mice (Bielawski et al. 2002). Importantly many genes are methylated from conception, and there are a substantial number of genes where only the maternal or the paternal copy of the gene is methylated (referred to as genomic imprinted genes). Disruption to the methylation status at known imprinting loci has been shown to have dramatic developmental effects (Monk et al. 2019). In this regard, studies have found reduced DNA methylation in imprinted gene loci *Snrpn*, *H19* and *Peg3* in sperm of chronic ethanol treated rodents (Knezovich and Ramsay 2012; Liang et al. 2014) similar to that reported in humans (Ouko et al. 2009). A reduction in methylation in the *BDNF* promoter in sperm of mice chronically exposed to ethanol vapour was also seen (Finegersh and Homanics 2014). Another study observed an increase in methylation of the dopamine transporter (DAT, *Slc6a3*) gene promoter in the sperm of sire mice, suggesting germline changes in the epigenetic methylation signature of DAT gene by ethanol exposure (Kim et al. 2014). There is also evidence that supports changes in offspring methylation, even in the absence of clear methylation changes in the paternal sperm. Knezovich and Ramsay noted trends but no significant DNA methylation changes in *H19* in the sperm but significant decreases in expression levels of the same gene in tail biopsies of ethanol-sired offspring (Knezovich and Ramsay 2012). Similar patterns with the paternally imprinted *Snrpn* gene found no methylation differences in the sperm but decreased methylation in the offspring cerebral cortex (Liang et al. 2014). Additionally, paternal voluntary alcohol consumption had no effect on sperm methylation but did alter hepatic fibrosis markers in offspring (Chang et al. 2017). Taken together, these suggest a small effect of methylation on cross-generational inheritance.

With respect to other epigenetic mechanisms, ethanol alters the integrity of sperm chromatin and DNA (Rahimipour et al. 2013). In another recent study, ethanol exposure was shown to elevate the levels of H3K4me3 (trimethylation of histone H3 at lysine 4) in sperm, leading to enhanced CTCF (CCCTC-binding factor) binding, which was associated with altered developmental programming (Bedi et al. 2022). Lastly, a third study found male alcohol consumption significantly decreased sperm histone epigenetic mark H3K4me3 in the treated group (Cambiasso et al. 2022). The lack of research looking at sperm chromatin profile is suggested to be due to the unique properties and minimal histone content; however, acute ethanol does increase histone acetylation in rat testes, which is important in developing germ cells (Kim and Shukla 2006). Ethanol is also a modulator of histone-modifying enzymes in the brain and liver (Pandey et al. 2008; You et al. 2008). Researchers note that as sperm chromatin profile continues to be detailed, it would lead to improved understanding of any epigenetic mechanisms that would explain environmental inheritance (Rompala and Homanics 2019).

Lastly, RNA modifications are considered essential to models of RNA-mediated inheritance and underlie sncRNA stability (Rompala and Homanics 2019). Multiple miRNAs and tDR (transfer RNA-derived RNA) in the sperm appear to be altered by chronic ethanol exposure (Rompala et al. 2018). One of those noted to be altered was miR10a, which is suggested to function as a posttranscriptional regulator for *BDNF* (Jiajie et al. 2017). The other prominent finding was the alteration of *tRNA-Gly-GCC*, which is associated with controlling retrotransposon-regulated gene networks in the fertilized oocyte (Sharma et al. 2016), and *tRNA-Glu-CTC*, as it makes up about half of the small RNA content in the sperm (Peng et al. 2012). Rompala et al. also found 2 altered posttranscriptional RNA modifications specific to mitochondrial tRNAs, which could be involved in sperm mitochondrial function (Rompala et al. 2018). Additionally, chronic alcohol use was shown to alter the ratio of tRNA-derived small RNAs to Piwi-interacting RNAs in sperm which was associated with late term foetal growth restriction (Bedi et al. 2022). While few studies have looked at alcohol, exposure of male mice to stress or paternal stress increases several sperm miRNAs (Rodgers et al. 2013, 2015). The same is seen with high-fat-diet-induced obese male mice showing altered miRNA levels (McPherson et al. 2015). As there are over 150 identified RNA modifications (Boccaletto et al. 2022), most of which have not been assessed, future studies will need to fully characterize the effects of alcohol on the epitranscriptome.

4.2.2 Gene Expression Changes in Offspring

We earlier discussed that the epigenetic changes in the sperm of fathers who consume alcohol may play a critical role in the phenotypical differences seen in PPAC offspring when compared to non-drinkers. We also pointed out that the methylation of a particular gene in the sperm does not coincide with dysregulation of the same gene in the offspring. Liang et al. (2014) found reduced DNA methylation in genes *H19* and *Peg3* in sperm of chronic ethanol-treated rodents but found no methylation differences of these in their offspring, instead finding decreased *Snrpn* methylation in the cerebral cortex, a gene that was not differentially methylated in the sperm. Similarly, a study found no significant DNA methylation of *H19* in the sperm but saw reduction in *H19* expression in the offspring (Knezovich and Ramsay 2012). Another study with comparable results found no changes in gene methylation in the sperm but found decreased expression of imprinted genes associated with foetal and placental growth in the offspring (Chang et al. 2017). Specifically, *Gnas*, *Grb10* (males) and *Sgce* (females) in the placenta, as well as the genes *Cdkn1c*, *Dcn*, *Gnas* (males), and *Meg3* in the foetal liver, are associated with liver fibrosis (Chang et al. 2017).

One of the common sets of genes with altered expression due to PPAC are neurotrophins, with a reduction in DNA methylation levels of the *BDNF* gene in the ventral tegmental area of male but not the female offspring of ethanol-exposed sires (Finegersh and Homanics 2014). The authors also reported no changes, in either sex, in *BDNF* methylation in the mPFC. On the other hand, increased *BDNF* and reduced NGF expression in the PFC, striatum, and olfactory lobes of offspring of ethanol-exposed mice as well as changes in Trk-A, which is a receptor for NGF, in the hippocampus have been found (Ceccanti et al. 2016). Additionally, there was an increase in DAT levels in the olfactory bulbs of the ethanol-sired offspring (Ceccanti et al. 2016). A recent rat study of PPAC found that while *BDNF* DNA methylation levels were lower in the NAc and greater in the mPFC of the ethanol sires, the same patterns of DNA methylation were not seen in the offspring despite the offspring showing aberrant *BDNF* DNA methylation patterns compared to control-sired offspring (Nieto et al. 2022).

Several other methylation patterns and gene expression changes as a result of PPAC have been noted. The offspring of ethanol-sires showed increased methylation in the *POMC* gene, which can lead to HPA (hypothalamic pituitary adrenal) axis dysregulation (Govorko et al. 2012). This possible HPA disruption was confirmed by another study, which found lower corticosterone levels in offspring of preconception ethanol animals compared to control offspring (Asimes 2018). The study also found lower mRNA expression of gonadotropin-releasing hormone, which is important in regulating the reproductive axis (Asimes 2018). Chronic paternal ethanol exposure from late adolescence was found to decrease DNMT1 and *MECP2* levels, as well as DAT expression in the mPFC and striatum of male offspring (Kim et al. 2014). Another study detailed the effects of ethanol exposure before conception on the offspring's hypothalamus and found differences in expression of genes that mediate neurogenesis and synaptic plasticity during neurodevelopment, genes important for directing chromatin remodelling, posttranslational modifications or transcription regulation and genes involved in regulation of obesity and reproductive function (Przybycien-Szymanska et al. 2014). Likewise, genome-wide DNA methylation patterns in the hypothalamus of offspring who had either one or both parents being exposed to ethanol preconception found methylation patterns were different when only one parent was exposed to ethanol compared to both and the patterns again differed whether it was the mother or father who was exposed (Asimes et al. 2017). All three cases did have different DNA methylation patterns when compared to controls, while the global methylations appeared to not be significantly different (Asimes et al. 2017).

To summarize, research has revealed a clear influence of paternal alcohol consumption on the expression levels of genes associated with neurotransmitter systems, synaptic plasticity, neurodevelopment, endocrine functions, transcription regulation and chromatin modifications. The studies also indicate the complexities of intergenerational epigenetic mechanisms, with methylation not being a direct predictor of RNA expression levels and the high sex-based and region-specific alterations seen thus far.

5 Transgenerational Impacts of PPAC

While the majority of research focuses on intergenerational non-genomic inheritance, it is important to also consider the effects across multiple generations. However, this inheritance appears slightly different when coming from males or females. For males, a phenotype being inherited from the father (F0 generation) to the offspring (F1 generation) is considered intergenerational, as the sperm that

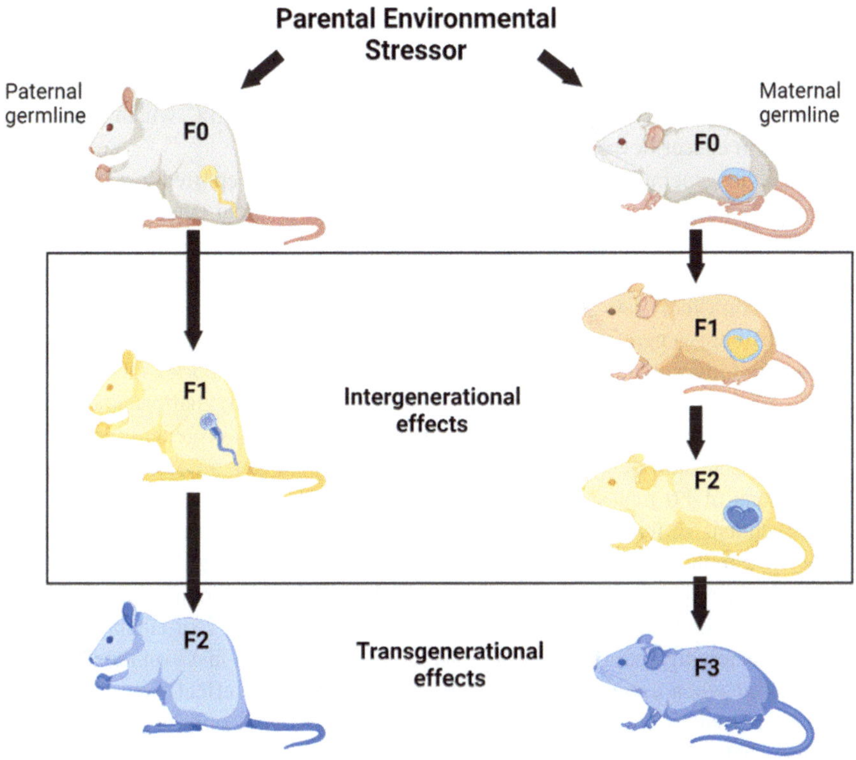

Fig. 3 Difference between intergenerational and transgenerational effects in male and female germline. Intergenerational and transgenerational inheritance of environmental stressors like alcohol consumption. In paternal exposure, only the sperm cells for the F1 generation are present and directly exposed to the stressor (intergenerational inheritance), whereas phenotypes that persist into F2 and beyond are transgenerational (blue) due to not having any direct contact with the stressor. Maternal exposure exposes both her and the developing foetus in utero (F1, orange) as well as the developing germ cells in the foetus, which later give rise to F2 (yellow), both of which are present at the time of environmental exposure (therefore, they both are intergenerational). The F3 generation cells (blue) are not present at the time of environmental exposure; if they carry phenotypes, they do so by transgenerational inheritance. (Created in BioRender. Ellenbroek (2025) https://BioRender.com/7r9nzy8)

produced the F1 is directly exposed to the environmental stressor. If the same phenotype persists in the F2 or subsequent generations, it can then be called transgenerational (Fig. 3). In females, as the F0 female already carries the F1 foetus and the already formed oocytes that will be used for the F2, a true transgenerational effect through the female germline would be reserved for the F3 generation (Perez and Lehner 2019).

The Swedish Överkalix study showed that paternal food availability and nutrition can affect the F2 grandchildren (Bygren et al. 2001). A further analysis of these cohorts found sex-dependent transgenerational links between paternal grandfather's food supply and mortality of grandsons, while paternal grandmother's food supply showed intergenerational effects and was associated with granddaughters' mortality (Pembrey et al. 2006). With regard to alcohol, a human study noted that ancestral exposure to alcohol can have enduring effects that impact epigenetic processes such as hypomethylation of *HRAS* and hypermethylation of *TP53*, which are genes involved in cancer development (Hill et al. 2017).

True transgenerational effects of PPAC have rarely been reported in rodents. One early study on guinea pigs indicated lasting effects of paternal ethanol on offspring mortality and deformations lasting up to 4 generations (Stockard and Papanicolaou 1916b). Other works have noted decreased testosterone levels and damage to sperm DNA in adult male rats sired by ethanol-exposed rats, which could lead to differences in future generations (Abel and Lee 1988; Cambiasso et al. 2022), although this was not investigated. One study did find a 5% reduction in cortex methylation in the F2 of ethanol-sired mice, although the study did not see any malformations in the animals at 8 weeks (Liang et al. 2014). Our lab looked at the transgenerational effects of PPAC and saw delay in reaching early developmental milestones in the F2 generations of F0 males exposed to ethanol (Hussain et al. 2025). We also saw sex-dependent differences in drinking behaviour with only the female F2 rats showing a reduction in ethanol consumption in a free-choice paradigm, while the male F2 consumed similar levels to controls. Motor coordination under the influence of ethanol also showed sex-dependent changes with female F2 of ethanol-exposed F0 males showing better performance on motor coordination tasks on the accelerating rotarod when compared to control females, while this effect was not seen in males (Hussain et al. 2025). These suggest that PPAC not only appears to cause changes that pass through multiple generations but also that these changes manifest in a sex-dependent manner.

To the best of our knowledge, no other studies have investigated truly transgenerational behavioural effects due to PPAC in animals. However, studies that have looked at prenatal alcohol exposure have seen *POMC* dysregulations caused by prenatal exposure to ethanol being passed down through the male germline for at least 2 further generations (Govorko et al. 2012; Sarkar 2016). While these changes are passed down through the males, they appear to not do so through the female germline (Sarkar 2016). Recently, another study looking at a transgenerational model of prenatal alcohol exposure saw abnormal sensorimotor processing, increased risk-taking behaviour, and increased depressive-like behaviours extending to the F3 generation in 20-day-old mice (Bottom et al. 2022). The study, however, only saw differences in phenotypes and could not see any significant differences in cortical connectivity after the directly exposed F1 generation. This is in line with two other studies that have observed maintenance of epigenetically encoded phenotypes despite loss of transmission of altered ncRNAs to offspring (Gapp et al. 2014; Radford et al. 2014). These studies clearly indicate that more research is needed to investigate whether the effects of PPAC extend beyond the F1 generation due to PPAC.

6 Conclusion

The chapter presents an overview of the current knowledge surrounding the intergenerational impacts of alcohol with a primary focus on paternal drinking. Research, while being heterogenous, has clearly shown that PPAC influences offspring behaviours and gene expression. Ethanol is shown to affect DNA methylation, ncRNAs and even histone modifications in germ cells, and these contribute to developmental and behavioural changes in the offspring in ways ranging from birth defects and increased mortality to altered sensitivity to alcohol and cognitive impairments. Interestingly, these changes appear to have the potential to be transmitted to future generations and highlight the complexity of non-Mendelian inheritance, although the research in this area has so far been very limited. Current research has shown that these intergenerational changes appear to be highly dependent on a range of factors, which include acute or chronic alcohol exposure, species of subject, sex of offspring and region analysed to name a few. Despite a clear effect of PPAC, the small number of studies and high variance within experiments indicate how early PPAC is in its research in terms of understanding the effects seen. Understanding these impacts is imperative for designing public health policy around alcohol use as it is increasingly showing to not only affect the health of the user but also any future offspring.

References

Abel E (1989) Paternal and maternal alcohol consumption: effects on offspring in two strains of rats. Alcohol Clin Exp Res 13(4):533–541. https://doi.org/10.1111/j.1530-0277.1989.tb00373.x

Abel E (1991a) Alarm substance emitted by rats in the forced-swim test is a low volatile pheromone. Physiol Behav 50(4):723–727. https://doi.org/10.1016/0031-9384(91)90009-D

Abel E (1991b) Paternal alcohol consumption affects grooming response in rat offspring. Alcohol (Fayetteville, NY) 8(1):21–23. https://doi.org/10.1016/0741-8329(91)91168-2

Abel E (1993) Paternal alcohol exposure and hyperactivity in rat offspring: effects of amphetamine. Neurotoxicol Teratol 15(6):445–449. https://doi.org/10.1016/0892-0362(93)90063-T

Abel E (1995) An update on incidence of FAS: FAS is not an equal opportunity birth defect. Neurotoxicol Teratol 17(4):437–443. https://doi.org/10.1016/0892-0362(95)00005-C

Abel E (1999) Was the fetal alcohol syndrome recognized by the Greeks and Romans? Alcohol Alcohol 34(6):868–872. https://doi.org/10.1093/alcalc/34.6.868

Abel E (2004) Paternal contribution to fetal alcohol syndrome. Addict Biol 9(2):127–133. https://doi.org/10.1080/13556210410001716980

Abel E, Lee J (1988) Paternal alcohol exposure affects offspring behavior but not body or organ weights in mice. Alcohol Clin Exp Res 12(3):349–355. https://doi.org/10.1111/j.1530-0277.1988.tb00205.x

Abel E, Tan SE (1988) Effects of paternal alcohol consumption on pregnancy outcome in rats. Neurotoxicol Teratol 10(3):187–192. https://doi.org/10.1016/0892-0362(88)90016-5

Armstrong EM (2017) Making sense of advice about drinking during pregnancy: does evidence even matter? J Perinat Educ 26(2):65–69. https://doi.org/10.1891/1058-1243.26.2.65

Aronson H, Gilbert A (1963) Preadolescent sons of male alcoholics: an experimental study of personality patterning. Arch Gen Psychiatry 8(3):235–241. https://doi.org/10.1001/archpsyc.1963.01720090023002

Asimes A (2018) Mechanisms and consequences of epigenetic inheritance following parental preconception alcohol exposure. Available from ProQuest One Academic Eastern Edition https://search.proquest.com/docview/2047585233

Asimes A, Torcaso A, Pinceti E, Kim CK, Zeleznik-Le NJ, Pak TR (2017) Adolescent binge-pattern alcohol exposure alters genome-wide DNA methylation patterns in the hypothalamus of alcohol-naïve male offspring. Alcohol (Fayetteville, NY) 60:179–189. https://doi.org/10.1016/j.alcohol.2016.10.010

Australian Bureau of Statistics (2021) Alcohol consumption. ABS. https://www.abs.gov.au/statistics/health/health-conditions-and-risks/alcohol-consumption/2020-21

Balhorn R (2007) The protamine family of sperm nuclear proteins. Genome Biol 8(9):227. https://doi.org/10.1186/gb-2007-8-9-227

Barbier E, Tapocik JD, Juergens N, Pitcairn C, Borich A, Schank JR, Sun H, Schuebel K, Zhou Z, Yuan Q, Vendruscolo LF, Goldman D, Heilig M (2015) DNA methylation in the medial prefrontal cortex regulates alcohol-induced behavior and plasticity. J Neurosci 35(15):6153–6164. https://doi.org/10.1523/JNEUROSCI.4571-14.2015

Barney TM, Vore AS, Deak T (2022) Acute ethanol challenge differentially regulates expression of growth factors and miRNA expression profile of whole tissue of the dorsal hippocampus. Front Neurosci 16:884197. https://doi.org/10.3389/fnins.2022.884197

Bedi Y, Wang H, Thomas KN, Basel A, Prunier J, Robert C, Golding MC (2022) Alcohol induced increases in sperm histone H3 lysine 4 trimethylation correlate with increased placental CTCF occupancy and altered developmental programming. Sci Rep 12(1):8839. https://doi.org/10.1038/s41598-022-12188-3

Beeler E, Nobile ZL, Homanics GE (2019) Paternal preconception every-other-Day ethanol drinking alters behavior and ethanol consumption in offspring. Brain Sci 9(3):56. https://doi.org/10.3390/brainsci9030056

Bell CG, Finer S, Lindgren CM, Wilson GA, Rakyan VK, Teschendorff AE, Akan P, Stupka E, Down TA, Prokopenko I, Morison IM, Mill J, Pidsley R, Deloukas P, Frayling TM, Hattersley AT, McCarthy MI, Beck S, Hitman GA (2010) Integrated genetic and epigenetic analysis identifies haplotype-specific methylation in the FTO type 2 diabetes and obesity susceptibility locus. PLoS One 5(11):e14040. https://doi.org/10.1371/journal.pone.0014040

Bielawski DM, Abel EL (1997) Acute treatment of paternal alcohol exposure produces malformations in offspring. Alcohol (Fayetteville, NY) 14(4):397–401. https://doi.org/10.1016/S0741-8329(97)87951-7

Bielawski DM, Zaher FM, Svinarich DM, Abel EL (2002) Paternal alcohol exposure affects sperm cytosine Methyltransferase messenger RNA levels. Alcohol Clin Exp Res 26(3):347–351. https://doi.org/10.1111/j.1530-0277.2002.tb02544.x

Biermann T, Reulbach U, Lenz B, Frieling H, Muschler M, Hillemacher T, Kornhuber J, Bleich S (2009) N-methyl-D-aspartate 2b receptor subtype (NR2B) promoter methylation in patients during alcohol withdrawal. J Neural Transm (Vienna, Austria: 1996) 116(5):615–622. https://doi.org/10.1007/s00702-009-0212-2

Billah MM, Khatiwada S, Morris MJ, Maloney CA (2022) Effects of paternal overnutrition and interventions on future generations. Int J Obes 46(5):901–917. https://doi.org/10.1038/s41366-021-01042-7

Blaseg E, Von Wald T, Hansen KA (2022) Vitamin D levels and human sperm DNA fragmentation: a prospective, cohort study. Basic Clin Androl 32(1):1–14. https://doi.org/10.1186/s12610-022-00166-8

Boccaletto P, Stefaniak F, Ray A, Cappannini A, Mukherjee S, Purta E, Kurkowska M, Shirvanizadeh N, Destefanis E, Groza P, Avşar G, Romitelli A, Pir P, Dassi E, Conticello SG, Aguilo F, Bujnicki JM (2022) MODOMICS: a database of RNA modification pathways. 2021 update. Nucleic Acids Res 50(D1):D231–D235. https://doi.org/10.1093/nar/gkab1083

Bohnsack JP, Teppen T, Kyzar EJ, Dzitoyeva S, Pandey SC (2019) The lncRNA BDNF-AS is an epigenetic regulator in the human amygdala in early onset alcohol use disorders. Transl Psychiatry 9(1):34. https://doi.org/10.1038/s41398-019-0367-z

Bönsch D, Lenz B, Reulbach U, Kornhuber J, Bleich S (2004) Homocysteine associated genomic DNA hypermethylation in patients with chronic alcoholism. J Neural Transm 111(12): 1611–1616. https://doi.org/10.1007/s00702-004-0232-x

Bönsch D, Lenz B, Fiszer R, Frieling H, Kornhuber J, Bleich S (2006) Lowered DNA methyltransferase (DNMT-3b) mRNA expression is associated with genomic DNA hypermethylation in patients with chronic alcoholism. J Neural Transm 113(9):1299–1304. https://doi.org/10.1007/s00702-005-0413-2

Bottom RT, Kozanian OO, Rohac DJ, Erickson MA, Huffman KJ (2022) Transgenerational effects of prenatal ethanol exposure in prepubescent mice. Front Cell Dev Biol 10:812429. https://doi.org/10.3389/fcell.2022.812429

Burchert S, Stammel N, Knaevelsrud C (2017) Transgenerational trauma in a post-conflict setting: effects on offspring PTSS/PTSD and offspring vulnerability in Cambodian families. Psychiatry Res 254:151–157. https://doi.org/10.1016/j.psychres.2017.04.033

Bygren LO, Kaati G, Edvinsson S (2001) Longevity determined by paternal ancestors' nutrition during their slow growth period. Acta Biotheor 49(1):53–59. https://doi.org/10.1023/A:1010241825519

Calhoun F, Warren K (2007) Fetal alcohol syndrome: historical perspectives. Neurosci Biobehav Rev 31(2):168–171. https://doi.org/10.1016/j.neubiorev.2006.06.023

Cambiasso MY, Gotfryd L, Stinson MG, Birolo S, Salamone G, Romanato M, Calvo JC, Fontana VA (2022) Paternal alcohol consumption has intergenerational consequences in male offspring. J Assist Reprod Genet 39(2):441–459. https://doi.org/10.1007/s10815-021-02373-0

Ceccanti M, Coccurello R, Carito V, Ciafrè S, Ferraguti G, Giacovazzo G, Mancinelli R, Tirassa P, Chaldakov GN, Pascale E, Ceccanti M, Codazzo C, Fiore M (2016) Paternal alcohol exposure in mice alters brain NGF and BDNF and increases ethanol-elicited preference in male offspring. Addict Biol 21(4):776–787. https://doi.org/10.1111/adb.12255

Chang RC, Skiles WM, Chronister SS, Wang H, Sutton GI, Bedi YS, Snyder M, Long CR, Golding MC (2017) DNA methylation-independent growth restriction and altered developmental programming in a mouse model of preconception male alcohol exposure. Epigenetics 12(10): 841–853. https://doi.org/10.1080/15592294.2017.1363952

Chang RC, Wang H, Bedi Y, Golding MC (2019) Preconception paternal alcohol exposure exerts sex-specific effects on offspring growth and long-term metabolic programming. Epigenetics Chromatin 12(1):9. https://doi.org/10.1186/s13072-019-0254-0

Chastain LG, Sarkar DK (2017) Alcohol effects on the epigenome in the germline: role in the inheritance of alcohol-related pathology. Alcohol (Fayetteville, NY) 60:53–66. https://doi.org/10.1016/j.alcohol.2016.12.007

Ciafrè S, Carito V, Ferraguti G, Greco A, Chaldakov GN, Fiore M, Ceccanti M (2019) How alcohol drinking affects our genes: an epigenetic point of view. Biochem Cell Biol 97(4):345–356. https://doi.org/10.1139/bcb-2018-0248

Conner KE, Bottom RT, Huffman KJ (2020) The impact of paternal alcohol consumption on offspring brain and behavioral development. Alcohol Clin Exp Res 44(1):125–140. https://doi.org/10.1111/acer.14245

Crossin R, Cleland L, Wilkins C, Rychert M, Adamson S, Potiki T, Pomerleau AC, MacDonald B, Faletanoai D, Hutton F (2023) The New Zealand drug harms ranking study: a multi-criteria decision analysis. J Psychopharmacol. https://doi.org/10.1177/02698811231182012

Dasgupta A (2016) Alcohol, drugs, genes and the clinical laboratory. Elsevier Science

Dugué P, Wilson R, Lehne B, Jayasekara H, Wang X, Jung C, Joo JE, Makalic E, Schmidt DF, Baglietto L, Severi G, Gieger C, Ladwig K, Peters A, Kooner JS, Southey MC, English DR, Waldenberger M, Chambers JC et al (2021) Alcohol consumption is associated with widespread changes in blood DNA methylation: analysis of cross-sectional and longitudinal data. Addict Biol 26(1):e12855. https://doi.org/10.1111/adb.12855

Easey KE, Sharp GC (2021) The impact of paternal alcohol, tobacco, caffeine use and physical activity on offspring mental health: a systematic review and meta-analysis. Reprod Health 18(1). https://doi.org/10.1186/s12978-021-01266-w

Ellenbroek B, Youn JU (2016) Gene-environment interactions in psychiatry: nature, nurture, neuroscience. Academic

Ervin CS, Little RE, Streissguth AP, Beck DE (1984) Alcoholic fathering and its relation to child's intellectual development: a pilot investigation. Alcohol Clin Exp Res 8(4):362–365. https://doi.org/10.1111/j.1530-0277.1984.tb05681.x

Feinberg AP (2007) Phenotypic plasticity and the epigenetics of human disease. Nature 447(7143): 433–440. https://doi.org/10.1038/nature05919

Ferraguti G, Codazzo C, Petrella C, Coccurello R, Ceccanti M, Fiore M (2020) Brainstem expression of SLC6A4, HTR2C, NGF, BDNF, TRKANGF, TRKBBDNF and P75NTR following paternal alcohol exposure in the male mouse. Bio-Med Rev (Varna) 31:75. https://doi.org/10.14748/bmr.v31.7707

Finegersh A, Homanics GE (2014) Paternal alcohol exposure reduces alcohol drinking and increases behavioral sensitivity to alcohol selectively in male offspring. PLoS One 9(6): e99078. https://doi.org/10.1371/journal.pone.0099078

Finegersh A, Rompala GR, Martin DIK, Homanics GE (2015) Drinking beyond a lifetime: new and emerging insights into paternal alcohol exposure on subsequent generations. Alcohol (Fayetteville, NY) 49(5):461–470. https://doi.org/10.1016/j.alcohol.2015.02.008

Fischer AG, Ullsperger M (2017) An update on the role of serotonin and its interplay with dopamine for reward. Front Hum Neurosci 11:484. https://doi.org/10.3389/fnhum.2017.00484

Gangisetty O, Sinha R, Sarkar DK (2019) Hypermethylation of proopiomelanocortin and period 2 genes in blood are associated with greater subjective and behavioral motivation for alcohol in humans. Alcohol Clin Exp Res 43(2):212–220. https://doi.org/10.1111/acer.13932

Gapp K, von Ziegler L, Tweedie-Cullen RY, Mansuy IM (2014) Early life epigenetic programming and transmission of stress-induced traits in mammals: how and when can environmental factors influence traits and their transgenerational inheritance? BioEssays 36(5):491–502. https://doi.org/10.1002/bies.201300116

Govorko D, Bekdash RA, Zhang C, Sarkar DK (2012) Male germline transmits fetal alcohol adverse effect on hypothalamic proopiomelanocortin gene across generations. Biol Psychiatry (1969) 72(5):378–388. https://doi.org/10.1016/j.biopsych.2012.04.006

Greeson KW, Crow KMS, Edenfield RC, Easley CA 4th (2023) Inheritance of paternal lifestyles and exposures through sperm DNA methylation. Nat Rev Urol 20(6):356–370. https://doi.org/10.1038/s41585-022-00708-9

Han J, Kwon H, Ha M, Paik K, Lim M, Gyu Lee S, Yoo S, Kim E (2015) The effects of prenatal exposure to alcohol and environmental tobacco smoke on risk for ADHD: a large population-based study. Psychiatry Res 225(1):164–168. https://doi.org/10.1016/j.psychres.2014.11.009

Harkness LL (1993) Transgenerational transmission of war-related trauma. In: International handbook of traumatic stress syndromes. Springer US, pp 635–643. https://doi.org/10.1007/978-1-4615-2820-3_53

Heijmans BT, Tobi EW, Stein AD, Putter H, Blauw GJ, Susser ES, Slagboom PE, Lumey LH (2008) Persistent epigenetic differences associated with prenatal exposure to famine in humans. Proc Natl Acad Sci – PNAS 105(44):17046–17049. https://doi.org/10.1073/pnas.0806560105

Henriksen TB, Hjollund NH, Jensen TK, Bonde JP, Andersson A, Kolstad H, Ernst E, Giwercman A, Skakkebæk NE, Olsen J (2004) Alcohol consumption at the time of conception and spontaneous abortion. Am J Epidemiol 160(7):661–667. https://doi.org/10.1093/aje/kwh259

Hill SY, Rompala G, Homanics GE, Zezza N (2017) Cross-generational effects of alcohol dependence in humans on HRAS and TP53 methylation in offspring. Epigenomics 9(9):1189–1203. https://doi.org/10.2217/epi-2017-0052

Hollander J, McNivens M, Pautassi RM, Nizhnikov ME (2019) Offspring of male rats exposed to binge alcohol exhibit heightened ethanol intake at infancy and alterations in T-maze performance. Alcohol (Fayetteville, NY) 76:65–71. https://doi.org/10.1016/j.alcohol.2018.07.013

Holoch D, Moazed D (2015) RNA-mediated epigenetic regulation of gene expression. Nat Rev Genet 16(2):71–84. https://doi.org/10.1038/nrg3863

Hottenga JJ, Ligthart L, Mbarek H, Boomsma DI (2018) Transancestral GWAS of alcohol dependence reveals common genetic underpinnings with psychiatric disorders. Nat Neurosci 21(12):1656–1669. https://doi.org/10.1038/s41593-018-0275-1

Hussain S, Lesscher HMD, Day DJ, Ellenbroek BA (2022) Genetics and epigenetics: paternal adolescent ethanol consumption in serotonin transporter knock-out rats and offspring sensitivity to ethanol. Psychopharmacology 239(10):3145–3159. https://doi.org/10.1007/s00213-022-06195-5

Hussain S, Day DJ, Ellenbroek BA (2025) Preconceptual paternal ethanol drinking induces sexually dimorphic Behavioural changes across 2 generations. Psychopharmacology. https://doi.org/10.1007/s00213-025-06807-w

Hutchinson D, Youssef GJ, McCormack C, Wilson J, Allsop S, Najman J, Elliott E, Burns L, Jacobs S, Honan I, Rossen L, Fiedler H, Teague S, Ryan J, Olsson CA, Mattick RP (2019) Prenatal alcohol exposure and infant gross motor development: a prospective cohort study. BMC Pediatr 19(1):149. https://doi.org/10.1186/s12887-019-1516-5

Jamerson PA, Wulser MJ, Kimler BF (2004) Neurobehavioral effects in rat pups whose sires were exposed to alcohol. Brain Res Dev Brain Res 149(2):103–111. https://doi.org/10.1016/j.devbrainres.2003.12.010

Jana K, Jana N, De DK, Guha SK (2010) Ethanol induces mouse spermatogenic cell apoptosis in vivo through over-expression of Fas/Fas-L, p53, and caspase-3 along with cytochrome c translocation and glutathione depletion. Mol Reprod Dev 77(9):820–833. https://doi.org/10.1002/mrd.21227

Jiajie T, Yanzhou Y, Hoi-Hung AC, Zi-Jiang C, Wai-Yee C (2017) Conserved miR-10 family represses proliferation and induces apoptosis in ovarian granulosa cells. Sci Rep 7(1):41304. https://doi.org/10.1038/srep41304

Joo K, Kwon Y, Myung S, Kim T (2012) The effects of smoking and alcohol intake on sperm quality: light and transmission electron microscopy findings. J Int Med Res 40(6):2327–2335. https://doi.org/10.1177/030006051204000631

Kaufman J, Yang B, Douglas-Palumberi H, Crouse-Artus M, Lipschitz D, Krystal JH, Gelernter J (2007) Genetic and environmental predictors of early alcohol use. Biol Psychiatry 61(11):1228–1234

Kim J, Shukla S (2006) Acute in vivo effect of ethanol (binge drinking) on histone H3 modifications in rat tissues. Alcohol Alcoholism (Oxford) 41(2):126–132. https://doi.org/10.1093/alcalc/agh248

Kim P, Choi CS, Park JH, Joo SH, Kim SY, Ko HM, Kim KC, Jeon SJ, Park SH, Han S, Ryu JH, Cheong JH, Han JY, Ko KN, Shin CY (2014) Chronic exposure to ethanol of male mice before mating produces attention deficit hyperactivity disorder-like phenotype along with epigenetic dysregulation of dopamine transporter expression in mouse offspring. J Neurosci Res 92(5):658–670. https://doi.org/10.1002/jnr.23275

Klassen RW, Persaud TVN (1976) Experimental studies on the influence of male alcoholism on pregnancy and progeny. Exp Pathol 12(1):38–45. https://doi.org/10.1016/S0014-4908(76)80031-2

Knezovich JG, Ramsay M (2012) The effect of preconception paternal alcohol exposure on epigenetic remodeling of the H19 and Rasgrf1 imprinting control regions in mouse offspring. Front Genet 3:10. https://doi.org/10.3389/fgene.2012.00010

Koabel J, McNivens M, McKee P, Pautassi R, Bordner K, Nizhnikov M (2021) The offspring of alcohol-exposed sires exhibit heightened ethanol intake and behavioral alterations in the elevated plus maze. Alcohol (Fayetteville, NY) 92:65–72. https://doi.org/10.1016/j.alcohol.2021.01.009

Krasteva M, Koycheva Y, Taseva T, Simeonova S (2020) Changes in the expression of DNA methylation related genes in leukocytes of persons with alcohol and drug dependence. Acta Medica Bulgarica 47(4):11–17. https://doi.org/10.2478/amb-2020-0039

Kruman II, Fowler A (2014) Impaired one carbon metabolism and DNA methylation in alcohol toxicity. J Neurochem 129(5):770–780. https://doi.org/10.1111/jnc.12677

Kyzar EJ, Zhang H, Pandey SC (2019) Adolescent alcohol exposure epigenetically suppresses amygdala arc enhancer RNA expression to confer adult anxiety susceptibility. Biol Psychiatry (1969) 85(11):904–914. https://doi.org/10.1016/j.biopsych.2018.12.021

Lanata CM, Chung SA, Criswell LA (2018) DNA methylation 101: what is important to know about DNA methylation and its role in SLE risk and disease heterogeneity. Lupus Sci Med 5(1): e000285. https://doi.org/10.1136/lupus-2018-000285

Ledig M, Simler S, Ciesielski L, Mandel P (1993) Alcohol exposure before pregnancy: effect on GABA levels and turnover in rat offspring. Alcohol Alcoholism (Oxford) 28(2):175–179. https://doi.org/10.1093/oxfordjournals.alcalc.a045358

Ledig M, Misslin R, Vogel E, Holownia A, Copin JC, Tholey G (1998) Paternal alcohol exposure: developmental and behavioral effects on the offspring of rats. Neuropharmacology 37(1):57–66. https://doi.org/10.1016/S0028-3908(97)00185-8

Lee HJ, Ryu JS, Choi NY, Park YS, Kim YI, Han DW, Ko KS, Shin CY, Hwng HS, Kang KS, Ko K (2013) Transgenerational effects of paternal alcohol exposure in mouse offspring. Anim Cells Syst 17(6):429–434. https://doi.org/10.1080/19768354.2013.865675

Lemoine P, Harousseau H, Borteyru JP, Menuet JC (2003) Children of alcoholic parents—observed anomalies: discussion of 127 cases. Ther Drug Monit 25(2):132–136. https://doi.org/10.1097/00007691-200304000-00002

Liang F, Diao L, Liu J, Jiang N, Zhang J, Wang H, Zhou W, Huang G, Ma D (2014) Paternal ethanol exposure and behavioral abnormities in offspring: associated alterations in imprinted gene methylation. Neuropharmacology 81:126–133. https://doi.org/10.1016/j.neuropharm.2014.01.025

Liang F, Diao L, Jiang N, Zhang J, Wang H, Zhou W, Huang G, Ma D (2015) Chronic exposure to ethanol in male mice may be associated with hearing loss in offspring. Asian J Androl 17(6): 985–990. https://doi.org/10.4103/1008-682X.160267

Liu N, Pan T (2015) RNA epigenetics. Transl Res 165(1):28–35. https://doi.org/10.1016/j.trsl.2014.04.003

Lohoff FW, Roy A, Jung J, Longley M, Rosoff DB, Luo A, O'Connell E, Sorcher JL, Sun H, Schwandt M, Hodgkinson CA, Goldman D, Momenan R, McIntosh AM, Adams MJ, Walker RM, Evans KL, Porteous D, Smith AK et al (2021) Epigenome-wide association study and multi-tissue replication of individuals with alcohol use disorder: evidence for abnormal glucocorticoid signaling pathway gene regulation. Mol Psychiatry 26(6):2224–2237. https://doi.org/10.1038/s41380-020-0734-4

Maeda K, Yamada H, Munetsuna E, Fujii R, Yamazaki M, Ando Y, Mizuno G, Ishikawa H, Ohashi K, Tsuboi Y, Hattori Y, Ishihara Y, Hashimoto S, Hamajima N, Suzuki K (2022) Association of drinking behaviors with TXNIP DNA methylation levels in leukocytes among the general Japanese population. Am J Drug Alcohol Abuse 48(3):302–310. https://doi.org/10.1080/00952990.2022.2037137

Mahesh PKB, Leung J, Senananayake S, Kumarapeli V, Ferdinando R, Jayawickrama WIU, Lakmini WD, Perera C, Thasleem S, McLaughlin D (2021) Effects of paternal alcoholism on the psycho-behavioural outcomes of the offspring: a systematic review and meta-analysis. J Coll Commun Phys Sri Lanka 27(3):442–461. https://doi.org/10.4038/jccpsl.v27i3.8398

Manzardo AM, Henkhaus RS, Butler MG (2012) Global DNA promoter methylation in frontal cortex of alcoholics and controls. Gene 498(1):5–12. https://doi.org/10.1016/j.gene.2012.01.096

McCormack C, Hutchinson D, Burns L, Youssef G, Wilson J, Elliott E, Allsop S, Najman J, Jacobs S, Rossen L, Olsson C, Mattick R (2018) Maternal and partner prenatal alcohol use and

infant cognitive development. Drug Alcohol Depend 185:330–338. https://doi.org/10.1016/j.drugalcdep.2017.12.038
McPherson NO, Owens JA, Fullston T, Lane M (2015) Preconception diet or exercise intervention in obese fathers normalizes sperm microRNA profile and metabolic syndrome in female offspring. Am J Physiol Endocrinol Metab 308(9):E805–E821. https://doi.org/10.1152/ajpendo.00013.2015
Meek LR, Myren K, Sturm J, Burau D (2007) Acute paternal alcohol use affects offspring development and adult behavior. Physiol Behav 91(1):154–160. https://doi.org/10.1016/j.physbeh.2007.02.004
Meng W, Sjöholm LK, Kononenko O, Tay N, Zhang D, Sarkisyan D, Geske JR, Ing A, Qiu W, Watanabe H, Almamoun R, Frieling H, Bleich S, Cui D, Biernacka JM, Mayfield RD, Dang Y, Karpyak VM, Schumann G et al (2021) Genotype-dependent epigenetic regulation of DLGAP2 in alcohol use and dependence. Mol Psychiatry 26(8):4367–4382. https://doi.org/10.1038/s41380-019-0588-9
Merrill SM, Moore SR, Gladish N, Giesbrecht GF, Dewey D, Konwar C, MacIssac JL, Kobor MS, Letourneau NL (2021) Paternal adverse childhood experiences: associations with infant DNA methylation. Dev Psychobiol 63(6):e22174. https://doi.org/10.1002/dev.22174
Ministry of Health (2022) Annual data explorer 2021/22: New Zealand Health Survey. Ministry of Health. https://minhealthnz.shinyapps.io/nz-health-survey-2021-22-annual-data-explorer/
Monk D, Mackay DJG, Eggermann T, Maher ER, Riccio A (2019) Genomic imprinting disorders: lessons on how genome, epigenome and environment interact. Nat Rev Genet 20(4):235–248. https://doi.org/10.1038/s41576-018-0092-0
Montesinos J, Pascual M, Rodríguez-Arias M, Miñarro J, Guerri C (2016) Involvement of TLR4 in the long-term epigenetic changes, rewarding and anxiety effects induced by intermittent ethanol treatment in adolescence. Brain Behav Immun 53:159–171. https://doi.org/10.1016/j.bbi.2015.12.006
Moore LD, Le T, Fan G (2013) DNA methylation and its basic function. Neuropsychopharmacology (New York, NY) 38(1):23–38. https://doi.org/10.1038/npp.2012.112
Muschler MAN, Hillemacher T, Kraus C, Kornhuber J, Bleich S, Frieling H (2010) DNA methylation of the POMC gene promoter is associated with craving in alcohol dependence. J Neural Transm 117(4):513–519. https://doi.org/10.1007/s00702-010-0378-7
Nieto SJ, Harding MJ, Nielsen DA, Kosten TA (2022) Paternal alcohol exposure has task- and sex-dependent behavioral effect in offspring. Alcohol Clin Exp Res 46(12):2191–2202. https://doi.org/10.1111/acer.14964
Niu Y, Zhao X, Wu Y, Li M, Wang X, Yang Y (2013) N6-methyl-adenosine (m6A) in RNA: an old modification with A novel epigenetic function. Genomics Proteomics Bioinformatics 11(1):8–17. https://doi.org/10.1016/j.gpb.2012.12.002
Olfson E, Edenberg HJ, Nurnberger J Jr, Agrawal A, Bucholz KK, Almasy LA, Chorlian D, Dick DM, Hesselbrock VM, Kramer JR (2014) An ADH 1B variant and peer drinking in progression to adolescent drinking milestones: evidence of a gene-by-environment interaction. Alcohol Clin Exp Res 38(10):2541–2549
ONS (2022) Drug misuse in England and Wales: year ending June 2022. ONS website: Office for National Statistics. https://www.ons.gov.uk/peoplepopulationandcommunity/crimeandjustice/articles/drugmisuseinenglandandwales/yearendingjune2022
Oppermann U (2013) Why is epigenetics important in understanding the pathogenesis of inflammatory musculoskeletal diseases? Arthritis Res Ther 15(2):209. https://doi.org/10.1186/ar4186
Orford J, Velleman R (1990) Offspring of parents with drinking problems: drinking and drug-taking as young adults. Br J Addict 85(Jun 90):779–794. https://search.proquest.com/docview/57759673
Otero NKH, Thomas JD, Saski CA, Xia X, Kelly SJ (2012) Choline supplementation and DNA methylation in the hippocampus and prefrontal cortex of rats exposed to alcohol during

development. Alcohol Clin Exp Res 36(10):1701–1709. https://doi.org/10.1111/j.1530-0277.2012.01784.x

Ouko LA, Shantikumar K, Knezovich J, Haycock P, Schnugh DJ, Ramsay M (2009) Effect of alcohol consumption on CpG methylation in the differentially methylated regions of H19 and IG-DMR in male gametes—implications for fetal alcohol Spectrum disorders. Alcohol Clin Exp Res 33(9):1615–1627. https://doi.org/10.1111/j.1530-0277.2009.00993.x

Ozkaragoz T, Satz P, Noble EP (1997) Neuropsychological functioning in sons of active alcoholic, recovering alcoholic, and social drinking fathers. Alcohol (Fayetteville, NY) 14(1):31–37. https://doi.org/10.1016/S0741-8329(96)00084-5

Painter RC, Rooij SRD, Bossuyt PM, Simmers TA, Osmond C, Barker DJ, Bleker OP, Roseboom TJ (2006) Early onset of coronary artery disease after prenatal exposure to the Dutch famine. Am J Clin Nutr 84(2):322–327. https://doi.org/10.1093/ajcn/84.2.322

Pandey SC, Ugale R, Zhang H, Tang L, Prakash A (2008) Brain chromatin remodeling: A novel mechanism of alcoholism. J Neurosci 28(14):3729–3737. https://doi.org/10.1523/JNEUROSCI.5731-07.2008

Passaro KT, Little RE, Savitz DA, Noss J (1998) Effect of paternal alcohol consumption before conception on infant birth weight. Teratology (Philadelphia) 57(6):294–301. https://doi.org/10.1002/(SICI)1096-9926(199806)57:63.0.CO;2-X

Pembrey ME, Bygren LO, Kaati G, Edvinsson S, Northstone K, Sjöström M, Golding J (2006) Sex-specific, male-line transgenerational responses in humans. Eur J Hum Genet: EJHG 14(2):159–166. https://doi.org/10.1038/sj.ejhg.5201538

Peng H, Shi J, Zhang Y, Zhang H, Liao S, Li W, Lei L, Han C, Ning L, Cao Y, Zhou Q, Chen Q, Duan E (2012) A novel class of tRNA-derived small RNAs extremely enriched in mature mouse sperm. Cell Res 22(11):1609–1612. https://doi.org/10.1038/cr.2012.141

Perez MF, Lehner B (2019) Intergenerational and transgenerational epigenetic inheritance in animals. Nat Cell Biol 21(2):143–151. https://doi.org/10.1038/s41556-018-0242-9

Philibert R, Dogan M, Noel A, Miller S, Krukow B, Papworth E, Cowley J, Knudsen A, Beach SRH, Black D (2018) Genome-wide and digital polymerase chain reaction epigenetic assessments of alcohol consumption. Am J Med Genet B Neuropsychiatr Genet 177(5):479–488. https://doi.org/10.1002/ajmg.b.32636

Ponomarev I, Wang S, Zhang L, Harris RA, Mayfield RD (2012) Gene Coexpression networks in human brain identify epigenetic modifications in alcohol dependence. J Neurosci 32(5):1884–1897. https://doi.org/10.1523/JNEUROSCI.3136-11.2012

Przybycien-Szymanska MM, Rao YS, Prins SA, Pak TR (2014) Parental binge alcohol abuse alters F1 generation hypothalamic gene expression in the absence of direct fetal alcohol exposure. PLoS One 9(2):e89320. https://doi.org/10.1371/journal.pone.0089320

Radford EJ, Ito M, Shi H, Corish JA, Yamazawa K, Isganaitis E, Seisenberger S, Hore TA, Reik W, Erkek S, Peters AHFM, Patti M, Ferguson-Smith AC (2014) In utero undernourishment perturbs the adult sperm methylome and intergenerational metabolism. Science (American Association for the Advancement of Science) 345(6198):785. https://doi.org/10.1126/science.1255903

Rahimipour M, Talebi AR, Anvari M, Sarcheshmeh AA, Omidi M (2013) Effects of different doses of ethanol on sperm parameters, chromatin structure and apoptosis in adult mice. Eur J Obstet Gynecol Reprod Biol 170(2):423–428. https://doi.org/10.1016/j.ejogrb.2013.06.038

Randall CL, Burling TA, Lochry EA, Sutker PB (1982) The effect of paternal alcohol consumption on fetal development in mice, vol 9. Elsevier Ireland Ltd. https://doi.org/10.1016/0376-8716(82)90028-X

Rathod RS, Ferguson C, Seth A, Baratta AM, Plasil SL, Homanics GE (2020) Effects of paternal preconception vapor alcohol exposure paradigms on behavioral responses in offspring. Brain Sci 10(9):658. https://doi.org/10.3390/brainsci10090658

Ricci E, Al Beitawi S, Cipriani S, Candiani M, Chiaffarino F, Viganò P, Noli S, Parazzini F (2017) Semen quality and alcohol intake: a systematic review and meta-analysis. Reprod Biomed Online 34(1):38–47. https://doi.org/10.1016/j.rbmo.2016.09.012

Rodgers AB, Morgan CP, Bronson SL, Revello S, Bale TL (2013) Paternal stress exposure alters sperm MicroRNA content and reprograms offspring HPA stress Axis regulation. J Neurosci 33(21):9003–9012. https://doi.org/10.1523/JNEUROSCI.0914-13.2013
Rodgers AB, Morgan CP, Leu NA, Bale TL (2015) Transgenerational epigenetic programming via sperm microRNA recapitulates effects of paternal stress. Proc Natl Acad Sci – PNAS 112(44): 13699–13704. https://doi.org/10.1073/pnas.1508347112
Rompala GR, Homanics GE (2019) Intergenerational effects of alcohol: A review of paternal preconception ethanol exposure studies and epigenetic mechanisms in the male germline. Alcohol Clin Exp Res 43(6):1032–1045. https://doi.org/10.1111/acer.14029
Rompala GR, Finegersh A, Homanics GE (2016) Paternal preconception ethanol exposure blunts hypothalamic-pituitary-adrenal axis responsivity and stress-induced excessive fluid intake in male mice. Alcohol (Fayetteville, NY) 53:19–25. https://doi.org/10.1016/j.alcohol.2016.03.006
Rompala GR, Finegersh A, Slater M, Homanics GE (2017) Paternal preconception alcohol exposure imparts intergenerational alcohol-related behaviors to male offspring on a pure C57BL/6J background. Alcohol (Fayetteville, NY) 60:169–177. https://doi.org/10.1016/j.alcohol.2016.11.001
Rompala GR, Mounier A, Wolfe CM, Lin Q, Lefterov I, Homanics GE (2018) Heavy chronic intermittent ethanol exposure alters small noncoding RNAs in mouse sperm and Epididymosomes. Front Genet 9:32. https://doi.org/10.3389/fgene.2018.00032
Roseboom TJ, Painter RC, van Abeelen AFM, Veenendaal MVE, de Rooij SR (2011) Hungry in the womb: what are the consequences? Lessons from the Dutch famine. Maturitas 70(2): 141–145. https://doi.org/10.1016/j.maturitas.2011.06.017
Rossetto IMU, Cagnon VHA, Lizarte FSN, Tirapelli LF, Tirapelli DPC, Arantes RMS, Chuffa LGA, Martinez FE, Martinez M (2019) Ethanol and caffeine consumption modulates the expression of miRNAs in the cerebellum and plasma of UChB rats. Life Sci 1973(229): 180–186. https://doi.org/10.1016/j.lfs.2019.05.016
Russo SJ, Mazei-Robison MS, Ables JL, Nestler EJ (2009) Neurotrophic factors and structural plasticity in addiction. Neuropharmacology 56(Suppl 1):73–82. https://doi.org/10.1016/j.neuropharm.2008.06.059
Sakharkar AJ, Kyzar EJ, Gavin DP, Zhang H, Chen Y, Krishnan HR, Grayson DR, Pandey SC (2019) Altered amygdala DNA methylation mechanisms after adolescent alcohol exposure contribute to adult anxiety and alcohol drinking. Neuropharmacology 157:107679. https://doi.org/10.1016/j.neuropharm.2019.107679
Sarkar DK (2016) Male germline transmits fetal alcohol epigenetic marks for multiple generations: a review. Addict Biol 21(1):23–34. https://doi.org/10.1111/adb.12186
Schandler S, Antick J, Cohen M, Brannock J, Caine K (1988) Visuospatial learning in elementary school children with and without a family history of alcoholism. J Stud Alcohol 49(1988): 538–545. https://search.proquest.com/docview/57766427
Scholte RHJ, Poelen EAP, Willemsen G, Boomsma DI, Engels RCME (2008) Relative risks of adolescent and young adult alcohol use: the role of drinking fathers, mothers, siblings, and friends. Addict Behav 33(1):1–14. https://doi.org/10.1016/j.addbeh.2007.04.015
Sharma U (2019) Paternal contributions to offspring health: role of sperm small RNAs in intergenerational transmission of epigenetic information. Front Cell Dev Biol 7:215. https://doi.org/10.3389/FCELL.2019.00215
Sharma U, Conine CC, Shea JM, Boskovic A, Derr AG, Bing XY, Belleannee C, Kucukural A, Serra RW, Sun F, Song L, Carone BR, Ricci EP, Li XZ, Fauquier L, Moore MJ, Sullivan R, Mello CC, Garber M, Rando OJ (2016) Biogenesis and function of tRNA fragments during sperm maturation and fertilization in mammals. Science (American Association for the Advancement of Science) 351(6271):391–396. https://doi.org/10.1126/science.aad6780
Shen HM, Chia SE, Ong CN (1999) Evaluation of oxidative DNA damage in human sperm and its association with male infertility. J Androl 20(6):718–723. https://doi.org/10.1002/j.1939-4640.1999.tb03376.x

Skinner MK (2016) Epigenetic transgenerational inheritance. Nat Rev Endocrinol 12(2):68–70. https://doi.org/10.1038/nrendo.2015.206

Stockard CR, Papanicolaou G (1916a) A further analysis of the hereditary transmission of degeneracy and deformities by the descendants of Alcoholized mammals. Am Nat 50(590):65–88. https://doi.org/10.1086/279525

Stockard CR, Papanicolaou G (1916b) A further analysis of the hereditary transmission of degeneracy and deformities by the descendants of Alcoholized mammals. II Am Naturalist 50(591): 144–177. https://doi.org/10.1086/279531

Tang Y, Banan A, Forsyth CB, Fields JZ, Lau CK, Zhang LJ, Keshavarzian A (2008) Effect of alcohol on miR-212 expression in intestinal epithelial cells and its potential role in alcoholic liver disease. Alcohol Clin Exp Res 32(2):355–364. https://doi.org/10.1111/j.1530-0277.2007.00584.x

Terracina S, Ferraguti G, Tarani L, Messina MP, Lucarelli M, Vitali M, De Persis S, Greco A, Minni A, Polimeni A, Ceccanti M, Petrella C, Fiore M (2022) Transgenerational abnormalities induced by paternal Preconceptual alcohol drinking: findings from humans and animal models. Curr Neuropharmacol 20(6):1158–1173. https://doi.org/10.2174/1570159X19666211101111430

Thomasson HR, Crabb DW, Edenberg HJ, Li T (1993) Alcohol and aldehyde dehydrogenase polymorphisms and alcoholism. Behav Genet 23(2):131–136. https://doi.org/10.1007/BF01067417

Thor S, Hemmingsson T, Danielsson A, Landberg J (2022) Fathers' alcohol consumption and risk of substance-related disorders in offspring. Drug Alcohol Depend 233:109354. https://doi.org/10.1016/j.drugalcdep.2022.109354

Tweedie-Cullen RY, Brunner AM, Grossmann J, Mohanna S, Sichau D, Nanni P, Panse C, Mansuy IM (2012) Identification of combinatorial patterns of Post-translational modifications on individual histones in the mouse brain. PLoS One 7(5):e36980. https://doi.org/10.1371/journal.pone.0036980

van Amsterdam J, Nutt D, Phillips L, van den Brink W (2015) European rating of drug harms. J Psychopharmacol (Oxford) 29(6):655–660. https://doi.org/10.1177/0269881115581980

van Amsterdam J, Ramaekers JG, Verkes R, Kuypers KP, Goudriaan AE, van den Brink W (2020) Alcohol-and drug-related public violence in Europe. Eur J Criminol 17(6):806–825

van der Zwaluw CS, Engels RC, Vermulst AA, Rose RJ, Verkes RJ, Buitelaar J, Franke B, Scholte RH (2010) A serotonin transporter polymorphism (5-HTTLPR) predicts the development of adolescent alcohol use. Drug Alcohol Depend 112(1–2):134–139

van Haaster LH, de Rooij DG (1993) Spermatogenesis is accelerated in the immature Djungarian and Chinese hamster and rat. Biol Reprod 49(6):1229–1235. https://doi.org/10.1095/biolreprod49.6.1229

Veenendaal MVE, Painter RC, de Rooij SR, Bossuyt PMM, van der Post JAM, Gluckman PD, Hanson MA, Roseboom TJ (2013) Transgenerational effects of prenatal exposure to the 1944-45 Dutch famine. BJOG: Int J Obstetr Gynaecol 120(5):548–553. https://doi.org/10.1111/1471-0528.12136

Verhulst B, Neale MC, Kendler KS (2015) The heritability of alcohol use disorders: a meta-analysis of twin and adoption studies. Psychol Med 45(5):1061–1072. https://doi.org/10.1017/S0033291714002165

Wang X, Feng J, Xue Y, Guan Z, Zhang D, Liu Z, Gong Z, Wang Q, Huang J, Tang C, Zou T, Yin P (2016) Structural basis of N6-adenosine methylation by the METTL3–METTL14 complex. Nature (London) 534(7608):575–578. https://doi.org/10.1038/nature18298

Wei J, Huang K, Yang C, Kang C (2017) Non-coding RNAs as regulators in epigenetics. Oncol Rep 37(1):3–9. https://doi.org/10.3892/or.2016.5236

Westman J, Wahlbeck K, Laursen TM, Gissler M, Nordentoft M, Hällgren J, Arffman M, Ösby U (2015) Mortality and life expectancy of people with alcohol use disorder in Denmark, Finland and Sweden. Acta Psychiatr Scand 131(4):297–306

WHO (2022) Alcohol. https://www.who.int/news-room/fact-sheets/detail/alcohol

WHO (2023) World health statistics 2023: monitoring health for the SDGs, Sustainable Development Goal. World Health Organization

Wozniak DF, Cicero TJ, Kettinger LI, Meyer ER (1991) Paternal alcohol consumption in the rat impairs spatial learning performance in male offspring. Psychopharmacologia 105(2):289–302. https://doi.org/10.1007/BF02244324

Xia R, Jin L, Li D, Liang H, Yang F, Chen J, Yuan W, Miao M (2018) Association between paternal alcohol consumption before conception and Anogenital distance of offspring. Alcohol Clin Exp Res 42(4):735–742. https://doi.org/10.1111/acer.13595

Yin X, Anwar A, Wang Y, Hu H, Liang G, Zhang C (2022) Paternal environmental exposure-induced spermatozoal small noncoding RNA alteration meditates the intergenerational epigenetic inheritance of multiple diseases. Front Med 16(2):176–184. https://doi.org/10.1007/s11684-021-0885-y

You M, Liang X, Ajmo JM, Ness GC (2008) Involvement of mammalian sirtuin 1 in the action of ethanol in the liver. Am J Physiol Gastrointest Liver Physiol 294(4):892–898. https://doi.org/10.1152/ajpgi.00575.2007

Young-Wolff KC, Enoch M, Prescott CA (2011) The influence of gene–environment interactions on alcohol consumption and alcohol use disorders: a comprehensive review. Clin Psychol Rev 31(5):800–816. https://doi.org/10.1016/j.cpr.2011.03.005

Youssef NA, Lockwood L, Su S, Hao G, Rutten BPF (2018) The effects of trauma, with or without PTSD, on the transgenerational DNA methylation alterations in human Offsprings. Brain Sci 8(5):83. https://doi.org/10.3390/brainsci8050083

Zhang S, Wang L, Yang T, Chen L, Zhao L, Wang T, Chen L, Ye Z, Zheng Z, Qin J (2020) Parental alcohol consumption and the risk of congenital heart diseases in offspring: an updated systematic review and meta-analysis. SAGE Publications. https://doi.org/10.1177/2047487319874530

Gestational Opioid Exposure: From Morphine to MOUD – How Opioids Impact Offspring Development and Maternal Care

Jecenia Duran, Sonia Khalid, Lauren Richardson, Kelly Bosse, and Susanne Brummelte

Contents

Abstract Opioid use disorder (OUD) continues to be a global problem, with particularly high opioid usage rates in the United States. One major contributor to

---

Jecenia Duran, Sonia Khalid, and Lauren Richardson contributed equally to this work.

J. Duran and L. Richardson
Department of Psychology, Wayne State University, Detroit, MI, USA

S. Khalid
Translational Neuroscience Program, Wayne State University, Detroit, MI, USA

K. Bosse
Translational Neuroscience Program, Wayne State University, Detroit, MI, USA

Research and Development Service, John D Dingell VA Medical Center, Detroit, MI, USA

Department of Psychiatry and Behavioral Neurosciences, Wayne State University School of Medicine, Detroit, MI, USA

S. Brummelte (✉)
Department of Psychology, Wayne State University, Detroit, MI, USA

Translational Neuroscience Program, Wayne State University, Detroit, MI, USA
e-mail: sbrummelte@wayne.edu

Curr Topics Behav Neurosci (2025) 75: 149–182
https://doi.org/10.1007/7854_2024_570
Published Online: 12 March 2025

this crisis has been the high rate of opioid prescriptions, which has increased access to opioids and contributed to many vulnerable individuals becoming dependent or addicted. Many of these affected people are women of reproductive age, which in turn results in many women using or abusing opioids during pregnancy and thus many infants being exposed to illicit opioids. OUD is typically treated with either methadone or buprenorphine (BUP), two effective opioid-based medications for OUD (MOUD). BUP has recently gained more attention and replaced methadone as the "gold standard" of treatment since its unique pharmacodynamic properties seem to result in better compliance, less withdrawal symptoms, and improved infant outcomes compared to methadone. However, the effects of BUP exposure on the long-term outcome of the offspring and mother-infant dyad are not fully understood. This chapter will review the current state of the literature regarding effects of gestational opioid exposure on offspring outcomes, focusing on morphine as a commonly used illicit substance and BUP as a widely used MOUD. Collectively, the literature reviewed here highlights the need for future research into the impact of gestational opioid use on mothers, their care behavior, and their subsequent mother-infant bonds.

Keywords Buprenorphine · Medication · Methadone · Morphine · Neonatal opioid withdrawal syndrome · Opioid use disorder · Pregnancy

1 Introduction

The United States continues to experience an "opioid crisis" brought on by an increase in opioid prescriptions and the expanded availability of illicit opioids (Kolodny et al. 2015). Potentially, this opioid epidemic may have worsened during the Covid-19 pandemic (National Center for Health Statistics 2001), partly due to a lack of access to medication for opioid use disorder (MOUD) (Haley and Saitz 2020). Of particular concern, the number of maternal opioid-related diagnoses at the time of delivery has increased by 131% between 2010 and 2017 (Hirai et al. 2021). Though updated data for the time during or after the pandemic are still missing, a 2019 study estimated that approximately 7% of women in the USA used prescription opioids during pregnancy and of these women, almost 21.2% reported misuse (i.e., using the drug illegally or for a purpose not consistent with medical guidelines) (Ko et al. 2020). Given the recent rise in fentanyl misuse, these numbers may currently be even higher. While it is already widely recognized and not surprising that the misuse of opioids during gestation can have adverse consequences for both mother and infant (CSAT 2006; Broussard et al. 2011; Whiteman et al. 2014; Pryor et al. 2017; Salihu et al. 2019), less is known about the long-term functional impact and specific implications of MOUD use during pregnancy on infants' brain development and the mother-infant dyad.

MOUDs are typically prescribed to anyone suffering from opioid use disorder (OUD) to diminish the negative effects of opioid misuse, such as recurring withdrawal symptoms, aiding individuals to overcome their addiction. However, many patients end up taking these medications for prolonged periods of time to mitigate the high risk of relapse after discontinuation. Thus, a woman may already be maintained on a MOUD when she becomes pregnant, or she may get started on a MOUD if she is diagnosed with OUD during pregnancy. In either case, the pharmacological treatment is often combined with counseling and behavioral therapies to be most effective, but there are no universal guidelines on which MOUD to use, dose or length of treatment, and thus there is a lot of variability in the clinical data on the possible outcomes for the offspring (Zhang et al. 2018; Martin et al. 2020).

The two most commonly prescribed opioids for OUD treatment are the synthetic opioid methadone and the semi-synthetic opioid buprenorphine (BUP). Both have been proven effective in reducing neonatal opioid withdrawal syndrome (NOWS) and other adverse consequences for the offspring when compared to continued opioid misuse during pregnancy (Klaman et al. 2017; Rodriguez and Klie 2019; Soyka 2021); see more details below. Pregnant women maintained on MOUDs are reported to have increased physiological and emotional stability compared to untreated/non-maintained mothers, which in turn is associated with improved neonatal outcomes. Further, MOUDs prevent the drastic cycling between opioid intoxication and withdrawal (Swain and Ho 2019), stabilizing opioid levels, which in turn helps to prevent relapse behavior. Recent systematic reviews and meta-analyses provide evidence that BUP may be a preferred option compared to methadone because BUP resulted in slightly better infant outcomes (for reviews see: Jones et al. 2010; Brogly et al. 2014; Farid et al. 2008). For example, a recent study published in the New England Journal of Medicine, utilizing a large cohort of over 2.5 Mil pregnancies in the USA, confirmed that BUP generally resulted in less NOWS, fewer preterm births, and fewer "small for gestational age" and low birth weight diagnoses when the outcomes of BUP (over 11 K patients) and methadone (~5 K patients) treatment in early or late pregnancy were compared (Suarez et al. 2022). These advantageous outcomes have contributed to BUP's reputation as the new "gold standard" for treating OUD during pregnancy (Wallin et al. 2021; Myers et al. 2023).

BUP and methadone differ quite drastically in their binding affinities at the various opioid receptors. Similar to morphine, methadone is a full agonist at the *mu*-opioid receptor (μOR) (methadone is also an NMDA receptor antagonist). Conversely, BUP is a partial agonist at μOR and an antagonist at both the *kappa*-opioid (κOR) and *delta*-opioid receptors (δOR) (Vuong et al. 2010; Walsh and Eissenberg 2003). These receptor binding characteristics may contribute to its effectiveness in treating withdrawal symptoms and improving infant outcomes (Suarez et al. 2022). However, BUP's distinctive pharmacologic profile may cause it to uniquely impact the exposed offspring and maternal physiology in the long-term, which is almost entirely undefined.

Though not much is known about the prenatal development of the endogenous opioid system in humans (or primates), in the rat brain opioid receptors are detectable very early during fetal development. Specially, μOR mRNA appears first in the neuroepithelium of the cortical plate around embryonic day (ED) 12, but can subsequently be found in other areas, such as the caudate putamen nucleus, an area involved in cognitive and motor functioning, shortly after (i.e., ~ED 13) (Tong et al. 2000). Similarly, brain expression of κORs have been reported as early as ED 12–16 (Tong et al. 2000; Clendeninn et al. 1976; Tan et al. 2018). By postnatal day (PD) 3, the presence of opioid receptors has been confirmed across extensive regions within the rodent brain; including the neostriatum, rostral midbrain, nucleus accumbens (NAc), thalamus, and hypothalamus (Kornblum et al. 1987; McDowell and Kitchen 1987). This widespread distribution of opioid receptors in the postnatal stage implies their likely importance in early brain development. It is tempting to speculate that opioid receptor development in primates and humans follows a similar pattern as reported for rodents, suggesting that human fetuses may start expressing opioid receptors sometime in the second trimester.

What is known in humans is that exogenous or endogenous opioids can cross the placental barrier (Ross et al. 2015) and that trophoblast cells, which form the outer layer of a blastocyst that serves as a precursor to the placenta, express opioid receptors (Rosenfeld 2022). Physiologically, activation of these receptors by endogenous opioids, such as β-endorphin, enkephalin, leucine enkephalin, and dynorphins, plays an essential role in placental function by increasing human chorionic gonadotropin and placental lactogen production (Rosenfeld 2022). Intriguingly, a study by Zharikova et al. (2006) also demonstrated that methadone or BUP can inhibit aromatase conversion of testosterone to estrogen in placental tissue, suggesting that these opioids could influence sexual development of the fetus. Taken together, these findings suggest that opioid exposure during gestation can have direct effects on the fetus but also indirectly impact placental structure and function as well as maternal physiology. However, there are some features of the relationship between opioids and the placenta that are generally ill-defined. For instance, given that placental opioid transport and metabolism could change throughout pregnancy and with maternal blood flow (Nanovskaya et al. 2008; Giroux et al. 1997), there is a lack of data on possible windows of vulnerability for transfer to the fetus. Also unknown is the capacity of opioids to be metabolized in the placenta and, thus, the possible impact of those opioid metabolites on the fetus, as well as understanding if semi-synthetic compared to synthetic MOUDs differ in their influence on placental and offspring development.

Importantly, the effects of maternal opioid exposure do not necessarily terminate with birth, as opioids can also be transmitted through breast milk. Mothers maintained on a MOUD are typically advised to breastfeed even if small amounts of opioids are transferred to the infant (Lindemalm et al. 2009) as the health benefits of breast milk for the infant outweigh the risks of opioid exposure (Welle-Strand et al. 2013; Yonke et al. 2019; Jansson et al. 2008). However, rates of breastfeeding are lower in MOUD moms than in non-opioid-using mothers (Jones et al. 2010; Yonke et al. 2019; Clark 2019) possibly due to societal stigma, health-related issues, or maternal concern regarding safety of breastfeeding while maintained on MOUDs (Pritham 2013; Yonke et al. 2020).

Taken together, the effects of gestational opioid exposure on infant brain development may be mediated through several pathways, including changes in placental function, maternal physiological changes, direct binding of opioids to opioid receptors in the developing infant's brain, or changes in maternal-infant interactions. This chapter will review reported short- and long-term outcomes of gestational opioid exposure, highlighting morphine as a common illicit substance and methadone and BUP as widely prescribed MOUDs, in humans and rodents. Additionally, this chapter will discuss the role of opioid-induced maternal changes in influencing offspring development.

2 Effects of Gestational Opioid Exposure on Neurodevelopmental Outcomes in Humans

The most widely recognized acute impact of gestational opioid exposure on the infant is NOWS. Typical symptoms include hyperactivity, tremors, irritability, high-pitched crying, restlessness, decreased sleep, and yawning (Wouldes and Lester 2023), with seizures and other neurological syndromes occurring more rarely (only 2.7%; Desai et al. 2015). NOWS is also associated with gastrointestinal dysfunction (vomiting and watery stools), respiratory and feeding difficulties, hypertonia, autonomic nervous system activation (fever and nasal stuffiness) and immunological outcomes, a hallmark of which is reduced white blood cell count implying vulnerability to future infections (Wouldes and Lester 2023; Desai et al. 2015; Miller et al. 2022). From 2010 to 2017, the estimated rate of NOWS increased from 4.0 to 7.3 per 1000 deliveries, representing a relative increase of 82% (Hirai et al. 2021). A population-based observational study discovered that a history of long-term use of opioids (prescribed or illicit) prior to pregnancy as well as later gestational opioid use increased the severity and likelihood of the infant developing NOWS (Desai et al. 2015). Polysubstance use is another common risk factor for NOWS severity (Desai et al. 2015). Though NOWS symptoms may be temporary, the presence and severity of NOWS has been linked to later neurodevelopmental outcomes (for review see: Yen and Davis 2022; Harder and Murphy 2019). Interestingly, the presence of NOWS, and the subsequent clinical treatment and support of the exposed infant, may even be protective for infant survival. A recent study by Leyenaar et al. (2021) demonstrated that opioid-exposed infants who were not diagnosed with NOWS had a higher mortality risk compared to a reference population, while opioid-exposed infants with a NOWS diagnosis did not differ in their odds ratio from controls.

Other short-term or acute effects of gestational opioid exposure include, but are not limited to, a higher risk for premature birth and low birth weight, as well as a smaller head circumference at birth (e.g., Graeve et al. 2022; Azuine et al. 2019). Smaller head circumference findings are particularly concerning as they suggest changes to brain development (Yen and Davis 2022). Indeed, accumulating evidence suggests opioid-exposed infants have delayed gray and white matter maturation. For instance, magnetic resonance imaging (MRI) studies have revealed white matter lesions (Merhar et al. 2019), as well as smaller relative volumes of the deep gray matter, bilateral insular white matter, bilateral thalamic ventrolateral nuclei, bilateral subthalamic nuclei, brainstem, and cerebrospinal fluid in opioid-exposed infants compared to unexposed controls (Merhar et al. 2021a). However, exposed infants also had enlargement in some brain areas (i.e., white matter of the right cingulate gyrus and left occipital lobe), suggesting that the influence of prenatal opioids on brain development is complex (Merhar et al. 2021a).

Complementary studies using resting state functional MRI showed significantly higher connectivity between the amygdala and medial prefrontal cortex in exposed infants compared to controls (Radhakrishnan et al. 2021), but smaller network volumes in the primary visual network (Merhar et al. 2021b). Alterations in white matter tract integrity and maturation have also been reported with diffusion tract imaging as indicated by reduced fractional anisotropy in opioid-exposed infants (Sikka et al. 2023). Importantly, most of these neurological studies had rather small sample sizes and subsequent studies with larger cohorts are lacking. Most clinical studies investigating short-term effects of opioid exposure also do not distinguish between prescribed and illicit opioids, and many infants may indeed have been exposed to both as well as additional substances such as nicotine or other psychoactive medications, which makes the interpretation of these findings rather difficult.

Notwithstanding, meta-analyses confirm that gestational opioid exposure in general is associated with a multitude of adverse outcomes later in life. For instance, a systematic review conducted in 2023, which assessed 79 sources across four databases (i.e., PubMed, Embase, PsycInfo, and the Web of Science), revealed that prenatal opioid exposure was consistently linked with neurodevelopmental delays in infancy to early childhood (birth – 6 years of age) across three dimensions: cognitive, behavioral/attention, and motor (Balalian et al. 2023). For infants and toddlers (<18 months old), 71% of studies evaluated by Balalian et al. (2023) reported lower cognitive scores in children with prenatal opioid exposure compared to unexposed children. Reduced cognitive scores seem to persist into early school age years (Balalian et al. 2023) and were most prominently observed in preschool aged boys (Davis and Templer 1988). In a similar manner, some studies indicate that children with NOWS as infants had significantly lower test scores in grades 3, 5, and 7 and were more likely to meet criteria for an educational disability (Oei et al. 2017; Fill et al. 2018). Grade school children (age ~8.5) also displayed reduced language and psycholinguistic ability scores, lower memory performance scores (Davis and Templer 1988; Guo et al. 1994), and had a lower average IQ (Nygaard et al. 2015). This is in line with an earlier meta-analysis report by Yeoh et al. (2019), who similarly found evidence linking lower cognitive and motor scores with prenatal opioid exposure in children up to 6 years of age. However, this relationship was only significant in the younger age group and not in 7- to 18-year-old exposed children. The lack of effect of developmental opioid exposure on these outcomes in the older group suggests that some deficits may be transient or addressable with proper interventions and support.

Behavioral problems are another commonly reported outcome in opioid-exposed children (for review see: Balalian et al. 2023). These issues often manifest in heightened levels of arousal, near-constant crying (Lesser-Katz 1982), reduced ability to orient to auditory and visual stimuli (Strauss et al. 1976; Walhovd et al. 2015), greater internalizing and externalizing behavior, increased attention problems and high scores on attention deficit hyperactivity disorder (ADHD) assessments compared to unexposed children. Moreover, problem behavior, such as impulsivity, aggression and inattentive ADHD (Davis and Templer 1988; Ornoy et al. 2001; Sandtorv et al. 2018), can, in turn, negatively impact cognitive and learning outcomes, indicating a multifaceted inter-relationship among disorder symptomologies. The observation that children with a NOWS diagnosis, and thus a history of opioid exposure, are more likely to be hospitalized for maltreatment, trauma, and mental and behavioral disorders in their first 13 years of life (Uebel et al. 2015), indicates this connection is made even more complex by environment and parental factors experienced by this population.

Along these lines, the relative scope of the impact of opioids on childhood development is often difficult to ascertain due to the heterogeneity inherent among the clinical population and differences that exists within adopted methodologies. Though a majority of studies document short- and/or long-term changes in cognitive, behavioral, and motor outcomes after gestational opioid exposure, there are numerous reports that cognitive performance or behavioral scores are not significantly altered between exposure groups (e.g., Walhovd et al. 2015; Messinger et al. 2004; Lifschitz et al. 1985; Wilson 1989). Maternal factors among these studies are often varied with regards to the timing of when opioid exposure occurred during pregnancy, whether polysubstance use was present, and if participants received MOUDs (Balalian et al. 2023). Consequently, to specifically parse out the effects of illicit opioid use would be next to impossible, plus, clinical studies that assess a single illicit opioid are not common and often lack the proper effect size when evaluated alone. Moreover, procedures to evaluate prenatal opioid exposure (i.e., maternal self-report, OUD diagnosis or NOWS as a marker) and the age at which outcomes were measured in human observational data often differ across studies. Due to these confounds and the often low sample/effect sizes of clinical reports, there is a dire need to extend and refine research efforts that more adequately control for these factors, particularly maternal polysubstance use and timing/type of opioid exposure.

2.1 *Effects of Gestational MOUD Exposure in Humans*

BUP was approved by the Food and Drug Administration (FDA) as a MOUD in 2002 and like most drugs it never received official clearance by the FDA for use during pregnancy. However, BUP has been used as a standard treatment for pregnant persons with opioid use disorder for over 20 years despite there being limited research on either its short- or long-term implications on offspring and maternal outcomes. The limited reported evidence does support more favorable outcomes with BUP compared to the other common MOUD, methadone. Though the overall rates or prevalence of NOWS (Jones et al. 2010; Welle-Strand et al. 2013; Kakko et al. 2008) and peak NOWS scores (Jones et al. 2010) may not always differ between the two treatments, many studies report that BUP use is associated with less severe NOWS symptoms compared to methadone-exposed infants, such that BUP-exposed infants received fewer total pharmacological interventions, had a shorter duration of treatment for NOWS (Jones et al. 2010; Kakko et al. 2008; Hall et al. 2016) and had a reduced length of neonatal hospital stay (Jones et al. 2010; Kakko et al. 2008). Further, some evidence suggests that gestational BUP exposure can also result in lower rates of NOWS compared to methadone (Lemon et al. 2018). For instance, a study by Suarez et al. (2022) found that NOWS occurred in 69% of infants exposed to methadone compared to 52% exposed to BUP within 30 days before delivery. The expression of NOWS is known to be dependent upon numerous factors, such as the type of opioid exposure, dosage, duration of gestational exposure, and maternal factors like physiology, postnatal care, and polysubstance use (Jansson et al. 2017; Kelty and Preen 2019; Kentner et al. 2019), making it difficult to isolate experimental variability attributable to just prenatal MOUD exposure.

Apart from its benefits compared to methadone related to NOWS, gestational BUP exposure has also been consistently linked with reductions in the prevalence of preterm birth, low birth weight, and small size for gestational age (Brogly et al. 2014; Suarez et al. 2022; Zedler et al. 2016). Further, BUP-exposed fetuses show better indices of fetal well-being than methadone-exposed fetuses, in that they had greater heart rate variability with more accelerations and better coupling between fetal movements and heart rate (Salisbury et al. 2012; Jansson et al. 2011). Regarding short-term neurobehavioral outcomes (i.e., within 30 days of birth), infants exposed to BUP showed less stress and hypertonia, lower excitability, and had better self-regulation compared to methadone-exposed infants (Jones et al. 2010; Coyle et al. 2012). While these studies do indicate that BUP results in more favorable outcomes relative to other MOUD treatments or continued illicit opioid use during pregnancy, it should be noted that BUP is not without risks. Compared to nonexposed infants, prenatal BUP exposure was associated with lower birth weight and body length as well as NOWS symptoms that required pharmacological treatment (Jansson et al. 2017). Further, most studies reporting the impact of perinatal opioid exposure on accompanying neurodevelopmental outcomes, such as reduced brain volume in the basal ganglia, thalamus, cerebellar white matter, brainstem, and deep gray matter (Merhar et al. 2021a; Sirnes et al. 2017), do not distinguish between illicit opioid and MOUD exposure, making it difficult to differentiate the extent to which methadone or BUP may play a role. One clinical study using MRI found that prenatal methadone exposure was associated with atypical white matter tract development, such that fiber tracts were more immature and less coherently organized compared to nonexposed controls (Monnelly et al. 2018). Future studies should investigate altered neonatal brain development and neurofunction following MOUD exposure to isolate their neurodevelopmental consequences from illicit opioid exposure in humans.

As previously mentioned, there is a relative lack of long-term outcome data in humans following gestational exposure to all opioids, including MOUDs. Also, as is the case with other opioids, findings for MOUDs in this context are often contradictory, revealing no clear pattern of long-term developmental effects. Generally, the limited evidence indicates gestational BUP and methadone exposure are associated with numerous neurobehavioral deficits, including lower self-regulation and positive affect, poorer attention, less interest in activities, and reduced responsiveness and involvement (Bakhireva et al. 2019; Konijnenberg et al. 2016; Salo et al. 2010; Wouldes and Woodward 2020) in exposed infants and toddlers compared to nonexposed children. For instance, a prospective, longitudinal study followed methadone-exposed infants and nonexposed controls from birth through 24 months of age and assessed neurobehavioral outcomes using the NICU Network Neurobehavioral Scale (NNNS), which uses various measures to assess neurologic function, behavior, and signs of stress abstinence. After controlling for factors like socioeconomic status and other psychoactive substance use, their findings indicate that methadone-exposed children exhibited nonoptimal reflexes, more hypertonicity, were more easily aroused, and had poorer behavioral regulation and attention compared to nonexposed children (Wouldes and Woodward 2020). However, many of the methadone-exposed children also exhibited more troublesome clinical outcomes at birth, as they were more likely to be born premature, required prolonged NOWS treatment, and longer hospitalization (Wouldes and Woodward 2020), which probably contributed to the long-term effects. This suggests that children who appear to be most at risk at birth after gestational opioid exposure, may benefit from targeted interventions and resources to prevent long-term deficits.

Further, one study investigating 4-year-old children found that gestational MOUD exposure resulted in lower performance on visual selective attention tasks but did not impact general attention parameters (Konijnenberg and Melinder 2015). For long-term cognitive effects, it has been shown that 9- to 11-year-old children prenatally exposed to MOUDs scored lower on cognitive development measures assessing visual-spatial skills, manipulation of visual stimuli, nonverbal problem-solving, and the ability to identify and articulate the relation between concepts compared to nonexposed children (Konijnenberg and Melinder 2020). This particular impact on visual processing and attention might be related to the smaller network volumes reported for the primary visual network in opioid-exposed children (Merhar et al. 2021b). Further, while the study by Konijnenberg and Melinder (Konijnenberg and Melinder 2020) found no differences in basal cortisol levels or reactivity between MOUD-exposed and unexposed children, they found an interesting relationship between cortisol levels and cognitive abilities among the MOUD-exposed group, where higher levels of cortisol at baseline were associated with better performance on the measure of visual information processing and abstract reasoning (e.g., matrix reasoning task). This would be in line with the two-hit model or match/mismatch hypothesis suggesting that early-life stress (i.e., here: gestational opioid exposure) can result in resilience to renewed stress later in life (i.e., better performance under stressful conditions) (Kentner et al. 2019; Santarelli et al. 2017). Overall, evidence suggests that MOUDs may have a subtle, unique influence on

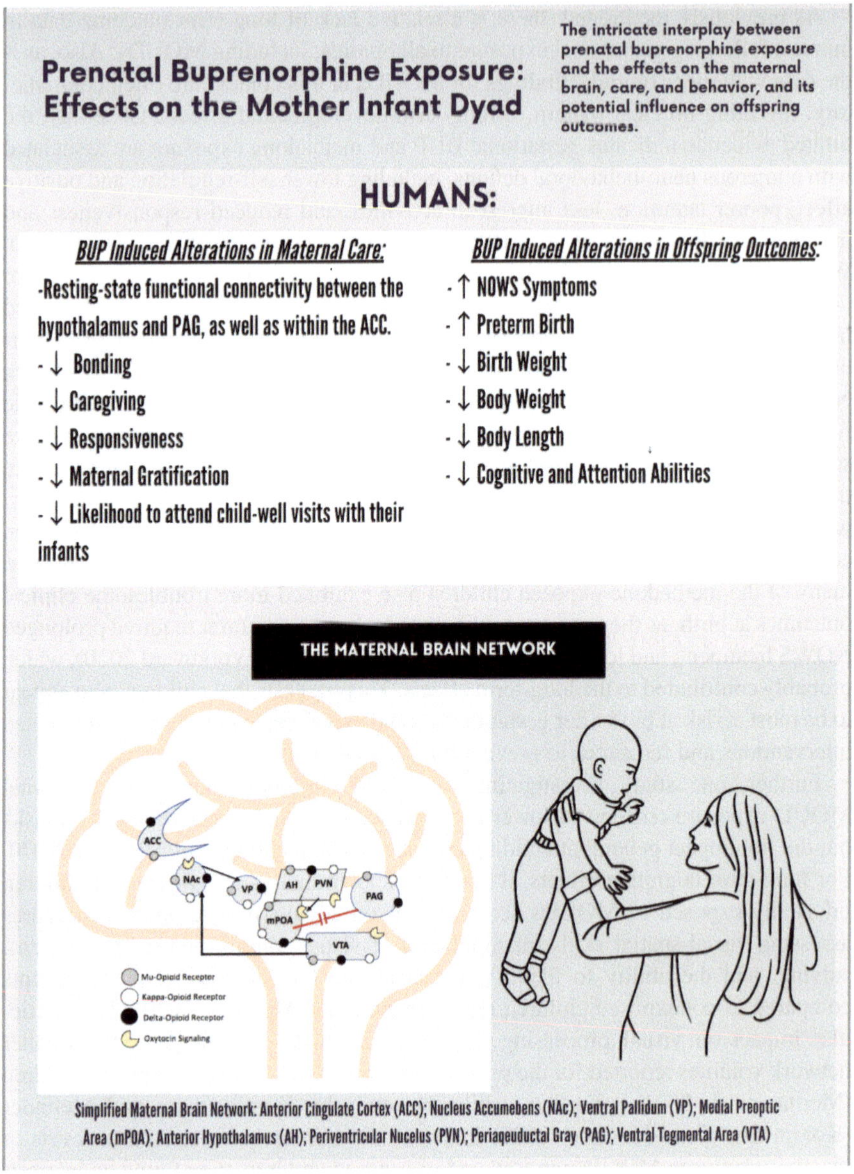

Fig. 1 Overview of effects of prenatal buprenorphine exposure for mother infant dyad

neurocognitive and neuroendocrine development that requires further attention (see Fig. 1). Despite the neurobehavioral findings from these studies, there are numerous others that find no impairments in physical and cognitive development among MOUD-exposed children (Bakhireva et al. 2019; Baldacchino et al. 2014; Kaltenbach et al. 2018). While no consensus has been reached regarding the long-term effects of gestational MOUD exposure, what all the aforementioned studies have in common is that they call for more research and long-term follow-ups to be conducted. Additionally, they highlight that exogenous factors like environmental influences and maternal care are difficult to disentangle in human-subjects research making it a challenge to determine the direct effects of MOUD exposure. Many suggest that differences reported may be an artifact of deviations in parental care as discussed further below.

3 Effects of Gestational Opioid Exposure on Neurodevelopmental Outcomes in Animals

For decades, researchers have utilized and refined animal models of developmental opioid exposure to evaluate the impact of opioids on offspring in a controlled setting, minimizing the influence of environmental, genetic and extraneous factors inherent in human studies. Through these efforts, the field has indeed gained a greater understanding of the neurodevelopmental consequences of opioids, as reviewed below. Of course, these findings are not without their own caveats, particularly with regards to validity. For example, many animal studies have isolated opioid administration to mid-term development (gestational day (GD) 11–18, second trimester equivalent), which is not very translationally relevant. Most pregnant women would not start using opioids midway through pregnancy then stop shortly before parturition. Further, opioid exposure patterns in animal studies can also widely vary among studies, ranging from acute (single dose), sub-chronic (1–2 weeks) and chronic (before conception or first week of gestation until pups are born/weaned) treatment and from low-dose (0.5–1.0 mg/kg) to high-dose exposure (75 mg/kg single bolus or 30–40 mg/kg daily injections), thus they do not fully circumvent the wide variability in opioid exposure patterns that make clinical data interpolation complicated. Despite these experimental variabilities and the short gestational period of rodents (21–22 days, comparable to a second trimester developing fetus), findings from preclinical studies generally map onto those reported in patient populations (Byrnes and Vassoler 2017), highlighting the translational value of animal models for development of clinical interventions and therapies.

3.1 Morphine: Short-Term Outcomes

Common findings related to neurodevelopmental outcomes in rodent models of gestational morphine exposure include decreases in absolute brain and cerebellar weights as well as reductions in levels of growth factors, neural progenitor proteins and neuronal cell markers, including cyclin-dependent kinase-5 (Cdk5), which is critical for neuronal migration and neurite outgrowth (Willner et al. 2014; Bhat et al.

2006). Indices of neuronal morphology, including dendritic length, branching, and spine density, were also found to be reduced with gestational morphine exposure, especially in pyramidal neurons in the visual and somatosensory cortices (Mei et al. 2009; Ricalde and Hammer 1990). Recent reports further indicate that developmental morphine administration may impact neuronal stability, connectivity and function. One avenue through which morphine could influence these outcomes is by altering myelin formation and growth, perhaps via direct activation of opioid receptors expressed on the myelin-producing oligodendrocyte cells. Indeed, offspring administered once-daily subcutaneous morphine injections from PD 1 to 14 (Borrelli et al. 2023) showed significant downregulation of myelin-associated transcripts in the NAc, though this was not associated with alterations in behavior. Alternatively, neuronal matrix proteins, such as perineuronal nets (PNNs), are increasingly recognized to play a critical role in the functional and structural stabilization of developing neurons. Intriguingly, preliminary data from the Bosse lab (unpublished) indicates PNN development in weanlings from morphine-exposed dams is developmentally delayed and current investigation is underway to isolate if these shifts are linked to mood and pain sensitivity impairments in offspring.

Functionally, prenatal morphine exposure in rodents has been shown to impact neuronal signaling specifically in the striatum, cortex, and hippocampus. Isolation of pre- and postsynaptic effects of gestational morphine exposure on excitatory and inhibitory neurons revealed changes in G-protein coupled receptor expression and calcium homeostasis. In the cortex, these effects manifested in reduced Gαs mRNA (Basheer et al. 1992) and altered distribution of calbindin expressing inhibitory neurons (Mithbaokar et al. 2016). Comparatively, striatal dopamine receptor 1-stimulated cAMP levels were elevated and greater electrically evoked norepinephrine release was detected in neocortical slices (De Vries et al. 1991). Additionally, long-term morphine exposure, starting in preconception and continuing until PD 7 in rats, led to altered kinetic properties of synaptic NMDA receptors in hippocampal CA1 pyramidal neurons detected at PD 14 (Yang et al. 2000). These findings, coupled with evidence for neurotransmitter imbalances among acetylcholine, dopamine (DA), serotonin (5HT), and glutamate in these brain regions (De Vries et al. 1991; Grecco and Atwood 2020), point to disrupted function across corticostriatal and integrated hippocampal circuits implicated in learning and memory formation within morphine-exposed offspring.

Studies defining the influence of gestational morphine exposure on the endogenous opioid system have reported dynamic shifts in these measures based on experimental design and time point of assessment. Studies that have restricted morphine administration from mid-to-late prenatal development report increased binding and expression of the striatal μOR in a sex-specific fashion (Vathy et al. 2003; Slamberova et al. 2005; Handelmann and Quirion 1983), whereas exposure to morphine from gestation to postpartum resulted in decreased μOR density in several regions, including the hippocampus, striatum, thalamus, and amygdala (Slamberova et al. 2003; Chiou et al. 2003). Similarly, alterations in the endogenous opioid peptide, met-enkephalin, was shown to vary based on developmental timing of morphine administration, such that high-dose exposure in the second trimester resulted in decreased levels (Tempel et al. 1995), but lower dose morphine access through gestation and postnatal development lead to higher levels (Tenconi et al. 1989). Intriguingly, both of these effects were isolated to the striatum, which together with the aforementioned studies characterizing μOR changes, suggests this brain region is particularly susceptible to the developmental impact of morphine on the endogenous opioid system. Given the regulatory influence of the endogenous opioid system in shaping brain maturation, further studies are warranted to establish the mediators that underlie these differential patterns of changes based on developmental stage at time of exposure and establish how other opioid peptides (i.e., endorphins and dynorphins) and receptor levels (i.e., δORs and κORs) are changed in this context.

Based on the regional prevalence of neural and molecular changes in cortical, striatal and hippocampal pathways following developmental opioid exposure, it is not surprising that prenatal morphine has behavioral effects on learning and memory as well as affective states, including sensitivity to pain and reward. For instance, rat pups (PD 22–31) exposed to morphine from GD 11 to 18 showed deficits in spatial memory and were found to have reduced long-term potentiation in the dentate gyrus region of the hippocampus (Niu et al. 2009). Regarding pain sensitivity, diminished pain responses were noticeable in early development, persistent even after pups were cross-fostered, suggesting these effects are more directly linked to the pharmacological effects of the drug than maternal/environmental factors (O'Callaghan and Holtzman 1976, 1977; Biglarnia et al. 2013). While these findings generally point to an early developmental impact of gestational opioid exposure on neurobehavioral outcomes in animal models, relatively few studies have characterized these effects in young compared to adult rodents. Thus, contrary to human studies, much more has been defined in terms of long-term outcomes in animal models as reviewed in the following section.

3.2 *Morphine: Long-Term Outcomes*

Preclinical rodent studies exploring the long-term impact of gestational morphine exposure have revealed effects on neurotransmitters, the endogenous opioid system, cognition, stress regulation, and drug responsiveness. For instance, prenatal exposure to morphine disrupted monoamine neurotransmission, reflected by alterations including enhanced 5HT and DA metabolism, in a region- and sex-dependent manner in rats aged from ~3 weeks to adulthood (Wu et al. 2009; Laborie et al. 2005; Siddiqui et al. 1997; Vathy and Katay 1992; Vathy et al. 1994, 1995). Persistent deficits in glutamate driven synaptic plasticity, receptor upregulation and neurotransmission were also observed in the cortex and striatum of opioid-exposed adult animals as reviewed by Grecco and Atwood (Grecco and Atwood 2020). Similar to monoamine changes, the directionality of long-term effects of prenatal morphine exposure on opioid receptors varied by region and sex. Tissue from adult female offspring, but not males, exposed to morphine in utero exhibited higher μOR density in the CA1 and CA3 regions of the hippocampus (Slamberova et al. 2003), though male offspring were shown to have greater μOR binding in the central amygdala and NAc (Vathy et al. 2003; Riley and Vathy 2006). Conversely, the medial preoptic area (mPOA) (Slamberova et al. 2005) and basolateral amygdala (BLA) (Vathy et al. 2003) of males displayed lower μOR binding. Though μOR density was reported to be lower in the striatum just after birth (Chiou et al. 2003), neither the striatum nor cerebral cortex showed altered μOR binding in male or female adults as a consequence of prenatal morphine exposure (Slamberova et al. 2003). This suggests that while some effects of gestational morphine exposure on ORs may be transient, this insult can drive long-lasting, sex-specific alterations in regions associated with motivation and learning/memory formation (Vathy et al. 2003; Slamberova et al. 2003, 2005; Rimanoczy et al. 2001).

Besides μOR, only a few studies have analyzed changes in κ- and δ-OR. Prenatal morphine exposure was found to block estrogen-induced elevations in κOR within the frontal cortex of female, but not male, adult offspring (Rimanoczy et al. 2001). Similarly, prenatal morphine impacted hormonal regulation of δORs in the frontal cortex and striatum, resulting in enhanced sensitivity of δORs to hormonal changes in males but decreased sensitivity in females (Vathy et al. 2000); however, no direct changes were seen in function or abundance of δORs in either region, nor in the hippocampus, of adult males (Slamberova et al. 2003; Vathy et al. 2000). In terms of endogenous opioids, differential changes were noted in hippocampal tissue from adult male offspring based on subregion and opioid peptide analyzed. The dentate gyrus showed lower proenkephalin mRNA and met-enkephalin levels, but higher prodynorphin mRNA and dynorphin levels, while a reduction in dynorphin-derived peptides was detected in the CA3 region (Villarreal et al. 2008; Slamberova et al. 2004).

In terms of behavioral, neuroendocrine and cognitive outcomes, previous literature has extensively described the long-term influence of prenatal morphine exposure on memory formation and stress regulation in rats (for review see: Byrnes and Vassoler 2018; Abu and Roy 2021). For instance, several studies have shown memory impairments that were sex, age, and morphine dose-dependent, as well as evidence for lower levels of brain-derived neurotrophic factor (BDNF) in the hippocampus, especially in adult females (Ahmadalipour et al. 2015; Nasiraei-Moghadam et al. 2013). Further, reports have consistently demonstrated alterations in long-term potentiation (LTP) and depression (LTD) in hippocampal slices that, in turn, are linked to learning deficits and increased impulsivity (Villarreal et al. 2008; Alaee et al. 2021; Tan et al. 2015). Prenatal morphine exposure was also shown to alter HPA axis regulation, such that adult rats exhibited dysregulated stress responses (Slamberova et al. 2004; Vathy 2002; Rimanoczy et al. 2003; Castellano and Ammassari-Teule 1984). Further, prenatal morphine exposure resulted in impaired acquisition of contextual fear memory, which could be overcome by enhancing the foot shock intensity used (Tan et al. 2015). This study also found evidence for enhanced maintenance as well as extinction deficits of contextual fear memory in morphine exposed compared with control animals, which is in line with the idea of increased vigilance in stressful situations. In other words, opioid-exposed animals remembered the fearful condition longer, which somewhat maps onto the human observation mentioned earlier demonstrating that higher cortisol levels in opioid-exposed children were associated with improved cognitive function and the two-hit hypothesis of early adversity preparing offspring for a challenging environment. This adaptation hypothesis could help explain various discrepancies in the literature with regards to alterations in mood behavior observed in adult rodents exposed to prenatal morphine, such that some report greater anxiety and depression-like behaviors (Ahmadalipour et al. 2015), while others indicate that anxiety-like behavior displayed in the elevated plus-maze and light/dark box test is reduced (Tan et al. 2015). Given that many preclinical studies do not report on maternal care deficits after opioid exposure or the concurrent circumstances of the testing environment (e.g., elevated plus maze is a more stressful test of anxiety-like behavior than the open field test) it is conceivable that the early-life (adverse) conditions or differences in testing environment could have influenced the behavioral performance of opioid-exposed animals and could thus help explain discrepancies between studies.

In general, gestational exposure to drugs of abuse has often been associated with increased risk for drug use later in life (for reviews see: Grecco and Atwood 2020; Abu and Roy 2021). Overall, prenatal opioid exposure has been shown to increase drug motivated behaviors in adult offspring, as observed by enhanced self-administration and consumption of drugs including morphine, augmented reward-driven behaviors in the conditioned place paradigm, and increased locomotor sensitization (Abu and Roy 2021). This enhanced motivation for drug taking and seeking seems to generalize to other drugs of abuse as well, as morphine-exposed rats also showed increased preference and sensitization for stimulant drugs, such as cocaine (Grecco and Atwood 2020; Vathy 2002). Of note, most studies only measured these behaviors in males and less is known about prenatal opioid exposure on drug-seeking behavior in females. The altered behavioral response to opioids and other drugs later in life suggests that morphine-exposed animals undergo persistent structural and/or functional changes in their brain reward pathways during neurodevelopment (Grecco and Atwood 2020). These changes could also explain the observed altered tolerance to morphine later in life (e.g., Grecco and Atwood 2020; Wallin et al. 2019; Wagner et al. 1986; Chiang et al. 2010).

Taken together, preclinical studies suggest that prenatal morphine exposure can affect brain development in multifaceted ways and promote long-lasting neurobehavioral disturbances. The region- and sex-dependent nature of these changes reflect the complexity of these adaptations involving intricate regulatory processes, which are not currently well understood. In addition, early-life adversity (such as exposure to reduced maternal care) or concurrent circumstances could further influence the outcome, highlighting the need for better translational models and increased research efforts.

3.3 MOUDs: Short-Term Outcomes

Prenatal exposure to MOUDs in rodent models can yield acute effects on the offspring, influencing various aspects of their development and well-being. These adverse effects predominantly include high mortality rates, NOWS, low birth weight, developmental delays, and neurological alterations. Concerning offspring mortality, several preclinical studies have reported significantly higher rates in pups exposed to high doses of BUP (ranging from 1 to 3 mg/kg) compared to unexposed counterparts (Wallin et al. 2019; Chen et al. 2015; Tiong and Olley 1988). One investigation found that subcutaneous BUP administration (1–2 mg/kg) prior to conception resulted in a 50–60% mortality rate by PD 5, despite cross-fostering (Tiong and Olley 1988). Utilizing a similar exposure procedure, research conducted by Wallin et al. (2019) found that some BUP-exposed (1 mg/kg) litters did not survive beyond PD 2. Notably, one study found an increase in mortality throughout the postpartum period after gestational BUP treatment, with a ~9% fatality rate at birth, followed by a 12% mortality rate into the postpartum period (Chen et al. 2015). However, some studies have not observed high mortality rates, possibly due to lower BUP doses (0.3 or 0.5 mg/kg), differences in route of administration (subcutaneous, intraperitoneal, or osmotic minipump), or differences in exposure time and length (Hung et al. 2013; Kongstorp et al. 2019; Hutchings et al. 1995; Evans et al. 1989; Belcheva et al. 1994).

Rodents also display several of the typical or similar NOWS symptoms to what can be observed and quantified in humans after gestational MOUD exposure, such as hyperactivity, tremors, high-pitched crying, yawning, and stretching. Research has shown that BUP exposed offspring display noticeable NOWS symptoms following birth (Wallin et al. 2019) as compared to controls. In severe instances, offspring may even experience seizures (Sutter et al. 2014), which generally has a low incidence of occurrence.

Due to methodological inconsistencies (i.e., route of administration, exposure time, dosing, cross-fostering paradigms, etc.), there are conflicting results within the literature regarding short-term neurodevelopmental outcomes from gestational BUP exposure. For example, some preclinical models have documented delayed developmental milestone achievement in prenatally exposed offspring (Wallin et al. 2019), while others have reported accelerated righting reflex (Robinson and Wallace 2001) as compared to drug-naive offspring. However, many preclinical investigations consistently report decreased weight and body length in litters exposed to BUP compared to controls (Wallin et al. 2019; Chiang et al. 2010; Chen et al. 2015; Tiong and Olley 1988; Hung et al. 2013; Robinson and Wallace 2001; Barron and Chung 1997; Wu et al. 2014, 2017). Lastly, research indicates that MOUD exposure during gestation can impose neurological alterations on the offspring, which could explain some of the altered neurobehavioral and cognitive outcomes discussed in the next section. These alterations include smaller brain weight, decreased expression of neuronal cell markers and growth factors, as well as changes in opioid receptor densities and binding (Hung et al. 2013; Wu et al. 2001, 2014; Hou et al. 2004). Given that most of these neurological findings have been reported for animals at or postweaning age, these results are discussed with the long-term effects.

3.4 MOUDs: Long-Term Outcomes

Compared to morphine, prenatal exposure to MOUDs, including BUP, has been consistently associated with long-term cognitive, behavioral, and neurological outcomes in offspring; however, studies investigating this are relatively few. Notably, offspring exposed to BUP (1–3 mg/kg) throughout the gestational period displayed heightened anxiety and depressive-like behaviors, along with impairments in learning and memory on PD 21 as compared to controls (Chen et al. 2015; Hung et al. 2013; Wu et al. 2014, 2017; Kongstorp et al. 2020).

Moreover, Hung et al. (2013) found that 21-day-old pups exposed to 1 mg/kg BUP from GD 7 to 21 spent significantly more time immobile in the forced swim test (FST). They also report evidence of oxidative stress, reduced plasma levels of BDNF and 5-HT as well as decreased phosphorylation levels of the BDNF receptor, tropomyosin receptor kinase B (TrkB), and the transcription factor, cAMP response element-binding protein (CREB), consistent with lower levels of the neurotrophic factors and, in turn, reduced brain mass and brain/body mass ratios observed. The same group later published additional evidence that prenatal BUP also affects neurogenesis and they confirmed their results in a parallel cell study that showed BUP had an adverse impact on cultured neurospheres, leading to decreased cell proliferation and differentiation as well as altered neurotrophic factors and signaling (Wu et al. 2014). These findings are in line with work from Nieto-Estévez et al. (2022) that suggests prenatal BUP (1 mg/kg, GD 11–21) interferes with interneuron migration into the developing prefrontal cortex and results in an altered neuron distribution across the cortical layers on PD 42. Interestingly, they also confirmed altered interneuron development and migration in human brain spheroids grown from fetal tissue obtained following BUP exposure. These limited studies are congruent with additional evidence for persistent structural and functional deficits in offspring brain development, including decreased abundance of cortical myelinated axons (Sanchez et al. 2008), lower expression of nerve growth factor within the striatum (Wu et al. 2001), delayed generation of cholinergic neurons (Robinson 2002), and alterations in myelination and oligodendrocyte formation (Sanchez et al. 2008). Overall, findings support the idea that prenatal BUP exposure can decrease neurotrophic factors and important developmental signaling pathways that have long-lasting impacts on brain maturation and later behavioral outcomes. In fact, several studies have linked these observed brain alterations following BUP, such as lower BDNF and 5-HT levels, with behavioral impairments, including augmented depressive-like behavior (Hung et al. 2013).

Of note, many of these findings are similar to reported outcomes after gestational morphine exposure. For instance, as mentioned above, Ahmadalipour et al. (2015), also reported lower BDNF levels in morphine exposed juvenile rats as well as increased immobility in the FST test. Intriguingly, the deficits seen after morphine exposure were somewhat reversible if pups grew up in an enriched environment. Similarly, Wu et al. (2017) reported that treadmill exercise could reverse the impact of gestational BUP on depressive-like behavior as well as normalize neurotrophin and oxidative stress measures. This again highlights the importance of postnatal circumstances (i.e., enriched environment, proper maternal care) in determining the long-term outcomes after gestational opioid exposure.

Taken together, evidence is accumulating that MOUDs, and in particular BUP, may have long-lasting adverse effects on offspring development (see Fig. 2), and

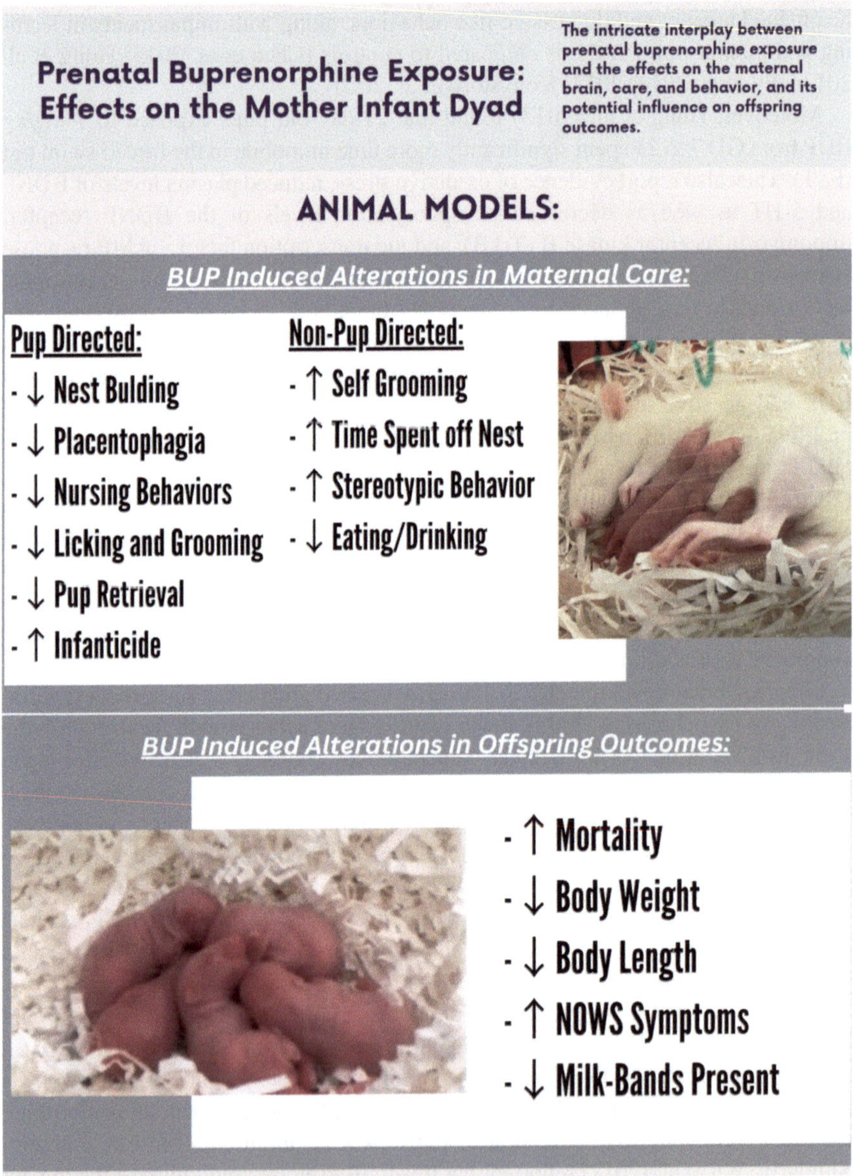

Fig. 2 Overview of effects of prenatal buprenorphine exposure for rodent dams and pups

most rodent studies do not support the idea of a preferable outcome in BUP- compared to morphine-exposed offspring. For instance, data from the Brummelte lab indicates that the mortality rate of BUP-exposed offspring is higher than morphine-exposed offspring, an effect that was attributed, at least partly, to maternal behavior deficits in BUP-exposed dams (Myers et al. 2024 and unpublished data). While animal models provide researchers with enhanced control over confounding variables associated with MOUD treatment compared to human studies, distinguishing between the direct effects of opioids and the impact of opioid-induced impaired maternal care on offspring mortality and morbidity remains challenging (as discussed below).

4 Maternal Brain and Mother-Infant Interactions After Gestational Opioid Exposure

For the most part, preclinical and clinical findings with opioids ranging from illicit compounds like morphine to opioid substitution therapies like BUP, provide evidence that gestational opioid exposure has the potential to influence short- and long-term offspring outcomes. However, it remains unclear what aspects of these effects are due to the direct impact of opioids on brain development and what may result from opioid-induced alterations in maternal brain, physiology, and care behavior. Thus, adopting a more holistic approach by discerning the effects of opioid use during pregnancy on maternal care may aid in understanding the underlying mechanisms of these offspring outcomes.

The Maternal Brain Network (MBN) is both intricate and integrative, functioning to increase interactions between the mother and child. Key structures within the MBN that subserve initiation and maintenance of maternal care include the mPOA, anterior hypothalamus, paraventricular nucleus, and regions specific to the mesolimbic DA reward circuitry, such as the ventral tegmental area, ventral pallidum, and NAc. Importantly, the MBN is rich in endogenous opioid receptors, including μOR, κOR, and δOR receptors, which are distributed extensively across the complex network. Endogenous opioids play a crucial role during the transition to motherhood by initiating or regulating many of the important changes in the MBN. Therefore, by binding to the same receptors, introduction of exogenous opioids during this pivotal phase has the potential to dysregulate the sensitive neurochemistry of the MBN network and interrupt development of maternal caregiving behaviors. Indeed, Wallin et al. (2021) extensively reviews how the MBN facilitates the maternal-infant bond via the transition from a “nulliparous brain” to a “maternal brain” and highlights that opioid or MOUD exposure may disrupt the reciprocal inhibition between the respective areas of the MBN (Wallin et al. 2021). In animal studies, this disruption may, in turn, result in diminished pup-directed behaviors and increased non-pup-directed behaviors (e.g., time off nest, self-grooming, defensive aggression) observed in dams during the postpartum period.

Indeed, preclinical studies have demonstrated decades ago that gestational morphine exposure can reduce maternal care behavior in rodents (Grimm and Bridges 1983; Mayer et al. 1985; Stafisso-Sandoz et al. 1998); or for review see Byrnes and Vassoler 2018). For instance, repeated administration of morphine in late pregnancy decreased active nursing behavior, pup grooming, and contact with pups (Yim et al. 2006; Lambert and Byrnes 2019). Further, morphine was found to significantly decrease preference for pup odors (Kinsley and Bridges 1990), delay pup retrieval (Bridges and Grimm 1982), and decrease maternal aggression in rodents (Kinsley and Bridges 1986). On the other hand, treatment with naloxone, a μOR antagonist, resulted in an increase in prolactin and oxytocin levels in the maternal brain (Russell and Brunton 2006; Brunton and Russell 2008), two hormones that are essential for proper maternal care and bonding. As such, these studies confirm that μORs homeostatically regulate processes involved in maternal bonding and care, such that their exogenous activation dysregulates maternal functions.

Despite the established beneficial role of maternal care behaviors and attachment on offspring development and their recognition as a fundamental element responsible for survival of mammalian offspring, there is a notable lack of comprehensive clinical studies defining the impact of MOUDs (and illicit opioid exposure) on the maternal brain and related behaviors. In particular, the unique pharmacodynamic properties of BUP as a partial μOR agonist and κOR antagonist increase concerns about its potential impact on the carefully orchestrated changes in the endogenous opioid system crucial for initiating maternal care behaviors during pregnancy (Myers et al. 2023), which could have important clinical implications for treatment-seeking mothers. The limited preclinical studies assessing the impact of MOUDs in this context report that these compounds, particularly BUP, have a detrimental impact on maternal care behaviors in dams (Wallin et al. 2019, 2021; Myers et al. 2023; Chen et al. 2015; Barron and Chung 1997; Schlagal et al. 2021) or for review see: Suarez et al. 2022; Wallin et al. 2021). Indeed, dam exposure to translational doses (1 mg/kg) of BUP from preconception to early postpartum manifested in severe maternal neglect that resulted in high mortality of offspring, while others showed significant reductions in time spent with pups as well as increased latencies to retrieve pups back to the nest (Myers et al. 2023; Wallin et al. 2019). However, not every study found drastic changes in maternal care behavior after gestational BUP exposure, which might be due to different routes of administration, doses, length, and time of exposure. Further, many studies simply did not measure or report on maternal behavior. For instance, a study using very low dose and intermittent BUP access (0.3 and 0.6 mg/kg, s.c., every other day from GD 6 to 20) described only subtle effects on maternal and pup interactions (Barron and Chung 1997). Chen et al. (2015) indicated that BUP dams had "a higher propensity of failing to lick or sniff their own pups and eat them instead compared with the dams of other treatment groups," which they associated with higher pup mortality, but then they made no further attempt to define or mention this behavior again throughout their paper. In other studies, pups are cross-fostered to drug-naive dams, preventing the influence of maternal behaviors to be reflected in offspring outcomes (Chiang et al. 2010; Hutchings et al. 1995, 1996); or for review see Byrnes and Vassoler 2018).

While human infants are, in some cases, removed from opioid-using mothers, the clinical goal of maintaining mothers on a MOUD is to keep the mother-infant dyad together and to strengthen the maternal bond, so there is a dire need to better understand how gestational MOUD exposure may impact the maternal brain and, subsequently, care and bonding. In the context of human studies, preliminary evidence suggests that BUP use during pregnancy may indeed disrupt the MBN, specifically altering the resting-state functional connectivity between the hypothalamus and periaqueductal gray, as well as within the anterior cingulate cortex (Swain et al. 2019; Swain and Ho 2023), which may be associated with infant bonding issues in this cohort. Though most human studies have focused on infant outcomes rather than on maternal behaviors, accumulating evidence suggests that the use of opioids (illicit and/or MOUDs) in pregnancy is associated with signs of impaired caregiving and reduced responsiveness to infant cues (Conradt et al. 2019; Wang et al. 2018). For mothers with OUD, it has been reported that they seem less attentive and/or responsive to their babies' needs and have reduced maternal gratification from infant cues (Rutherford et al. 2011, 2021; Rutherford and Mayes 2017). For instance, one study found that mothers with OUD displayed lower sensitivity, expressed less positive affect and engaged in fewer activities with their children at 12 months and 4 years, and their children also displayed lower positive affect, less interest in activities, and less involvement with people around them (Konijnenberg et al. 2016). Given the overlap between areas of the MBN that regulate sensory and emotional processing of infant cues and the reward circuitry in the brain involved in drug-seeking behaviors, it is tempting to spectulate that opioid use during pregnancy may interfere with typical infant-cued activation of these areas and thus have negative impacts on maternal caregiving (Landi et al. 2011). In line with this idea, clinical studies indicate that individuals with OUD may perceive opioids as more rewarding than engaging with their infants (Swain et al. 2019; Cataldo et al. 2019; Brancato and Cannizzaro 2018). This altered behavioral selection may result in a shift towards drug-seeking behaviors rather than prioritizing the well-being of the offspring (Cataldo et al. 2019). Somewhat in line with this, one study found that mothers receiving MOUD are less likely to attend child-well visits with their infants compared to controls (Camden et al. 2023).

A mechanism through which exogenous opioids could influence the mother-infant dyad is by interfering with oxytocin signaling, since it is an essential lactation and contraction hormone and plays a role in mediating maternal bonding. Indeed, morphine administration suppressed the release of oxytocin in mothers during breastfeeding (Lindow et al. 1999), indicating that opioid-exposed mothers may also have problems with the milk production or "let down", while BUP dose-dependently (0.045–0.45 mg/kg) repressed suckling-induced oxytocin release in rats (Clarke and Wright 1984). During pregnancy, oxytocin neurons develop a tolerance and dependence on endogenous opioids, which, in turn, inhibit their release. To initiate the process of childbirth, this inhibitory influence of opioids is removed, triggering a surge in oxytocin and other related hormones. This event is believed to hasten the onset of maternal behaviors and override pup avoidance (Brunton 2019; Russell et al. 2003). Moreover, this surge is theorized to offer protective and anti-inflammatory benefits to the developing fetal brain (Kingsbury and Bilbo 2019). As such, exposure to exogenous opioids risks interrupting these critical maternal-infant developmental interactions.

Maternal nutrition and food intake represent another potential pathway through which gestational opioid or MOUD exposure could impact maternal physiology and, consequently, offspring outcomes. Some of BUP's side effects include constipation, nausea or feeling sick, which may result in inadequate nutritional intake during pregnancy. A low quality maternal diet has, in turn, been linked to alterations in fetal brain development and cognitive deficits (Cortes-Albornoz et al. 2021), similar to those found in children exposed in utero to opioids, including BUP (Yeoh et al. 2019; Conradt et al. 2019; Lee et al. 2020). This is just one more example demonstrating the complexity of trying to tease apart the direct impact of gestational opioid exposure on offspring from opioid-induced changes to maternal physiology and behavior. Unfortunately, most human studies have additional confounding variables, such as maternal polydrug use, relapse behavior, food insecurities or other environmental challenges, which can contribute to adverse infant outcomes. In fact, a recent review by Conradt et al. (2023) suggests that "the factors that increase risk for opioid use [..] may be more likely to affect child neurodevelopment than the opioid exposure itself".

5 Conclusion

Our chapter summarized the consequences of gestational morphine or MOUD exposure on offspring short- and long-term outcomes through a review of human, as well as rodent, literature. Overall, illicit opioids are associated with worse outcomes compared to MOUD treatment in humans, but both common MOUD options, methadone and BUP, have also each been linked to adverse neurodevelopmental outcomes in infants. Of note, BUP treatment seems to result in more favorable infant outcomes compared to methadone, but most clinical studies have not investigated the long-term consequences or the impact of MOUD treatment on maternal behavior and well-being. Of concern, some preclinical studies have reported high offspring mortality rates with gestational BUP exposure, which has been associated with deficits in maternal care behaviors in BUP-exposed dams. Better translational animal models that start opioid treatment before conception, for example, and use proper dosing and administration routes are desperately needed to increase our understanding of the impact of exogenous opioids on brain development and maternal physiology. Further, there is a dearth of knowledge about drug interactions during pregnancy. Considering that most pregnant women suffering from substance use disorders do not limit their drug intake to just one specific drug, more animal studies are needed to investigate the effects of polydrug use during pregnancy. Another way to improve current animal models is to ensure studies measure and account for maternal care behaviors. The extent to which mothers suffering from OUD, who are often already stigmatized by society, bond with their infants is a complex and multifaceted issue. Considering the crucial role that the endogenous opioid system plays during the transition to motherhood, it is imperative that we illuminate how opioid use and, particularly MOUD treatment, may interfere with the fine-tuned brain chemistry that mediates mother-infant bonding. Achieving a better understanding of whether the bonding challenges recognized in mothers receiving treatment for OUDs may be attributable to the pharmacological effects of the MOUD medication itself, would reframe prejudicial discourse surrounding maternal bonding in this context and increase support for affected mothers and foster healthier outcomes for both mother and the child.

References

Abu Y, Roy S (2021) Prenatal opioid exposure and vulnerability to future substance use disorders in offspring. Exp Neurol 339:113621. Epub 2021/02/01. https://doi.org/10.1016/j.expneurol.2021.113621

Ahmadalipour A, Sadeghzadeh J, Vafaei AA, Bandegi AR, Mohammadkhani R, Rashidy-Pour A (2015) Effects of environmental enrichment on behavioral deficits and alterations in hippocampal BDNF induced by prenatal exposure to morphine in juvenile rats. Neuroscience 305:372–383. Epub 2015/08/15. https://doi.org/10.1016/j.neuroscience.2015.08.015

Alaee E, Moazen P, Pattij T, Semnanian S, Azizi H (2021) Prenatal exposure to morphine impairs attention and impulsivity in adult rats. Psychopharmacology (Berl) 238(10):2729–2741. Epub 2021/08/19. https://doi.org/10.1007/s00213-021-05888-7

Azuine RE, Ji Y, Chang HY, Kim Y, Ji H, DiBari J, Hong X, Wang G, Singh GK, Pearson C, Zuckerman B, Surkan PJ, Wang X (2019) Prenatal risk factors and perinatal and postnatal outcomes associated with maternal opioid exposure in an urban, low-income, Multiethnic US population. JAMA Netw Open 2(6):e196405. Epub 2019/06/30. https://doi.org/10.1001/jamanetworkopen.2019.6405

Bakhireva LN, Holbrook BD, Shrestha S, Leyva Y, Ashley M, Cano S, Lowe J, Stephen JM, Leeman L (2019) Association between prenatal opioid exposure, neonatal opioid withdrawal syndrome, and neurodevelopmental and behavioral outcomes at 5-8 months of age. Early Hum Dev 128:69–76. Epub 2018/12/17. https://doi.org/10.1016/j.earlhumdev.2018.10.010

Balalian AA, Graeve R, Richter M, Fink A, Kielstein H, Martins SS, Philbin MM, Factor-Litvak P (2023) Prenatal exposure to opioids and neurodevelopment in infancy and childhood: a systematic review. Front Pediatr 11:1071889. Epub 2023/03/11. https://doi.org/10.3389/fped.2023.1071889

Baldacchino A, Arbuckle K, Petrie DJ, McCowan C (2014) Neurobehavioral consequences of chronic intrauterine opioid exposure in infants and preschool children: a systematic review and meta-analysis. BMC Psychiatry 14:104. Epub 2014/04/09. https://doi.org/10.1186/1471-244X-14-104

Barron S, Chung VM (1997) Prenatal buprenorphine exposure and sexually dimorphic nonreproductive behaviors in rats. Pharmacol Biochem Behav 58(2):337–343. Epub 1997/09/23. https://doi.org/10.1016/s0091-3057(97)00260-8

Basheer R, Yang J, Tempel A (1992) Chronic prenatal morphine treatment decreases G alpha s mRNA levels in neonatal frontal cortex. Brain Res Dev Brain Res 70(1):145–148. Epub 1992/11/20. https://doi.org/10.1016/0165-3806(92)90113-b

Belcheva MM, Dawn S, Barg J, McHale RJ, Ho MT, Ignatova E, Coscia CJ (1994) Transient down-regulation of neonatal rat brain mu-opioid receptors upon in utero exposure to buprenorphine. Brain Res Dev Brain Res 80(1–2):158–162. Epub 1994/07/15. https://doi.org/10.1016/0165-3806(94)90100-7

Bhat R, Chari G, Rao R, Wirtshafter D (2006) Prenatal cocaine and morphine alter brain cyclin-dependent kinase 5 (Cdk5) activity in rat pups. Neurotoxicol Teratol 28(5):625–628. Epub 2006/09/12. https://doi.org/10.1016/j.ntt.2006.06.006

Biglarnia M, Karami M, Hafshejani ZK (2013) Differences in morphine-induced antinociception in male and female offspring born of morphine exposed mothers. Indian J Pharmacol 45(3): 227–231. Epub 2013/07/09. https://doi.org/10.4103/0253-7613.111904

Borrelli KN, Wingfield KK, Yao EJ, Zamorano CA, Sena KD, Beierle JA, Roos MA, Zhang H, Wachman EM, Bryant CD (2023) Decreased myelin-related gene expression in the nucleus accumbens during spontaneous neonatal opioid withdrawal in the absence of long-term behavioral effects in adult outbred CFW mice. Neuropharmacology 240:109732. Epub 2023/09/30. https://doi.org/10.1016/j.neuropharm.2023.109732

Brancato A, Cannizzaro C (2018) Mothering under the influence: how perinatal drugs of abuse alter the mother-infant interaction. Rev Neurosci 29(3):283–294. https://doi.org/10.1515/revneuro-2017-0052

Bridges RS, Grimm CT (1982) Reversal of morphine disruption of maternal behavior by concurrent treatment with the opiate antagonist naloxone. Science 218(4568):166–168. https://doi.org/10.1126/science.7123227

Brogly SB, Saia KA, Walley AY, Du HM, Sebastiani P (2014) Prenatal buprenorphine versus methadone exposure and neonatal outcomes: systematic review and meta-analysis. Am J Epidemiol 180(7):673–686. Epub 2014/08/26. https://doi.org/10.1093/aje/kwu190

Broussard CS, Rasmussen SA, Reefhuis J, Friedman JM, Jann MW, Riehle-Colarusso T, Honein MA, National Birth Defects Prevention Study (2011) Maternal treatment with opioid analgesics and risk for birth defects. Am J Obstet Gynecol 204(4):314 e1–314 11. Epub 2011/02/25. https://doi.org/10.1016/j.ajog.2010.12.039

Brunton PJ (2019) Endogenous opioid signalling in the brain during pregnancy and lactation. Cell Tissue Res 375(1):69–83. Epub 2018/11/12. https://doi.org/10.1007/s00441-018-2948-1

Brunton PJ, Russell JA (2008) The expectant brain: adapting for motherhood. Nat Rev Neurosci 9(1):11–25. https://doi.org/10.1038/nrn2280

Byrnes EM, Vassoler FM (2017. Epub 2017/10/03) Modeling prenatal opioid exposure in animals: current findings and future directions. Front Neuroendocrinol. https://doi.org/10.1016/j.yfrne.2017.09.001

Byrnes EM, Vassoler FM (2018) Modeling prenatal opioid exposure in animals: current findings and future directions. Front Neuroendocrinol 51:1–13. Epub 20170928. https://doi.org/10.1016/j.yfrne.2017.09.001

Camden A, To T, Gomes T, Ray J, Bai L, Guttmann A (2023) Prenatal opioid exposure and well-child care in the first 2 years of life: population-based cohort study. Arch Dis Child 108(9): 754–761. Epub 20230612. https://doi.org/10.1136/archdischild-2022-325029

Castellano C, Ammassari-Teule M (1984) Prenatal exposure to morphine in mice: enhanced responsiveness to morphine and stress. Pharmacol Biochem Behav 21(1):103–108. Epub 1984/07/01. https://doi.org/10.1016/0091-3057(84)90138-2

Cataldo I, Azhari A, Coppola A, Bornstein MH, Esposito G (2019) The influences of drug abuse on mother-infant interaction through the lens of the biopsychosocial model of health and illness: a review. Front Public Health:7. ARTN 45. https://doi.org/10.3389/fpubh.2019.00045

Chen HH, Chiang YC, Yuan ZF, Kuo CC, Lai MD, Hung TW, Ho IK, Chen ST (2015) Buprenorphine, methadone, and morphine treatment during pregnancy: behavioral effects on the offspring in rats. Neuropsychiatr Dis Treat 11:609–618. Epub 2015/04/04. https://doi.org/10.2147/ndt.S70585

Chiang YC, Hung TW, Lee CW, Yan JY, Ho IK (2010) Enhancement of tolerance development to morphine in rats prenatally exposed to morphine, methadone, and buprenorphine. J Biomed Sci 17(1):46. Epub 2010/06/10. https://doi.org/10.1186/1423-0127-17-46

Chiou LC, Yeh GC, Fan SH, How CH, Chuang KC, Tao PL (2003) Prenatal morphine exposure decreases analgesia but not K+ channel activation. Neuroreport 14(2):239–242. Epub 2003/02/25. https://doi.org/10.1097/00001756-200302100-00016

Clark RRS (2019) Breastfeeding in women on opioid maintenance therapy: a review of policy and practice. J Midwifery Womens Health 64(5):545–558. Epub 20190711. https://doi.org/10.1111/jmwh.12982

Clarke G, Wright DM (1984) A comparison of analgesia and suppression of oxytocin release by opiates. Br J Pharmacol 83(3):799–806. https://doi.org/10.1111/j.1476-5381.1984.tb16235.x

Clendeninn NJ, Petraitis M, Simon EJ (1976) Ontological development of opiate receptors in rodent brain. Brain Res 118(1):157–160. https://doi.org/10.1016/0006-8993(76)90852-0

Conradt E, Flannery T, Aschner JL, Annett RD, Croen LA, Duarte CS, Friedman AM, Guille C, Hedderson MM, Hofheimer JA, Jones MR, Ladd-Acosta C, McGrath M, Moreland A, Neiderhiser JM, Nguyen RHN, Posner J, Ross JL, Savitz DA, Ondersma SJ, Lester BM (2019) Prenatal opioid exposure: neurodevelopmental consequences and future research priorities. Pediatrics 144(3). https://doi.org/10.1542/peds.2019-0128

Conradt E, Camerota M, Maylott S, Lester BM (2023) Annual research review: prenatal opioid exposure – a two-generation approach to conceptualizing neurodevelopmental outcomes. J Child Psychol Psychiatry 64(4):566–578. Epub 20230207. https://doi.org/10.1111/jcpp.13761

Cortes-Albornoz MC, Garcia-Guaqueta DP, Velez-van-Meerbeke A, Talero-Gutierrez C (2021) Maternal nutrition and neurodevelopment: a scoping review. Nutrients 13(10) Epub 2021/10/24. https://doi.org/10.3390/nu13103530

Coyle MG, Salisbury AL, Lester BM, Jones HE, Lin H, Graf-Rohrmeister K, Fischer G (2012) Neonatal neurobehavior effects following buprenorphine versus methadone exposure. Addiction 107(Suppl 1):63–73. Epub 2012/11/01. https://doi.org/10.1111/j.1360-0443.2012.04040.x

Center for Substance Abuse Treatment (2006) Substance abuse: Clinical issues in intensive outpatient treatment. Substance Abuse and Mental Health Services Administration (US), Rockville (MD); 2006. Report No.: (SMA) 06-4182. PMID: 22514853.

Davis DD, Templer DI (1988) Neurobehavioral functioning in children exposed to narcotics in utero. Addict Behav 13(3):275–283. Epub 1988/01/01. https://doi.org/10.1016/0306-4603(88)90054-8

De Vries TJ, Van Vliet BJ, Hogenboom F, Wardeh G, Van der Laan JW, Mulder AH, Schoffelmeer AN (1991) Effect of chronic prenatal morphine treatment of mu-opioid receptor-regulated adenylate cyclase activity and neurotransmitter release in rat brain slices. Eur J Pharmacol 208(2):97–104. Epub 1991/10/14. https://doi.org/10.1016/0922-4106(91)90059-q

Desai RJ, Huybrechts KF, Hernandez-Diaz S, Mogun H, Patorno E, Kaltenbach K, Kerzner LS, Bateman BT (2015) Exposure to prescription opioid analgesics in utero and risk of neonatal abstinence syndrome: population based cohort study. BMJ 350:h2102. Epub 2015/05/16. PubMed PMID: 25975601; https://doi.org/10.1136/bmj.h2102

Evans RG, Olley JE, Rice GE, Abrahams JM (1989) Effects of subacute opioid administration during late pregnancy in the rat on the initiation, duration and outcome of parturition and maternal levels of oxytocin and arginine vasopressin. Clin Exp Pharmacol Physiol 16(3):169–178. Epub 1989/03/01. https://doi.org/10.1111/j.1440-1681.1989.tb01541.x

Farid WO, Dunlop SA, Tait RJ, Hulse GK (2008) The effects of maternally administered methadone, buprenorphine and naltrexone on offspring: review of human and animal data. Curr Neuropharmacol 6(2):125–150. Epub 2009/03/24. https://doi.org/10.2174/157015908784533842

Fill MA, Miller AM, Wilkinson RH, Warren MD, Dunn JR, Schaffner W, Jones TF (2018) Educational disabilities among children born with neonatal abstinence syndrome. pediatrics. 142(3) Epub 2018/09/01. https://doi.org/10.1542/peds.2018-0562

Giroux M, Teixera MG, Dumas JC, Desprats R, Grandjean H, Houin G (1997) Influence of maternal blood flow on the placental transfer of three opioids – fentanyl, alfentanil, sufentanil. Biol Neonate 72(3):133–141. https://doi.org/10.1159/000244476

Graeve R, Balalian AA, Richter M, Kielstein H, Fink A, Martins SS, Philbin MM, Factor-Litvak P (2022) Infants' prenatal exposure to opioids and the association with birth outcomes: a systematic review and meta-analysis. Paediatr Perinat Epidemiol 36(1):125–143. Epub 2021/11/11. https://doi.org/10.1111/ppe.12805

Grecco GG, Atwood BK (2020) Prenatal opioid exposure enhances responsiveness to future drug reward and alters sensitivity to pain: a review of preclinical models and contributing mechanisms. eNeuro 7(6) Epub 2020/10/17. https://doi.org/10.1523/ENEURO.0393-20.2020

Grimm CT, Bridges RS (1983) Opiate regulation of maternal behavior in the rat. Pharmacol Biochem Behav 19(4):609–616. https://doi.org/10.1016/0091-3057(83)90336-2

Guo X, Spencer JW, Suess PE, Hickey JE, Better WE, Herning RI (1994) Cognitive brain potential alterations in boys exposed to opiates: in utero and lifestyle comparisons. Addict Behav 19(4): 429–441. Epub 1994/07/01. https://doi.org/10.1016/0306-4603(94)90065-5

Haley DF, Saitz R (2020) The opioid epidemic during the COVID-19 pandemic. JAMA 324(16): 1615–1617. https://doi.org/10.1001/jama.2020.18543

Hall ES, Isemann BT, Wexelblatt SL, Meinzen-Derr J, Wiles JR, Harvey S, Akinbi HT (2016) A cohort comparison of buprenorphine versus methadone treatment for neonatal abstinence syndrome. J Pediatr 170:39–44.e1. Epub 2015/12/26. https://doi.org/10.1016/j.jpeds.2015.11.039

Handelmann GE, Quirion R (1983) Neonatal exposure to morphine increases mu opiate binding in the adult forebrain. Eur J Pharmacol 94(3–4):357–358. Epub 1983/10/28. https://doi.org/10.1016/0014-2999(83)90429-6

Harder HJ, Murphy AZ (2019) Early life opioid exposure and potential long-term effects. Neurobiol Stress 10:100156. Epub 2019/07/25. https://doi.org/10.1016/j.ynstr.2019.100156

Hirai AH, Ko JY, Owens PL, Stocks C, Patrick SW (2021) Neonatal abstinence syndrome and maternal opioid-related diagnoses in the US, 2010-2017. JAMA 325(2):146–155. https://doi.org/10.1001/jama.2020.24991

Hou Y, Tan Y, Belcheva MM, Clark AL, Zahm DS, Coscia CJ (2004) Differential effects of gestational buprenorphine, naloxone, and methadone on mesolimbic mu opioid and ORL1 receptor G protein coupling. Brain Res Dev Brain Res 151(1–2):149–157. Epub 2004/07/13. https://doi.org/10.1016/j.devbrainres.2004.05.002

Hung CJ, Wu CC, Chen WY, Chang CY, Kuan YH, Pan HC, Liao SL, Chen CJ (2013) Depression-like effect of prenatal buprenorphine exposure in rats. PLoS One 8(12):e82262. Epub 2013/12/25. https://doi.org/10.1371/journal.pone.0082262

Hutchings DE, Zmitrovich AC, Hamowy AS, Liu PY (1995) Prenatal administration of buprenorphine using the osmotic minipump: a preliminary study of maternal and offspring toxicity and growth in the rat. Neurotoxicol Teratol 17(4):419–423. Epub 1995/07/01. https://doi.org/10.1016/0892-0362(94)00079-s

Hutchings DE, Hamowy AS, Williams EM, Zmitrovich AC (1996) Prenatal administration of buprenorphine in the rat: effects on the rest-activity cycle at 22 and 30 days of age. Pharmacol Biochem Behav 55(4):607–613. Epub 1996/12/01. https://doi.org/10.1016/s0091-3057(96)00287-0

Jansson LM, Choo R, Velez ML, Harrow C, Schroeder JR, Shakleya DM, Huestis MA (2008) Methadone maintenance and breastfeeding in the neonatal period. Pediatrics 121(1):106–114. https://doi.org/10.1542/peds.2007-1182

Jansson LM, Dipietro JA, Velez M, Elko A, Williams E, Milio L, O'Grady K, Jones HE (2011) Fetal neurobehavioral effects of exposure to methadone or buprenorphine. Neurotoxicol Teratol 33(2):240–243. Epub 2010/09/28. https://doi.org/10.1016/j.ntt.2010.09.003

Jansson LM, Velez M, McConnell K, Spencer N, Tuten M, Jones HE, King VL, Gandotra N, Milio LA, Voegtline K, DiPietro JA (2017) Maternal buprenorphine treatment and fetal neurobehavioral development. Am J Obstet Gynecol 216(5):529 e1–529 e8. Epub 2017/02/12. https://doi.org/10.1016/j.ajog.2017.01.040

Jones HE, Kaltenbach K, Heil SH, Stine SM, Coyle MG, Arria AM, O'Grady KE, Selby P, Martin PR, Fischer G (2010) Neonatal abstinence syndrome after methadone or buprenorphine exposure. N Engl J Med 363(24):2320–2331. Epub 2010/12/15. https://doi.org/10.1056/NEJMoa1005359

Kakko J, Heilig M, Sarman I (2008) Buprenorphine and methadone treatment of opiate dependence during pregnancy: comparison of fetal growth and neonatal outcomes in two consecutive case

series. Drug Alcohol Depend 96(1–2):69–78. Epub 2008/03/22. https://doi.org/10.1016/j.drugalcdep.2008.01.025
Kaltenbach K, O'Grady KE, Heil SH, Salisbury AL, Coyle MG, Fischer G, Martin PR, Stine S, Jones HE (2018) Prenatal exposure to methadone or buprenorphine: early childhood developmental outcomes. Drug Alcohol Depend 185:40–49. Epub 2018/02/08. https://doi.org/10.1016/j.drugalcdep.2017.11.030
Kelty E, Preen DB (2019) Risk factors associated with the occurrence of neonatal opioid withdrawal syndrome: a review. CNS Drugs 33(11):1113–1120. Epub 2019/11/07. https://doi.org/10.1007/s40263-019-00681-9
Kentner AC, Cryan JF, Brummelte S (2019) Resilience priming: translational models for understanding resiliency and adaptation to early life adversity. Dev Psychobiol. 61(3):350-375 (Epub 2018/10/13) https://doi.org/10.1002/dev.21775
Kingsbury MA, Bilbo SD (2019) The inflammatory event of birth: how oxytocin signaling may guide the development of the brain and gastrointestinal system. Front Neuroendocrinol 55: 100794. Epub 2019/09/29. https://doi.org/10.1016/j.yfrne.2019.100794
Kinsley CH, Bridges RS (1986) Opiate involvement in postpartum aggression in rats. Pharmacol Biochem Behav 25(5):1007–1011. https://doi.org/10.1016/0091-3057(86)90078-x
Kinsley CH, Bridges RS (1990) Morphine treatment and reproductive condition alter olfactory preferences for pup and adult male odors in female rats. Dev Psychobiol 23(4):331–347. https://doi.org/10.1002/dev.420230405
Klaman SL, Isaacs K, Leopold A, Perpich J, Hayashi S, Vender J, Campopiano M, Jones HE (2017) Treating women who are pregnant and parenting for opioid use disorder and the concurrent care of their infants and children: literature review to support national guidance. J Addict Med 11(3): 178–190. https://doi.org/10.1097/adm.0000000000000308
Ko JY, D'Angelo DV, Haight SC, Morrow B, Cox S, Salvesen von Essen B, Strahan AE, Harrison L, Tevendale HD, Warner L, Kroelinger CD, Barfield WD (2020) Vital signs: prescription opioid pain reliever use during pregnancy – 34 U.S. Jurisdictions, 2019. Morb Mortal Wkly Rep 69(28):897–903. Epub 20200717. https://doi.org/10.15585/mmwr.mm6928a1
Kolodny A, Courtwright DT, Hwang CS, Kreiner P, Eadie JL, Clark TW, Alexander GC (2015) The prescription opioid and heroin crisis: a public health approach to an epidemic of addiction. Annu Rev Public Health 36:559–574. Epub 20150112. https://doi.org/10.1146/annurev-publhealth-031914-122957
Kongstorp M, Bogen IL, Stiris T, Andersen JM (2019) High accumulation of methadone compared with buprenorphine in fetal rat brain after maternal exposure. J Pharmacol Exp Ther 371(1): 130–137. Epub 2019/07/31. https://doi.org/10.1124/jpet.119.259531
Kongstorp M, Bogen IL, Stiris T, Andersen JM (2020) Prenatal exposure to methadone or buprenorphine impairs cognitive performance in young adult rats. Drug Alcohol Depend 212: 108008. Epub 2020/05/14. https://doi.org/10.1016/j.drugalcdep.2020.108008
Konijnenberg C, Melinder A (2015) Executive function in preschool children prenatally exposed to methadone or buprenorphine. Child Neuropsychol 21(5):570–585. Epub 2014/10/31. https://doi.org/10.1080/09297049.2014.967201
Konijnenberg C, Melinder A (2020) Salivary cortisol levels relate to cognitive performance in children prenatally exposed to methadone or buprenorphine. Dev Psychobiol 62(3):409–418. Epub 2019/09/30. https://doi.org/10.1002/dev.21921
Konijnenberg C, Sarfi M, Melinder A (2016) Mother-child interaction and cognitive development in children prenatally exposed to methadone or buprenorphine. Early Hum Dev 101:91–97. Epub 2016/09/11. https://doi.org/10.1016/j.earlhumdev.2016.08.013
Kornblum HI, Hurlbut DE, Leslie FM (1987) Postnatal development of multiple opioid receptors in rat brain. Brain Res 465(1–2):21–41. Epub 1987/12/15. https://doi.org/10.1016/0165-3806(87)90226-4
Laborie C, Dutriez-Casteloot I, Montel V, Dickes-Coopman A, Lesage J, Vieau D (2005) Prenatal morphine exposure affects sympathoadrenal axis activity and serotonin metabolism in adult

male rats both under basal conditions and after an ether inhalation stress. Neurosci Lett 381(3): 211–216. Epub 2005/05/18. https://doi.org/10.1016/j.neulet.2005.01.083

Lambert KG, Byrnes EM (2019) Challenges to the parental brain: neuroethological and translational considerations. Front Neuroendocrinol 53:100747. Epub 20190417. https://doi.org/10.1016/j.yfrne.2019.04.004

Landi N, Montoya J, Kober H, Rutherford HJ, Mencl WE, Worhunsky PD, Potenza MN, Mayes LC (2011) Maternal neural responses to infant cries and faces: relationships with substance use. Front Psych 2:32. Epub 20110615. https://doi.org/10.3389/fpsyt.2011.00032

Lee SJ, Bora S, Austin NC, Westerman A, Henderson JMT (2020) Neurodevelopmental outcomes of children born to opioid-dependent mothers: a systematic review and meta-analysis. Acad Pediatr 20(3):308–318. Epub 2019/11/18. https://doi.org/10.1016/j.acap.2019.11.005

Lemon LS, Caritis SN, Venkataramanan R, Platt RW, Bodnar LM (2018) Methadone versus buprenorphine for opioid use dependence and risk of neonatal abstinence syndrome. Epidemiology 29(2):261–268. Epub 2017/11/08. https://doi.org/10.1097/EDE.0000000000000780

Lesser-Katz M (1982) Some effects of maternal drug addiction on the neonate. Int J Addict 17(5): 887–896. Epub 1982/07/01. https://doi.org/10.3109/10826088209056335

Leyenaar JK, Schaefer AP, Wasserman JR, Moen EL, O'Malley AJ, Goodman DC (2021) Infant mortality associated with prenatal opioid exposure. JAMA Pediatr 175(7):706–714. Epub 2021/04/13. PubMed PMID: 33843963; https://doi.org/10.1001/jamapediatrics.2020.6364.

Lifschitz MH, Wilson GS, Smith EO, Desmond MM (1985) Factors affecting head growth and intellectual function in children of drug addicts. Pediatrics 75(2):269–274. Epub 1985/02/01. PMID: 3969327

Lindemalm S, Nydert P, Svensson JO, Stahle L, Sarman I (2009) Transfer of buprenorphine into breast milk and calculation of infant drug dose. J Hum Lact 25(2):199–205. Epub 2009/01/13. https://doi.org/10.1177/0890334408328295

Lindow SW, Hendricks MS, Nugent FA, Dunne TT, van der Spuy ZM (1999) Morphine suppresses the oxytocin response in breast-feeding women. Gynecol Obstet Invest 48(1):33–37. Epub 1999/07/08. https://doi.org/10.1159/000010130

Martin CE, Shadowen C, Thakkar B, Oakes T, Gal TS, Moeller FG (2020) Buprenorphine dosing for the treatment of opioid use disorder through pregnancy and postpartum. Curr Treat Options Psychiatry 7(3):375–399. Epub 20200728. https://doi.org/10.1007/s40501-020-00221-z

Mayer AD, Faris PL, Komisaruk BR, Rosenblatt JS (1985) Opiate antagonism reduces placentophagia and pup cleaning by parturient rats. Pharmacol Biochem Behav 22(6): 1035–1044. https://doi.org/10.1016/0091-3057(85)90314-4

McDowell J, Kitchen I (1987) Development of opioid systems: peptides, receptors and pharmacology. Brain Res 434(4):397–421. Epub 1987/11/01. https://doi.org/10.1016/0165-0173(87)90006-3

Mei B, Niu L, Cao B, Huang D, Zhou Y (2009) Prenatal morphine exposure alters the layer II/III pyramidal neurons morphology in lateral secondary visual cortex of juvenile rats. Synapse 63(12):1154–1161. Epub 2009/08/12. https://doi.org/10.1002/syn.20694

Merhar SL, Parikh NA, Braimah A, Poindexter BB, Tkach J, Kline-Fath B (2019) White matter injury and structural anomalies in infants with prenatal opioid exposure. Am J Neuroradiol 40(12):2161–2165. Epub 2019/10/19. https://doi.org/10.3174/ajnr.A6282

Merhar SL, Kline JE, Braimah A, Kline-Fath BM, Tkach JA, Altaye M, He L, Parikh NA (2021a) Prenatal opioid exposure is associated with smaller brain volumes in multiple regions. Pediatr Res 90(2):397–402. Epub 2020/11/13. https://doi.org/10.1038/s41390-020-01265-w

Merhar SL, Jiang W, Parikh NA, Yin W, Zhou Z, Tkach JA, Wang L, Kline-Fath BM, He L, Braimah A, Vannest J, Lin W (2021b) Effects of prenatal opioid exposure on functional networks in infancy. Dev Cogn Neurosci 51:100996. Epub 2021/08/14. https://doi.org/10.1016/j.dcn.2021.100996

Messinger DS, Bauer CR, Das A, Seifer R, Lester BM, Lagasse LL, Wright LL, Shankaran S, Bada HS, Smeriglio VL, Langer JC, Beeghly M, Poole WK (2004) The maternal lifestyle study: cognitive, motor, and behavioral outcomes of cocaine-exposed and opiate-exposed infants through three years of age. Pediatrics 113(6):1677–1685. Epub 2004/06/03. https://doi.org/10.1542/peds.113.6.1677

Miller NW, Seman BG, Akers SM, Povroznik JM, Brundage K, Fang W, Robinson CM (2022) The impact of opioid exposure during pregnancy on the human neonatal immune profile. Pediatr Res 92(6):1566–1574. Epub 2022/03/16. https://doi.org/10.1038/s41390-022-02014-x

Mithbaokar P, Fiorito F, Della Morte R, Maharajan V, Costagliola A (2016) Chronic maternal morphine alters calbindin D-28k expression pattern in postnatal mouse brain. Synapse 70(1): 15–23. Epub 2015/09/30. https://doi.org/10.1002/syn.21866

Monnelly VJ, Anblagan D, Quigley A, Cabez MB, Cooper ES, Mactier H, Semple SI, Bastin ME, Boardman JP (2018) Prenatal methadone exposure is associated with altered neonatal brain development. Neuroimage Clin 18:9–14. Epub 2018/01/13. https://doi.org/10.1016/j.nicl.2017.12.033

Myers AM, Bowen SE, Brummelte S (2023) Maternal care behavior and physiology moderate offspring outcomes following gestational exposure to opioids. Dev Psychobiol 65(8):e22433. https://doi.org/10.1002/dev.22433

Myers AM, Wallin CM, Richardson LM, Duran J, Neole SR, Kulaglic N, Davidson C, Perrine SA, Bowen SE, Brummelte S (2024) The effects of buprenorphine and morphine during pregnancy: Impact of exposure length on maternal brain behavior and offspring neurodevelopment Neuropharmacology 257, 110060. Epub 2024 Jul 1. https://doi.org/10.1016/j.neuropharm.2024.110060

Nanovskaya TN, Nekhayeva IA, Hankins GD, Ahmed MS (2008) Transfer of methadone across the dually perfused preterm human placental lobule. Am J Obstet Gynecol 198(1):126e1–126e4. https://doi.org/10.1016/j.ajog.2007.06.073

Nasiraei-Moghadam S, Sherafat MA, Safari MS, Moradi F, Ahmadiani A, Dargahi L (2013) Reversal of prenatal morphine exposure-induced memory deficit in male but not female rats. J Mol Neurosci 50(1):58–69. Epub 2012/08/07. https://doi.org/10.1007/s12031-012-9860-z

National Center for Health Statistics (2001) Drug overdose deaths in the U.S. Top 100,000 Annually https://www.cdc.gov/nchs/pressroom/nchs_press_releases/2021/20211117.htm: Centers for Disease Control and Prevention

Nieto-Estevez V, Donegan JJ, McMahon CL, Elam HB, Chavera TA, Varma P, Berg KA, Lodge DJ, Hsieh J (2022) Buprenorphine exposure alters the development and migration of interneurons in the cortex. Front Mol Neurosci 15:889922. Epub 2022/05/24. https://doi.org/10.3389/fnmol.2022.889922

Niu L, Cao B, Zhu H, Mei B, Wang M, Yang Y, Zhou Y (2009) Impaired in vivo synaptic plasticity in dentate gyrus and spatial memory in juvenile rats induced by prenatal morphine exposure. Hippocampus 19(7):649–657. Epub 2008/12/31. https://doi.org/10.1002/hipo.20540

Nygaard E, Moe V, Slinning K, Walhovd KB (2015) Longitudinal cognitive development of children born to mothers with opioid and polysubstance use. Pediatr Res 78(3):330–335. Epub 2015/05/16. https://doi.org/10.1038/pr.2015.95

O'Callaghan JP, Holtzman SG (1976) Prenatal administration of morphine to the rat: tolerance to the analgesic effect of morphine in the offspring. J Pharmacol Exp Ther 197(3):533–544. Epub 1976/06/01

O'Callaghan JP, Holtzman SG (1977) Prenatal administration of levorphanol or dextrorphan to the rat: analgesic effect of morphine in the offspring. J Pharmacol Exp Ther 200(2):255–262. Epub 1977/02/01

Oei JL, Melhuish E, Uebel H, Azzam N, Breen C, Burns L, Hilder L, Bajuk B, Abdel-Latif ME, Ward M, Feller JM, Falconer J, Clews S, Eastwood J, Li A, Wright IM (2017) Neonatal abstinence syndrome and high school performance. Pediatrics 139(2) Epub 2017/01/18. https://doi.org/10.1542/peds.2016-2651

Ornoy A, Segal J, Bar-Hamburger R, Greenbaum C (2001) Developmental outcome of school-age children born to mothers with heroin dependency: importance of environmental factors. Dev Med Child Neurol 43(10):668–675. Epub 2001/10/23. https://doi.org/10.1017/s0012162201001219

Pritham UA (2013) Breastfeeding promotion for management of neonatal abstinence syndrome. J Obstet Gynecol Neonatal Nurs 42(5):517–526. Epub 20130904. https://doi.org/10.1111/1552-6909.12242

Pryor JR, Maalouf FI, Krans EE, Schumacher RE, Cooper WO, Patrick SW (2017) The opioid epidemic and neonatal abstinence syndrome in the USA: a review of the continuum of care. Arch Dis Child Fetal Neonatal Ed 102(2):F183–F1F7. Epub 20170110. https://doi.org/10.1136/archdischild-2015-310045

Radhakrishnan R, Elsaid NMH, Sadhasivam S, Reher TA, Hines AC, Yoder KK, Saykin AJ, Wu YC (2021) Resting state functional MRI in infants with prenatal opioid exposure-a pilot study. Neuroradiology 63(4):585–591. Epub 2020/09/27. https://doi.org/10.1007/s00234-020-02552-3

Ricalde AA, Hammer RP Jr (1990) Perinatal opiate treatment delays growth of cortical dendrites. Neurosci Lett 115(2–3):137–143. Epub 1990/07/31. https://doi.org/10.1016/0304-3940(90)90444-e

Riley MA, Vathy I (2006) Mid- to late gestational morphine exposure does not alter the rewarding properties of morphine in adult male rats. Neuropharmacology 51(2):295–304. Epub 2006/05/16. https://doi.org/10.1016/j.neuropharm.2006.03.022

Rimanoczy A, Slamberova R, Vathy I (2001) Prenatal morphine exposure alters estrogen regulation of kappa receptors in the cortex and POA of adult female rats but has no effects on these receptors in adult male rats. Brain Res 894(1):154–156. Epub 2001/03/14. https://doi.org/10.1016/s0006-8993(00)03326-6

Rimanoczy A, Slamberova R, Riley MA, Vathy I (2003) Adrenocorticotropin stress response but not glucocorticoid-negative feedback is altered by prenatal morphine exposure in adult male rats. Neuroendocrinology 78(6):312–320. Epub 2003/12/23. https://doi.org/10.1159/000074884

Robinson SE (2002) Effects of perinatal buprenorphine and methadone exposures on striatal cholinergic ontogeny. Neurotoxicol Teratol 24(2):137–142. Epub 2002/04/12. https://doi.org/10.1016/s0892-0362(01)00185-4

Robinson SE, Wallace MJ (2001) Effect of perinatal buprenorphine exposure on development in the rat. J Pharmacol Exp Ther 298(2):797–804. Epub 2001/07/17

Rodriguez CE, Klie KA (2019) Pharmacological treatment of opioid use disorder in pregnancy. Semin Perinatol 43(3):141–148. Epub 20190121. https://doi.org/10.1053/j.semperi.2019.01.003

Rosenfeld CS (2022) The placenta as a target of opioid drugsdagger. Biol Reprod 106(4):676–686. https://doi.org/10.1093/biolre/ioac003

Ross EJ, Graham DL, Money KM, Stanwood GD (2015) Developmental consequences of fetal exposure to drugs: what we know and what we still must learn. Neuropsychopharmacology 40(1):61–87. Epub 20140618. https://doi.org/10.1038/npp.2014.147

Russell JA, Brunton PJ (2006) Neuroactive steroids attenuate oxytocin stress responses in late pregnancy. Neuroscience 138(3):879–889. Epub 20051128. https://doi.org/10.1016/j.neuroscience.2005.09.009

Russell JA, Leng G, Douglas AJ (2003) The magnocellular oxytocin system, the fount of maternity: adaptations in pregnancy. Front Neuroendocrinol 24(1):27–61. Epub 2003/03/01. https://doi.org/10.1016/s0091-3022(02)00104-8

Rutherford HJ, Mayes LC (2017) Parenting and addiction: neurobiological insights. Curr Opin Psychol 15:55–60. Epub 20170220. https://doi.org/10.1016/j.copsyc.2017.02.014

Rutherford HJ, Williams SK, Moy S, Mayes LC, Johns JM (2011) Disruption of maternal parenting circuitry by addictive process: rewiring of reward and stress systems. Front Psych 2:37. Epub 20110706. https://doi.org/10.3389/fpsyt.2011.00037

Rutherford HJ, Kim S, Yip SW, Potenza MN, Mayes LC, Strathearn L (2021) Parenting and addictions: current insights from human neuroscience. Curr Addict Rep 8(3):380–388. Epub 20210709. https://doi.org/10.1007/s40429-021-00384-6

Salihu HM, Salinas A, Medina I, Krishnaswami J, Aliyu MH (2019) Biopsychosocial determinants of opioid use disorder (OUD) and implications for maternal and child health research: a scoping review. J Opioid Manag 15(1):77–91. https://doi.org/10.5055/jom.2019.0488

Salisbury AL, Coyle MG, O'Grady KE, Heil SH, Martin PR, Stine SM, Kaltenbach K, Weninger M, Jones HE (2012) Fetal assessment before and after dosing with buprenorphine or methadone. Addiction 107(Suppl 1):36–44. Epub 2012/11/01. https://doi.org/10.1111/j.1360-0443.2012.04037.x

Salo S, Politi J, Tupola S, Biringen Z, Kalland M, Halmesmäki E, Kahila H, Kivitie-Kallio S (2010) Early development of opioid-exposed infants born to mothers in buprenorphine-replacement therapy. J Reprod Infant Psychol 28(2):161–179. https://doi.org/10.1080/02646830903219109

Sanchez ES, Bigbee JW, Fobbs W, Robinson SE, Sato-Bigbee C (2008) Opioid addiction and pregnancy: perinatal exposure to buprenorphine affects myelination in the developing brain. Glia 56(9):1017–1027. Epub 2008/04/03. https://doi.org/10.1002/glia.20675

Sandtorv LB, Fevang SKE, Nilsen SA, Boe T, Gjestad R, Haugland S, Elgen IB (2018) Symptoms associated with attention deficit/hyperactivity disorder and autism spectrum disorders in school-aged children prenatally exposed to substances. Subst Abus 12:1178221818765773. Epub 2018/04/06. https://doi.org/10.1177/1178221818765773

Santarelli S, Zimmermann C, Kalideris G, Lesuis SL, Arloth J, Uribe A, Dournes C, Balsevich G, Hartmann J, Masana M, Binder EB, Spengler D, Schmidt MV (2017) An adverse early life environment can enhance stress resilience in adulthood. Psychoneuroendocrinology 78:213–221. Epub 2017/02/22. https://doi.org/10.1016/j.psyneuen.2017.01.021

Schlagal CR, Dunn TJ, Xu P, Felsing DE, Merritt CR, Manja S, Fox RG, Buffington SA, Saade G, Dineley KT, Yu Y, Cunningham KA, Wu P (2021) Maternal opioid exposure culminates in perturbed murine neurodevelopment and hyperactive phenotype in adolescence. Neuroscience 463:272–287. Epub 2021/04/04. https://doi.org/10.1016/j.neuroscience.2021.03.014

Siddiqui A, Haq S, Shah BH (1997) Perinatal exposure to morphine disrupts brain norepinephrine, ovarian cyclicity, and sexual receptivity in rats. Pharmacol Biochem Behav 58(1):243–248. Epub 1997/09/01. https://doi.org/10.1016/s0091-3057(97)00012-9

Sikka P, Madan N, Yen E (2023) Early white matter tract changes in neonates with prenatal opioid exposure: a pilot study. J Perinatol 43(3):390–391. Epub 2022/06/19. https://doi.org/10.1038/s41372-022-01427-0

Sirnes E, Oltedal L, Bartsch H, Eide GE, Elgen IB, Aukland SM (2017) Brain morphology in school-aged children with prenatal opioid exposure: a structural MRI study. Early Hum Dev 106–107:33–39. Epub 2017/02/12. https://doi.org/10.1016/j.earlhumdev.2017.01.009

Slamberova R, Rimanoczy A, Bar N, Schindler CJ, Vathy I (2003) Density of mu-opioid receptors in the hippocampus of adult male and female rats is altered by prenatal morphine exposure and gonadal hormone treatment. Hippocampus 13(4):461–471. Epub 2003/07/03. https://doi.org/10.1002/hipo.10076

Slamberova R, Rimanoczy A, Riley MA, Vathy I (2004) Hypothalamo-pituitary-adrenal axis-regulated stress response and negative feedback sensitivity is altered by prenatal morphine exposure in adult female rats. Neuroendocrinology 80(3):192–200. Epub 2004/12/08. https://doi.org/10.1159/000082359

Slamberova R, Rimanoczy A, Cao D, Schindler CJ, Vathy I (2005) Alterations of prenatal morphine exposure in mu-opioid receptor density in hypothalamic nuclei associated with sexual behavior. Brain Res Bull 65(6):479–485. Epub 2005/05/03. https://doi.org/10.1016/j.brainresbull.2005.02.030

Soyka M (2021) Transition from full mu opioid agonists to buprenorphine in opioid dependent patients – a critical review. Front Pharmacol 12:718811. Epub 20211123. https://doi.org/10.3389/fphar.2021.718811

Stafisso-Sandoz G, Polley D, Holt E, Lambert KG, Kinsley CH (1998) Opiate disruption of maternal behavior: morphine reduces, and naloxone restores, c-fos activity in the medial preoptic area of lactating rats. Brain Res Bull 45(3):307–313. https://doi.org/10.1016/s0361-9230(97)00375-4

Strauss ME, Starr RH, Ostrea EM, Chavez CJ, Stryker JC (1976) Behavioural concomitants of prenatal addiction to narcotics. J Pediatr 89(5):842–846. Epub 1976/11/01. https://doi.org/10.1016/s0022-3476(76)80822-0

Suarez EA, Huybrechts KF, Straub L, Hernandez-Diaz S, Jones HE, Connery HS, Davis JM, Gray KJ, Lester B, Terplan M, Mogun H, Bateman BT (2022) Buprenorphine versus methadone for opioid use disorder in pregnancy. N Engl J Med 387(22):2033–2044. https://doi.org/10.1056/NEJMoa2203318

Sutter MB, Leeman L, Hsi A (2014) Neonatal opioid withdrawal syndrome. Obstet Gynecol Clin North Am 41(2):317–334. Epub 2014/05/23. https://doi.org/10.1016/j.ogc.2014.02.010

Swain JE, Ho SS (2019) Early postpartum resting-state functional connectivity for mothers receiving buprenorphine treatment for opioid use disorder: a pilot study. J Neuroendocrinol 31(9):e12770. Epub 20190729. https://doi.org/10.1111/jne.12770

Swain JE, Ho SS (2023) Brain circuits for maternal sensitivity and pain involving anterior cingulate cortex among mothers receiving buprenorphine treatment for opioid use disorder. J Neuroendocrinol 35(7):e13316. Epub 20230725. https://doi.org/10.1111/jne.13316

Swain JE, Ho SS, Fox H, Garry D, Brummelte S (2019) Effects of opioids on the parental brain in health and disease. Front Neuroendocrinol 54:100766. Epub 20190522. https://doi.org/10.1016/j.yfrne.2019.100766

Tan JW, Duan TT, Zhou QX, Ding ZY, Jing L, Cao J, Wang LP, Mao RR, Xu L (2015) Impaired contextual fear extinction and hippocampal synaptic plasticity in adult rats induced by prenatal morphine exposure. Addict Biol 20(4):652–662. Epub 2014/06/07. https://doi.org/10.1111/adb.12158

Tan KZ, Cunningham AM, Joshi A, Oei JL, Ward MC (2018) Expression of kappa opioid receptors in developing rat brain – implications for perinatal buprenorphine exposure. Reprod Toxicol 78: 81–89. Epub 20180407. https://doi.org/10.1016/j.reprotox.2018.04.006

Tempel A, Yang J, Basheer R (1995) Prenatal morphine exposure differentially alters expression of opioid peptides in striatum of newborns. Brain Res Mol Brain Res 33(2):227–232. Epub 1995/11/01. https://doi.org/10.1016/0169-328x(95)00127-e

Tenconi B, La Croix R, Donadoni ML, Di Giulio AM, Mantegazza P, Gorio A (1989) The developmental pattern of met-enkephalin innervation is altered in animals perinatally exposed to morphine. Pharmacol Res 21(Suppl 1):61–62. Epub 1989/11/01. https://doi.org/10.1016/s1043-6618(89)80052-0

Tiong GK, Olley JE (1988) Effects of exposure in utero to methadone and buprenorphine on enkephalin levels in the developing rat brain. Neurosci Lett 93(1):101–106. Epub 1988/10/31. https://doi.org/10.1016/0304-3940(88)90020-1

Tong Y, Chabot JG, Shen SH, O'Dowd BF, George SR, Quirion R (2000) Ontogenic profile of the expression of the mu opioid receptor gene in the rat telencephalon and diencephalon: an in situ hybridization study. J Chem Neuroanat 18(4):209–222. https://doi.org/10.1016/s0891-0618(00)00043-0

Uebel H, Wright IM, Burns L, Hilder L, Bajuk B, Breen C, Abdel-Latif ME, Feller JM, Falconer J, Clews S, Eastwood J, Oei JL (2015) Reasons for rehospitalization in children who had neonatal abstinence syndrome. Pediatrics 136(4):e811–e820. Epub 2015/09/16. https://doi.org/10.1542/peds.2014-2767

Vathy I (2002) Prenatal opiate exposure: long-term CNS consequences in the stress system of the offspring. Psychoneuroendocrinology 27(1–2):273–283. Epub 2001/12/26. https://doi.org/10.1016/s0306-4530(01)00049-x

Vathy I, Katay L (1992) Effects of prenatal morphine on adult sexual behavior and brain catecholamines in rats. Brain Res Dev Brain Res 68(1):125–131. Epub 1992/07/24. https://doi.org/10.1016/0165-3806(92)90254-t

Vathy I, Rimanoczy A, Eaton RC, Katay L (1994) Modulation of catecholamine turnover rate in brain regions of rats exposed prenatally to morphine. Brain Res 662(1–2):209–215. Epub 1994/10/31. https://doi.org/10.1016/0006-8993(94)90814-1

Vathy I, Rimanoczy A, Eaton RC, Katay L (1995) Sex dimorphic alterations in postnatal brain catecholamines after gestational morphine. Brain Res Bull 36(2):185–193. Epub 1995/01/01. https://doi.org/10.1016/0361-9230(94)00192-4

Vathy I, Rimanoczy A, Slamberova R (2000) Prenatal exposure to morphine differentially alters gonadal hormone regulation of delta-opioid receptor binding in male and female rats. Brain Res Bull 53(6):793–800. Epub 2001/02/17. https://doi.org/10.1016/s0361-9230(00)00409-3

Vathy I, Slamberova R, Rimanoczy A, Riley MA, Bar N (2003) Autoradiographic evidence that prenatal morphine exposure sex-dependently alters mu-opioid receptor densities in brain regions that are involved in the control of drug abuse and other motivated behaviors. Prog Neuropsychopharmacol Biol Psychiatry 27(3):381–393. Epub 2003/04/15. https://doi.org/10.1016/S0278-5846(02)00355-X

Villarreal DM, Derrick B, Vathy I (2008) Prenatal morphine exposure attenuates the maintenance of late LTP in lateral perforant path projections to the dentate gyrus and the CA3 region in vivo. J Neurophysiol 99(3):1235–1242. Epub 2008/01/18. https://doi.org/10.1152/jn.00981.2007

Vuong C, Van Uum SH, O'Dell LE, Lutfy K, Friedman TC (2010) The effects of opioids and opioid analogs on animal and human endocrine systems. Endocr Rev 31(1):98–132. Epub 20091110. https://doi.org/10.1210/er.2009-0009

Wagner GC, Jarvis MF, Gottesfeld DS, Giorlando M, Rabii J (1986) The effects of prenatal morphine on the responsiveness to morphine and amphetamine. Pharmacol Biochem Behav 24(3):757–759. Epub 1986/03/01. https://doi.org/10.1016/0091-3057(86)90586-1

Walhovd KB, Bjornebekk A, Haabrekke K, Siqveland T, Slinning K, Nygaard E, Fjell AM, Due-Tonnessen P, Bjornerud A, Moe V (2015) Child neuroanatomical, neurocognitive, and visual acuity outcomes with maternal opioid and polysubstance detoxification. Pediatr Neurol 52(3):326–32 e1–3. Epub 2015/01/18. https://doi.org/10.1016/j.pediatrneurol.2014.11.008

Wallin CM, Bowen SE, Roberge CL, Richardson LM, Brummelte S (2019) Gestational buprenorphine exposure: effects on pregnancy, development, neonatal opioid withdrawal syndrome, and behavior in a translational rodent model. Drug Alcohol Depend 205:107625. Epub 20191017. https://doi.org/10.1016/j.drugalcdep.2019.107625

Wallin CM, Bowen SE, Brummelte S (2021) Opioid use during pregnancy can impair maternal behavior and the maternal brain network: a literature review. Neurotoxicol Teratol 86:106976. Epub 20210401. https://doi.org/10.1016/j.ntt.2021.106976

Walsh SL, Eissenberg T (2003) The clinical pharmacology of buprenorphine: extrapolating from the laboratory to the clinic. Drug Alcohol Depend 70(2 Suppl):S13–S27. Epub 2003/05/10. https://doi.org/10.1016/s0376-8716(03)00056-5

Wang AL, Lowen SB, Elman I, Shi Z, Fairchild VP, Bouril A, Gur RC, Langleben DD (2018) Sustained opioid antagonism modulates striatal sensitivity to baby schema in opioid use disorder. J Subst Abuse Treat 85:70–77. Epub 20171018. https://doi.org/10.1016/j.jsat.2017.10.007

Welle-Strand GK, Skurtveit S, Jansson LM, Bakstad B, Bjarko L, Ravndal E (2013) Breastfeeding reduces the need for withdrawal treatment in opioid-exposed infants. Acta Paediatr 102(11):1060–1066. Epub 20130902. https://doi.org/10.1111/apa.12378

Whiteman VE, Salemi JL, Mogos MF, Cain MA, Aliyu MH, Salihu HM (2014) Maternal opioid drug use during pregnancy and its impact on perinatal morbidity, mortality, and the costs of medical care in the United States. J Pregnancy 2014:906723. Epub 20140828. https://doi.org/10.1155/2014/906723

Willner D, Cohen-Yeshurun A, Avidan A, Ozersky V, Shohami E, Leker RR (2014) Short term morphine exposure in vitro alters proliferation and differentiation of neural progenitor cells and promotes apoptosis via mu receptors. PLoS One 9(7):e103043. Epub 2014/07/30. https://doi.org/10.1371/journal.pone.0103043

Wilson GS (1989) Clinical studies of infants and children exposed prenatally to heroin. Ann N Y Acad Sci 562:183–194. Epub 1989/01/01. https://doi.org/10.1111/j.1749-6632.1989.tb21017.x

Wouldes TA, Lester BM (2023) Opioid, methamphetamine, and polysubstance use: perinatal outcomes for the mother and infant. Front Pediatr 11:1305508. Epub 2024/01/22. https://doi.org/10.3389/fped.2023.1305508

Wouldes TA, Woodward LJ (2020) Neurobehavior of newborn infants exposed prenatally to methadone and identification of a neurobehavioral profile linked to poorer neurodevelopmental outcomes at age 24 months. PLoS One 15(10):e0240905. Epub 2020/10/17. https://doi.org/10.1371/journal.pone.0240905

Wu VW, Mo Q, Yabe T, Schwartz JP, Robinson SE (2001) Perinatal opioids reduce striatal nerve growth factor content in rat striatum. Eur J Pharmacol 414(2–3):211–214. Epub 2001/03/10. https://doi.org/10.1016/s0014-2999(01)00807-x

Wu LY, Chen JF, Tao PL, Huang EY (2009) Attenuation by dextromethorphan on the higher liability to morphine-induced reward, caused by prenatal exposure of morphine in rat offspring. J Biomed Sci 16(1):106. Epub 2009/11/26. https://doi.org/10.1186/1423-0127-16-106

Wu CC, Hung CJ, Shen CH, Chen WY, Chang CY, Pan HC, Liao SL, Chen CJ (2014) Prenatal buprenorphine exposure decreases neurogenesis in rats. Toxicol Lett 225(1):92–101. Epub 2013/12/11. https://doi.org/10.1016/j.toxlet.2013.12.001

Wu CC, Hung CJ, Lin SY, Wang YY, Chang CY, Chen WY, Liao SL, Raung SL, Yang CP, Chen CJ (2017) Treadmill exercise alleviated prenatal buprenorphine exposure-induced depression in rats. Neurochem Int 110:91–100. Epub 2017/10/01. https://doi.org/10.1016/j.neuint.2017.09.012

Yang SN, Yang JM, Wu JN, Kao YH, Hsieh WY, Chao CC, Tao PL (2000) Prenatal exposure to morphine alters kinetic properties of NMDA receptor-mediated synaptic currents in the hippocampus of rat offspring. Hippocampus 10(6):654–662. Epub 2001/01/12. https://doi.org/10.1002/1098-1063(2000)10:6<654::AID-HIPO1003>3.0.CO;2-T

Yen E, Davis JM (2022) The immediate and long-term effects of prenatal opioid exposure. Front Pediatr 10:1039055. Epub 2022/11/25. https://doi.org/10.3389/fped.2022.1039055

Yeoh SL, Eastwood J, Wright IM, Morton R, Melhuish E, Ward M, Oei JL (2019) Cognitive and motor outcomes of children with prenatal opioid exposure: a systematic review and meta-analysis. JAMA Netw Open 2(7):e197025. Epub 2019/07/13. https://doi.org/10.1001/jamanetworkopen.2019.7025

Yim AJ, Miranda-Paiva CM, Florio JC, Oliveira CA, Nasello AG, Felicio LF (2006) A comparative study of morphine treatment regimen prior to mating and during late pregnancy. Brain Res Bull 68(5):384–391. Epub 20051012. https://doi.org/10.1016/j.brainresbull.2005.09.014

Yonke N, Maston R, Weitzen S, Leeman L (2019) Breastfeeding intention compared with breastfeeding postpartum among women receiving medication-assisted treatment. J Hum Lact 35(1):71–79. Epub 20180503. https://doi.org/10.1177/0890334418769637

Yonke N, Jimenez EY, Leeman L, Leyva Y, Ortega A, Bakhireva LN (2020) Breastfeeding motivators and barriers in women receiving medications for opioid use disorder. Breastfeed Med 15(1):17–23. Epub 20191106. https://doi.org/10.1089/bfm.2019.0122

Zedler BK, Mann AL, Kim MM, Amick HR, Joyce AR, Murrelle EL, Jones HE (2016) Buprenorphine compared with methadone to treat pregnant women with opioid use disorder: a systematic review and meta-analysis of safety in the mother, fetus and child. Addiction 111(12):2115–2128. Epub 2016/05/26. https://doi.org/10.1111/add.13462

Zhang H, Kalluri HV, Bastian JR, Chen H, Alshabi A, Caritis SN, Venkataramanan R (2018) Gestational changes in buprenorphine exposure: a physiologically-based pharmacokinetic analysis. Br J Clin Pharmacol 84(9):2075–2087. Epub 20180621. https://doi.org/10.1111/bcp.13642

Zharikova OL, Deshmukh SV, Nanovskaya TN, Hankins GD, Ahmed MS (2006) The effect of methadone and buprenorphine on human placental aromatase. Biochem Pharmacol 71(8):1255–1264. Epub 20060207. https://doi.org/10.1016/j.bcp.2005.12.035

The Effects of THC and Nicotine on Attention: A Narrative Review

Kennedy Oleszak, Lily Freeman Striegel, Nicole Roeder, Patrick Mohr, Samantha Penman, Lorraine Collins, Danielle M. Smith, and Panayotis K. Thanos

Contents

Kennedy Oleszak and Lily Freeman Striegel contributed equally to this work.

K. Oleszak, L. F. Striegel, P. Mohr, and S. Penman
Behavioral Neuropharmacology and Neuroimaging Laboratory on Addictions (BNNLA), Research Institute on Addictions, Department of Pharmacology and Toxicology, Jacobs School of Medicine and Biomedical Sciences, University at Buffalo, Buffalo, NY, USA

N. Roeder
Behavioral Neuropharmacology and Neuroimaging Laboratory on Addictions (BNNLA), Research Institute on Addictions, Department of Pharmacology and Toxicology, Jacobs School of Medicine and Biomedical Sciences, University at Buffalo, Buffalo, NY, USA

Department of Psychology, University at Buffalo, Buffalo, NY, USA

L. Collins
Department of Community Health and Health Behavior, University at Buffalo, Buffalo, NY, USA

D. M. Smith
Department of Health Behavior, Roswell Park Comprehensive Cancer Center, Buffalo, NY, USA

P. K. Thanos (✉)
Behavioral Neuropharmacology and Neuroimaging Laboratory on Addictions (BNNLA), Research Institute on Addictions, Department of Pharmacology and Toxicology, Jacobs School of Medicine and Biomedical Sciences, University at Buffalo, Buffalo, NY, USA

Department of Psychology, University at Buffalo, Buffalo, NY, USA

Department of Exercsie and Nutrition Sciences, University at Buffalo, Buffalo, NY, USA

Department of Molecular Biology and Adelson School of Medicine, Ariel University, Ariel, Israel
e-mail: thanos@buffalo.edu

Curr Topics Behav Neurosci (2025) 75: 183–238
https://doi.org/10.1007/7854_2024_568
Published Online: 07 March 2025

Abstract Since cannabis and nicotine are two of most commonly used substances and are often used together, this paper will review the effects of cannabis (specifically THC, or tetrahydrocannabinol) and nicotine on selective attention, sustained attention, visuospatial attention, attentional bias, and attentional disorders. This review includes preclinical and clinical findings throughout all periods of development and adulthood. Selective attention is directly impacted by cannabis use, while reaction time is dependent on the timing of the last cannabis exposure. Among individuals who use cannabis, there is an attentional bias that reduces anxiety and increases focus on cannabis-related cues. Preclinical studies show that cannabis induces attention deficits that persist even after an abstinence period. Preclinical and clinical studies of prenatal cannabis exposure (PCE) provide evidence that offspring will have an increased risk for drug-seeking behavior, attention deficits, and impulsivity, which may lead to attentional disorders such as attention deficit hyperactivity disorder (ADHD) and schizophrenia. Nicotine has a dose-dependent effect on attention in adults, though preclinical studies have shown mixed results, possibly due to differences in experimental design. Prenatal nicotine exposure (PNE) impairs attentional networks by increasing one's risk for ADHD, oppositional defiant disorder, and conduct disorder. Additionally, maternal secondhand smoke exposure is linked to ADHD/conduct disorder risk in offspring. Preclinical studies on prenatal nicotine exposure suggest that there may be sex differences in which males are affected more so than females with PNE. *Summary*: Overall, cannabis/ THC impairs attention, and nicotine enhances attention; however, both substances impair attention when individuals are exposed prenatally.

Keywords Addiction · Attention · Cannabis · Delta-9-tetrahydrocannabinol · Nicotine · Tobacco

1 Introduction

The increasing prevalence of cannabis has prompted research to better understand its effects on the brain and behavior. Cannabinoids, such as delta-9-tetrahydrocannabinol (THC), within cannabis bind to cannabinoid receptors which are found primarily in the neocortex of the brain (Herkenham et al. 1990; Eggan and Lewis 2007).

Receptor binding causes attentional disturbances, euphoria, relaxation, and psychedelic effects (Böcker et al. 2010). These effects may lead to many of the reported positive experiences among those who consume cannabis (e.g., improved mood and sleep) (Babson et al. 2017; Matheson et al. 2020), but may also heighten the development of attention deficit hyperactivity disorder (ADHD) and schizophrenia in those with genetic high risk (Hollis et al. 2008). Cannabis has a causal relationship with psychotic-like behaviors like those viewed in schizophrenia (Arseneault et al. 2004; Fergusson et al. 2006). In terms of prenatal exposure to cannabis, further research is warranted since cannabis is becoming more widely used, even among pregnant women (Nashed et al. 2021; Ko et al. 2015; Mark et al. 2021; Odom et al. 2020; Campolongo et al. 2010). Cannabis alters all aspects of attention such as selective attention (Böcker et al. 2010), attentional bias (O'Neill et al. 2020), attentional processing dysfunction (Rentzsch et al. 2016), and divided attention (Newmeyer et al. 2017). Since the brain and body undergo critical developmental stages during fetal life, prenatal cannabis exposure (PCE) negatively impacts these development stages, leading to alterations in children's cognition (including attention), sleep, birth weight, gestational age at birth, body mass index (BMI), brain structure, social behavior, and intelligence quotient (IQ) (Paul et al. 2021; Goldschmidt et al. 2008; El Marroun et al. 2009).

Smoking tobacco is one of the most preventable causes of death and disease in our society (Evans and Drobes 2009; Pomerleau et al. 2003; Glautier 2004). Nicotine, the major addictive substance in tobacco, makes quitting extremely difficult. Smoking tobacco is both positively and negatively reinforcing on the brain and body (Pomerleau et al. 2003; Glautier 2004). Evidence indicates that maternal cigarette smoking during pregnancy (SDP) leads to undesirable effects on offspring including low birth weight, preterm birth, and various birth defects (Drake 2018). In 2016, the United States reported that 7.2% of women who gave birth smoked cigarettes during their pregnancy, and the prevalence of SDP was highest among women ages 20–24 (10.7%), and lowest among women aged 45 and over (2.0%) (Drake 2018). Nicotine activates virtually all neurotransmitter and neuromodulator systems by activation of nicotinic acetylcholine receptors (nAchRs) (Evans and Drobes 2009). Nicotine-dependent smokers experience attentional deficits during nicotine withdrawal since tobacco is quickly removed from circulation (Evans and Drobes 2009). Smoking temporarily improves attention and recall by enhancing cognitive function and decreasing tiredness (Evans and Drobes 2009). Despite these effects, cognitive enhancement is not a large factor in the initiation of smoking nicotine (Evans and Drobes 2009). Prenatal nicotine exposure (PNE) also remains a growing concern in many areas of the world, including developed countries. SDP is harmful to the fetus and causes significant developmental and cognitive issues that persist into adulthood such as low IQ, attention problems, and reduced academic achievement (Clifford et al. 2012). Electronic cigarette (e-cig) use has increased among adolescents despite a lack of research on e-cig usage's long-term effects (Breland et al. 2017). More research is required to determine e-cigarettes' effects, especially since e-cigs were introduced to the public with little or no regulation (Breland et al. 2017).

1.1 Cannabis

Cannabis comes from the plant *Cannabis sativa* L. and is one of the most widely used federally illicit drugs in the Western world (Böcker et al. 2010). With an increase in legalization, cannabis is becoming more widely accepted by the public. Approximately 49.5% of people ages 12 and older in the United States have reported using cannabis in their lifetime (Rangel-Pacheco et al. 2021). Intake affects cognitive functioning, including attention (Böcker et al. 2010). Adolescents are particularly vulnerable to cannabis' long-term cognitive effects (Nicholls et al. 2015). In rats, long-term cannabinoid exposure resulted in impairment of working memory and executive function (Pattij et al. 2008). Cannabis contains THC, which produces relaxing, euphoric, and psychedelic effects. Chronic use of THC at high doses is associated with the early onset of psychological disorders (i.e., schizophrenia and ADHD) and relapse of psychotic disorders (Hollis et al. 2008; Power et al. 2014). Cannabis use is also associated with depressive and anxiety symptoms, violent offending, neurocognitive and social impairments, and medical comorbidities (O'Neill et al. 2020). Cannabis additionally induces acute deficits in motor skills, verbal expression, mathematical abilities, and time perception (Jager et al. 2006). Some neurocognitive impairments with cannabis are also viewed in patients with schizophrenia (Rentzsch et al. 2016). Long-term use of cannabis in individuals with and without schizophrenia had differences in attentional processing, which may reflect alterations in the endocannabinoid system in patients with schizophrenia (Rentzsch et al. 2016). Cannabis use in adolescents with a genetic high risk of developing schizophrenia is associated with worse mental health problems (Hollis et al. 2008). As concluded by a 2008 review paper, regular cannabis use is associated with an increased risk of developing schizophrenia (Hall and Degenhardt 2008), but there is conflicting evidence of whether cannabis worsens cognitive impairments in those with schizophrenia. A recent review concluded that for individuals with schizophrenia, cannabis use may be linked with improved learning and memory (National Academies of Sciences Engineering and Medicine 2017). However, during active intoxication with cannabis, cognitive impairments in attention and memory are observed (Turner et al. 2023). Heavy cannabis use is associated with reduced function of the attentional/executive system, but it is unclear whether these cognitive impairments are permanent (Lundqvist 2005). Cannabis has different mechanisms of action in the brain depending on route of administration (ROA), which creates a more complex understanding of cannabis binding in the brain; therefore, impairments with cannabis should not be associated with blood cannabinoid concentrations (Newmeyer et al. 2017). Despite this, cannabis has many medical applications, including the potential to treat multiple sclerosis-related symptoms of muscle spasms and tremors, stimulate appetite and prevent vomiting in cancer patients, and relieve pain in patients with neuropathic pain (Jager et al. 2006).

1.2 Cannabis Mechanism of Action

The main psychoactive cannabinoid extracted from cannabis, THC, functions by binding to and activating cannabinoid-1 receptors (CB1Rs) and cannabinoid-2 receptors (CB2Rs) (Chayasirisobhon 2020). Cannabis alters cognitive processing in those who use cannabis, which is correlated to functional changes in areas of the brain that are rich in CB1Rs (Kim et al. 2019). CB1Rs are particularly concentrated in areas of the brain involved with executive functioning, reward processing, attention, and memory (Nicholls et al. 2015). The brain areas involved in these processes include the prefrontal cortex (PFC), anterior cingulate cortex (ACC), basal ganglia, hippocampus, and cerebellum (Herkenham et al. 1990; Burns et al. 2007). In the attentional network model, there are three systems working together: networks for alerting, orienting, and executive control (Petersen and Posner 2012). The alerting network is most important for sustained attention, and the other two networks are most involved with selective attention (Nicholls et al. 2015). The orienting network is involved with the connection between the frontal and parietal lobes and has associations with the top-down control of visuospatial attention (Petersen and Posner 2012; Corbetta and Shulman 2002). The executive control network is mainly involved with top-down attentional control and involves the frontoparietal network of the brain, which includes the dorsolateral PFC (dLPFC), inferior parietal areas, and cingulo-opercular system (ACC, and anterior insula/frontal operculum) (Petersen and Posner 2012; Dosenbach et al. 2008). The areas in the PFC (ACC and dLPFC) have been linked to the initiation and maintenance of cannabis use behaviors due to their high density of CB1Rs (Yucel et al. 2007). It is possible that CB1Rs in the PFC are associated with functional changes to the executive control network with cannabis use, such that there is a reduced volume in the orbitofrontal cortex and greater functional connectivity.

1.3 Nicotine

Nicotine is a widely used drug: approximately 90% of adult smokers start smoking cigarettes before age 18, and more than 70% of young adults reported that they smoked a cigarette at least once in their lives (Currie et al. 1999; Cengelli et al. 2012). In clinical observations, the age of first cigarette use is critical in determining the individual's dependence on tobacco since those who start earlier, such as in their teens, have greater difficulty quitting (Cengelli et al. 2012; Kendler et al. 2013). Adolescence is a sensitive period for the maturation of brain circuits that regulate cognition and emotion, so nicotine exposure may cause long-term changes in behaviors and neural circuits such as deficits in serotonergic signaling, cognitive function, and attention span, and enhanced impulsivity, anxiety, and fear (Yuan et al. 2015; Goriounova and Mansvelder 2012; Slotkin and Seidler 2009; Counotte et al. 2011, 2009; Smith et al. 2006). In those who use tobacco, nicotine increases heart rate and blood pressure and results in dysphoria and pleasure (Kleykamp et al. 2005).

Nicotine functions by binding to and activating nAChRs, leading to the release of neurotransmitters like dopamine (Tiwari et al. 2020). Dopamine discharge in response to nicotine is associated with drug reinforcement and dependence (Tiwari et al. 2020). Nicotine is a potent, centrally acting cholinergic agonist that enhances cognition (Wignall and de Wit 2011). Nicotine's enhancing properties are seen in smokers and psychiatric patients with Alzheimer's disease, schizophrenia, and ADHD, but nicotine's cognitive-enhancing effects are less understood for healthy nonsmokers (Wignall and de Wit 2011). Nicotine enhances activation and functional connectivity between certain regions of the brain involved in attention (Jacobsen et al. 2004a). As a result, individuals who have attentional problems, such as patients with schizophrenia, may use nicotine-containing products to mitigate this adverse symptom, resulting in a therapeutic effect (Jacobsen et al. 2004a). For those with attentional problems and healthy people alike, nicotine's cognitive-enhancing properties include improved attentional processing (Semenova et al. 2007), improved working memory in both humans and animals (Valentine and Sofuoglu 2018), and increased concentration (Lawrence et al. 2002). These effects contribute to the development of nicotine dependence (ND) especially in people with attentional deficits, since nicotine attenuates these deficits (Guimarães et al. 2017). Today, there are many different methods to use nicotine including cigarettes, e-cigs, transdermal nicotine patches, and oral nicotine products, many of which contain additives other than nicotine and tobacco. Originally e-cigs were marketed as a smoking cessation tool, but currently, adolescents use them instead of cigarettes due to flavorants (Breland et al. 2017). In the United States and Canada, e-cig use peaked in 2019, declined in 2020 during the pandemic, and nearly returned to pre-pandemic levels in 2022 (Hammond and Reid 2023).

1.4 Nicotine Mechanism of Action

The PFC is important for executive functioning and attentional performance, but it is one of the last brain regions to fully develop, so nicotine can cause deficits or incomplete maturation of the PFC (Goriounova and Mansvelder 2012). Nicotine may also lead to higher levels of dependence in adolescence by exerting neurotoxic effects in the PFC, which interferes with cognitive development, executive functioning, and inhibitory control (DeBry and Tiffany 2008). In adolescence, there is a developmental window in which the brain is sensitive to novel experiences, which coincides with an increased vulnerability to drug misuse (Bernheim et al. 2013). Nicotine use during adolescence may produce deficits in the PFC that persist long-term (Goriounova and Mansvelder 2012). During sustained attention tasks such as the rapid visual information processing (RVIP) task, the frontal, parietal, occipital, thalamic, and cerebellar regions were activated in smokers and nonsmokers, as evidenced by functional magnetic resonance imaging (fMRI) (Lawrence et al. 2002). Smokers had increased activation in the parietal cortex, occipital cortex, thalamus, and caudate during task performance when given a transdermal nicotine patch (Lawrence et al. 2002). Nicotine administration also deactivated specific regions, suggesting that nicotine facilitates the focusing of attentional resources during attention-demanding tasks (Lawrence et al. 2002). In adolescent smokers, nicotine significantly reduced activation in the PFC compared to nonsmokers (Musso et al. 2007). Chronic nicotine use during adolescence can lead to long-term effects on brain function due to reduced presynaptic mGluR2 protein with nicotine use (Counotte et al. 2011).

2 Cannabis/THC Effects on Attention

Cannabis use effects on attention and specifically, selective attention (Böcker et al. 2010), attentional bias (O'Neill et al. 2020), attentional processing (Rentzsch et al. 2016), and divided attention (see Table 1) (Newmeyer et al. 2017). There are three different processing systems/networks (see Table 2) that join to form attention within all individuals. When smoking cannabis, one or more of these processes may be altered, resulting in attentional dysfunction. Studies show a correlation between cannabis and psychological disorders that affect attention, such as ADHD and schizophrenia (Hollis et al. 2008; Rentzsch et al. 2016; Groenman et al. 2013; Wallace et al. 2019).

Research showed differences in resting state functional connectivity (RSFC) within the dorsal (DAN) and ventral (VAN) attention networks in adolescent and young adults who use cannabis after a two-week abstinence period (Harris et al. 2022). THC activates the endogenous cannabinoid system (ECB) and binds to CB1Rs (Chayasirisobhon 2020). CB1Rs become desensitized or downregulated after repeated cannabis exposure (González et al. 2006). The ECB develops throughout adolescence and is involved in sustained attention, so when adolescents use cannabis regularly, they experience poor sustained and selective attention and increased attention lapses (Lisdahl et al. 2014; Jacobsen et al. 2004b; Abdullaev et al. 2010). Looking back to the DAN and VAN, it is theorized that these two frontoparietal areas are segregated but complementary attentional networks that are

Table 1 Attention types

Attention type	Definition	Reference
Selective attention	The ability to focus only on information that is important to an individual	Böcker et al. (2010)
Attentional bias	The ability to only see one perspective and not another	O'Neill et al. (2020)
Attentional processing	Depending on the individual, attentional processing dysfunction can mean a lack of focus, lack of organization, and impaired executive function	Rentzsch et al. (2016)
Divided attention	Divided attention allows the brain to focus on many stimuli at one time	Newmeyer et al. (2017)

Table 2 Attentional processing networks

Attention network	Definition	Reference(s)
Alerting	The process of attentional alert maintains a state of alertness, arousal, or focus	Mezzacappa (2004)
Orienting	Orienting attention is the ability to shift attention from one stimulus to another, which allows an individual to disengage and reengage focus to different stimuli	Mezzacappa (2004)
Executive control	Executive control includes the execution of goal-directed behaviors such as planning actions, anticipating consequences, selecting among competing demands, and monitoring outcomes of behavior and/or interrupting or modifying behavior accordingly	Mezzacappa (2004)
Dorsal attention network (DAN)	The DAN is responsible for top-down visual attentional control and voluntary directed attention in the frontoparietal area. DAN becomes activated with top-down processing, which is when there is prior knowledge or expectation that a stimulus will be present at a certain time. DAN associates relevant stimuli to certain responses and becomes active during the preparation of a task	Corbetta and Shulman (2002), Corbetta et al. (2008, 2000)
Ventral attention network (VAN)	The VAN is responsible for bottom-up processing in the frontoparietal area. Bottom-up processing allows one to attend to new and unexpected sensory stimuli, shift attention, and have goal-directed behaviors	Corbetta and Shulman (2002), Harris et al. (2022)

involved in alerting, orienting, and executive control (Table 2) (Corbetta and Shulman 2002; Xu et al. 2015). Harris and colleagues found that there was lower functional connectivity in the DAN network in individuals who use cannabis compared to the control group (Harris et al. 2022). This was associated with heavier lifetime use patterns, greater cannabis use disorder symptoms, and a shorter length of cannabis abstinence.

In addition, cannabis' effect on the visual attention network and cerebellar hypoactivation have been demonstrated (Chang et al. 2006). Attention and memory deficits are correlated with heavy cannabis use; however, prolonged abstinence may reverse these deficits. This study performed BOLD (blood oxygen level-dependent) fMRI in individuals who use cannabis (active and abstinent) and controls and found decreased activation in the right prefrontal, medial, dorsal parietal, and medial cerebellar regions and greater activation in many frontal, parietal, and occipital regions during visual attention tasks in those who use cannabis. The results indicated that the right frontal and medial cerebellar regions normalized compared to controls with the duration of abstinence. Altered BOLD signals in the attention network and hypoactivation of the cerebellum indicate neuroadaptation, altered brain development, and downregulation of CB1Rs in those who chronically use cannabis.

2.1 *Prenatal Effects of Cannabis and THC on Attention*

Clinical and preclinical research has examined the prenatal THC and/or cannabis exposure's effects on attention in humans and rodents (Table 3).

2.1.1 Clinical Studies on Cannabis/THC and Attention

Many prenatal cannabis studies show a correlation with attention deficits (Cioffredi et al. 2022), social problems (Paul et al. 2021), and a lower IQ (Newmeyer et al. 2017). Additionally, many children experience difficulties with cognition, sleep, birth weight, gestational age at birth, BMI, and brain structure (Paul et al. 2021).

Using The Child Behavior Checklist for Toddlers (CBCL 1 ½–5 years), which focuses on anxiety/depression, attention, and aggression, it was found that girls who were exposed to cannabis during gestation experience aggression and attention problems at 18 months old (El Marroun et al. 2011). These effects were observed in female but not male children. Researchers also found that aggressive behavior was only associated with maternal, but not paternal, cannabis intake (El Marroun et al. 2011).

Another clinical study of PCE observed attention and social problems in the offspring using the CBCL and data from the Adolescent Brain Cognitive Development (ABCD) study (Paul et al. 2021). Paul and colleagues observed associated problems in the offspring with PCE prior to and after maternal knowledge of pregnancy. There was an increased association and risk for psychopathology with PCE both before and after maternal knowledge of pregnancy. However, outcomes in children exposed to cannabis prior to maternal knowledge of pregnancy did not significantly differ from children who were never exposed (Table 3).

Another study focused on prenatal cocaine/polydrug exposure (PCP)'s cognitive effects (Singer et al. 2018). The results showed that PCP increased vulnerability to substance use disorders (SUDs) and produced deficits in perceptual reasoning IQ and visual attention in adolescents.

PCE is also associated with an increased risk of children developing ADHD. Roncero and colleagues conducted a systematic review of the relationship between maternal cannabis use and the risk of developing psychiatric disorders, including ADHD (Roncero et al. 2020). This review mentioned that preclinical studies supported an association between prenatal cannabinoid exposure and central nervous system (CNS)/cognition/attention deficits (Table 3).

Clinical studies examining cannabis and attention vary depending on what attention the study focused on. The following subsections will review, in more depth, the cognitive effects of cannabis on selective attention, sustained attention, visuospatial attention, attentional bias, and attentional disorders. Please see Table 4 for a more comprehensive summary of the literature.

Table 3 Prenatal THC and attention

Subject	Age	ROA	THC dosage	Length of exposure	Attention test	Results and conclusions	Reference(s)
No exposure ($n = 10{,}834$), exposure prior to maternal knowledge of pregnancy only ($n = 413$), and exposure after maternal knowledge of pregnancy ($n = 242$, with or without exposure prior to maternal knowledge)	Children born between 2005 and 2009	N/A	N/A	N/A	The Prodromal Questionnaire–Brief Child Version, Child Behavior Checklist, The National Institutes of Health Toolbox Cognition Battery–Total Cognition Composite, The Sleep Disturbance Scale for Children	• Cannabis exposure before and after maternal knowledge of pregnancy ↑ offspring psychopathology characteristics, sleep problems, and BMI, as well as ↓ cognition and gray matter volume compared to no exposure • Exposure after knowledge of pregnancy ↓ birth weight, total intracranial volume, and white matter volumes relative to no exposure and exposure only before knowledge	Paul et al. (2021)
Cannabis exposure in pregnancy ($n = 88$), only tobacco exposure in early pregnancy ($n = 435$), only tobacco exposure throughout pregnancy ($n = 276$), no tobacco or cannabis exposure ($n = 3278$)	18 months	N/A	N/A	N/A	Child Behavior Check List for Toddlers	• Prenatal exposure to cannabis is associated with behavioral problems in early childhood only in girls and in the areas of aggression and attention problems	El Marroun et al. (2011)

Prenatal cocaine/polydrug exposure (PCE) (including tobacco, alcohol, and marijuana) ($n = 354$ total), PCE ($n = 180$), non-cocaine/polydrug (NCE) ($n = 174$)	Birth – 15.5 years	N/A	N/A	Duration of pregnancy	Wechsler Intelligence Scales for Children-IV (WISC-IV) (performed from birth), Integrated Visual and Auditory Continuous Performance Test (IVA/CPT) (performed at 15.5 years of age)	• Adolescents with PCE had deficits in Perceptual Reasoning IQ and visual attention. • PCE ↑ vulnerability to substance use disorders and produced cognitive deficits. Specifically, prenatal and perinatal exposure to cannabinoids alters the dopaminergic activity of the corpus striatum in rats, leading to ADHD and changes in locomotion • In humans, PCE is associated with mood and behavioral changes, alterations in executive functions like visuospatial attention/reasoning, increased impulsiveness and hyperactivity, lower attention capacity, and a higher prevalence of delinquent behavior • PCE leads to detrimental effects on social interaction, ↑ risk-taking/impulsivity, and an early onset of drug use or drug-seeking behavior	El Marroun et al. (2011), Singer et al. (2018), Tirado-Muñoz et al. (2020), Trezza et al. (2012)

(continued)

Table 3 (continued)

Subject	Age	ROA	THC dosage	Length of exposure	Attention test	Results and conclusions	Reference(s)
See review	See review	See review	See review	See review	See review	• Low doses of cannabinoid compounds result in atypical locomotor activity, cognitive impairments, altered emotional behavior, and an enhanced sensitivity to the drugs • Rats exposed to cannabis during pregnancy or lactation have an ↑ likelihood of developing motor hyperactivity and impulsivity, analogous to ADHD in humans	Campolongo et al. (2010), Roncero et al. (2020)
See review	See review	See review	See review	See review	See review	• Prenatal cannabis exposure leads to detrimental effects on social interaction, ↑ risk-taking/impulsivity, and an early onset of drug use or drug-seeking behavior	Tirado-Muñoz et al. (2020)
Male Sprague-Dawley rats, control ($n = 12$), 2 mg/kg THC ($n = 10$), 4 mg/kg THC ($n = 10$)	9 weeks old	Subcutaneous injection	0, 2, 4 mg/kg	Dosed daily for 28 days	Operant Visual Attention Task, Novel Object Recognition	• Paternal-only THC exposure can cause persisting behavioral and cognitive deficits in offspring; these effects may be related to THC-induced alterations in the methylation of rat sperm DNA	Holloway et al. (2020), Murphy et al. (2018)

Table 4 Clinical THC and attention

Subjects	Age	ROA	THC dosage	Length of exposure	Attention test	Results and conclusions	Reference
Males ($n = 24$)	18–33	Joint (300 mg cannabis, 700 mg tobacco)	LD 29.3 mg THC, MD 49.2 mg THC, HD 69.4 mg THC, Placebo <0.003% THC	Joints smoked within ~22 min	Spatial Visual Attention Task	• ERPs measured neural activity-related cognitive and sensory processing in the brain • Accuracy and reaction times are dose-related: as THC concentration ↑, reaction times ↑ and accuracy ↓ • HD THC ↓ reaction times, and ↑ inaccuracy • Acute cannabis and/or THC intake interferes with attention-related stimulus processing, stimulus categorizing, and context updating	Böcker et al. (2010)
Males ($n = 30$)	18–32	N/A	N/A	Abstinent for 24 h prior to testing	Flanker go/no-go Task	• Cannabis caused ↓ reaction times and ↓ performance on the component of the task related to executive control, but not on those related to the alerting or orienting networks	Nicholls et al. (2015)
Males ($n = 26$), Females ($n = 22$)	~32–33	N/A	N/A	Abstinence on test days	Visual Selective Attention Task	• Theta oscillations were observed using magnetoencephalography imaging while individuals performed an arrow-based flanker interference task • Results indicated a robust difference in occipital theta oscillations and fronto-	Rangel-Pacheco et al. (2021)

(continued)

Table 4 (continued)

Subjects	Age	ROA	THC dosage	Length of exposure	Attention test	Results and conclusions	Reference
						occipital connectivity during the flanker task in individuals who use cannabis compared to nonsmoker controls • Participants who use cannabis exhibited performance within the normal range which indicates a possible compensatory mechanism being used to selectively attend to stimuli	
Males ($n = 14$), Females ($n = 6$)	~22	N/A	N/A	Abstinence for at least a week prior to testing	Visuo-Auditory Selective Attention Task	• The SAT involved the detection of tones with higher and lower pitches than baseline as well as the detection of dots that had a smaller or larger size than baseline • No significant differences were found between the cannabis and control groups regarding the SAT, and there were no differences between active brain regions when performing the attention task	Jager et al. (2006)
See review	See review	See review	See review	See review	Pictorial stimuli, word stimuli, and non-cannabis stimuli tasks	• The studies included in this review used three types of attentional bias tasks including pictorial stimuli, word stimuli, and non-cannabis stimuli tasks	O'Neill et al. (2020)

						• ↑ attentional bias was observed towards cannabis pictures and words as well as both types of stimuli overall in the participants who use cannabis • Suggest automatic orienting rather than controlled attentional processing • If studies employ a ↓ exposure duration it could validate attentional bias as a treatment for cannabis use disorder	
Males ($n = 16$), Females ($n = 24$)	18–27	N/A	N/A	Abstinence for >8 h prior to study	Dot-Probe Task	• Bias to fearful faces was measured by the P1, N170, and N2pc ERP components • There was ↓ attentional bias in those who use cannabis, similar to low-anxiety individuals • Cannabis use was associated with ↑ N2pc, an ERP component linked to selective attention, which indicates a ↑ spatial orientation of attention • Overall, cannabis reduced attention toward fearful faces	Torrence et al. (2018)

(continued)

Table 4 (continued)

Subjects	Age	ROA	THC dosage	Length of exposure	Attention test	Results and conclusions	Reference
Males ($n = 49$), Females ($n = 31$)	18–48	N/A	N/A	Abstinence for 28 days prior to study	Auditory three-stimulus oddball task	• Schizophrenic patients exhibited a global deficit in attentional processing • Individuals who chronically used cannabis with schizophrenia had a combined deficit in frontal automatic/bottom-up attentional processing as well as in top-down context updating • Chronic cannabis use produced attentional deficits in healthy controls and schizophrenic patients, but may lead to different cognitive sequelae in patients with schizophrenia due to differences in the endocannabinoid system with schizophrenia	Rentzsch et al. (2016)
Males ($n = 70$), Females ($n = 63$)	14–21	N/A	N/A	N/A	Schizotypal Personality Questionnaire, Strengths and difficulties questionnaire, Global assessment of function	• Young people with a high genetic risk of schizophrenia are more vulnerable to mental health issues associated with cannabis use • More research is needed to determine the causal relationship between cannabis and mental health in genetically vulnerable individuals	Hollis et al. (2008)

Males ($n = 38$), Females ($n = 34$)	16–26	N/A	N/A	Abstinence for 3 weeks prior to study	Ruff 2&7, CPT-II, WAIS-III LNS, D-KEFS Color-Word Interference	• Individuals who use cannabis demonstrated ↓ reaction times on attention tasks, indicative of general inattentiveness • Subclinical ADHD symptoms did not independently predict or moderate cannabis' impact on cognitive outcomes, meaning that attention deficits in those who use cannabis are more related to the substance than ADHD symptomatology	Wallace et al. (2019)
Males ($n = 10$), Females ($n = 8$)	21–35	One-gram marijuana cigarette	0%, 1.8%, or 3.9% THC	Three 5 s inhalations of cigarette before each session	Neuropsychological assessment battery	• The results indicated that there were no significant differences in response accuracy on attentional tasks • Sustained attention significantly ↑ after the high dose of THC (3.9%) in chronic daily cannabis smokers	Hart (2001)
Males ($n = 22$)	22–34	Intravenous THC	2.5 mg THC	1 ml/min pulses over 5 min	Digit Span	• THC ↓ performance on the Digit Span • Suggests that injected THC produces significant impairments in attention and concentration for those who use cannabis infrequently	Morrison et al. (2009)
Males ($n = 28$), Females ($n = 42$)	13–18	N/A	N/A	Abstinence for 12 h prior to interview	RVIP, Digit Span, and Symbol and Digit Modalities Test	• 70 adolescents who smoke cannabis, including 34 who used it regularly (more than once per week) and 36 who used it non-regularly	Harvey et al. (2007)

(continued)

Table 4 (continued)

Subjects	Age	ROA	THC dosage	Length of exposure	Attention test	Results and conclusions	Reference
						(once per week or less) were recruited • Those who used cannabis regularly had ↓ performance than non-regular users on the RVIP task, Digit Span, and the Symbol and Digit Modalities Test • Suggests that regular cannabis use impairs sustained attention in adolescents	
Males ($n = 22$), Females ($n = 19$)	18–35	N/A	N/A	Abstinence for 24 h prior to study	Three-stimuli standard oddball paradigm	• Participants included 18 individuals who began using cannabis at an early and 23 controls • Motion stimuli induced a negative peak from occipitotemporal and parietal cortical areas during the EEG • For individuals who use cannabis, N2 latency was longer for target and standard stimuli as compared to the control group • Those who use cannabis, especially those who started at an early age, experience ↓ neural activity when attending to motion stimuli	Mikulskaya and Martin (2018)

2.1.2 Selective Attention

Böcker and colleagues observed cannabis' effects on perception and visual selective attention in 24 male individuals who reported smoking at least 2–18 cannabis cigarettes a month (Böcker et al. 2010). Two hours after smoking cannabis, six event-related potential (ERP) components were recorded at various latencies and scalp positions while individuals performed a visual selective attention task. Participants smoked one of four joints on test days, and as THC concentration increased, reaction times increased and accuracy decreased.

Nicholls and colleagues observed visual selective attention in individuals who use cannabis using a flanker go/no-go task (Nicholls et al. 2015). ERP components were used when individuals performed an attention-related task. Selected participants used cannabis at least weekly in the month prior to the study. Individuals who use cannabis abstained from cannabis for 24 h prior to testing. Similar to Böcker et al. 2010, the results of the flanker go/no-go task indicated that cannabis slowed reaction times relative to controls, but Nicholls and colleagues did not find a difference in the accuracy of responses. The individuals who used cannabis also showed poorer performance on the component of the task related to executive control, but not on those related to the alerting or orienting networks (Nicholls et al. 2015).

Another study regarding visual selective attention in individuals who use cannabis regularly looked at fronto-occipital connectivity using magnetoencephalography (MEG) imaging during a visual selective attention task (SAT) (Rangel-Pacheco et al. 2021). Those who did not use cannabis had stronger theta activity during certain portions of the task, which was the opposite for those who did. This difference between the groups suggests an altered response to the initial stimuli since theta oscillations are implicated in visual selective attention. The participants who used cannabis exhibited performance within the normal range of the task; therefore, individuals who use cannabis may have a compensatory mechanism to selectively attend to stimuli (Rangel-Pacheco et al. 2021).

Jager and colleagues used a visuo-auditory SAT and fMRI imaging to observe selective attention and active brain regions (Jager et al. 2006). To exclude acute pharmacological effects of cannabis use, subjects abstained from cannabis for at least a week before testing. The individuals who frequently used cannabis and controls displayed comparable levels and patterns of overall brain activity in brain regions involved with selective attention during the SAT. Abstinence from cannabis for a week may have caused participants' brain functioning to return to a level comparable to the controls (Jager et al. 2006).

Lundqvist and colleagues explored cannabis use's cognitive consequences in comparison to using stimulants or heroin (Lundqvist 2005). The studies within this review paper looked at long-term and short-term cannabis use and its effects on attention and executive functioning. One included study found that long-term cannabis use causes attentional impairments that continue past the period of intoxication (Solowij and Grenyer 2002). Additionally, attention may worsen with increasing years of regular cannabis use (Solowij and Grenyer 2002).

2.1.3 Sustained Attention

Hart and colleagues explored acute smoked marijuana's effects on sustained attention (Hart 2001). Eighteen participants, who used cannabis chronically and daily, completed baseline computerized cognitive tasks and were given one marijuana cigarette (0%, 1.8%, or 3.9% THC). After smoking the cigarette, participants completed a neuropsychological assessment battery. The results indicated that performance on a tracking task, a measure of sustained attention, significantly improved after the high dose of THC (3.9%). This suggests that chronic daily cannabis smokers may actually have improved sustained attention after smoking a high dose of THC.

Another study investigated synthetic THC's acute effect on cognitive functioning (Morrison et al. 2009). Twenty-two healthy male subjects, who were infrequently used cannabis, were intravenously given 2.5 mg THC or a placebo. Psychological assessments were administered before the injection and at 30, 80, and 120 min following the injection. Morrison and colleagues found that THC decreased performance on the Digit Span, which measures attention. This suggests that for those with infrequent cannabis use, injected THC produces significant impairments in attention and concentration.

Harvey and colleagues explored how cannabis affects cognition in 70 adolescents who use cannabis (Harvey et al. 2007). After abstaining from cannabis for 12 h, participants completed a two-hour interview which included a cognitive test battery. The results indicated that participants who regularly used cannabis had poorer performance than non-regular users on the RVIP task, Digit Span, and the Symbol and Digit Modalities Test. This finding suggests that regular cannabis use impairs sustained attention in adolescents.

2.1.4 Visuospatial Attention

One study examined visual attention to motion stimuli in early-onset cannabis use and non-cannabis use (Mikulskaya and Martin 2018). Individuals who used cannabis abstained from cannabis 24 h prior to data collection. To measure visual attention, a three-stimuli standard oddball paradigm was used, and EEG activity was recorded continuously. Imaging suggested that those who use cannabis, especially those who began using it at an early age, experience reduced neural activity when attending to motion stimuli.

2.1.5 Attentional Bias

O'Neill and colleagues compiled literature regarding attentional bias towards cannabis cues in individuals who use cannabis (O'Neill et al. 2020). The results showed that there was a greater attentional bias towards cannabis cues in those who use cannabis compared to controls. Experienced individuals who use drugs also displayed an automatic selective attentional orientation toward cannabis-related stimuli that occurs outside of their awareness.

Torrence and colleagues examined cannabis use's effect on attentional bias towards fearful faces (Torrence et al. 2018). ERPs were measured during a dot-probe task with fearful and neutral facial expressions to observe cannabis' residual effects on attentional bias. A reduction in attentional bias was found in those who use cannabis, which is similar to findings in low-anxiety individuals. The decrease in attentional bias could potentially be a mechanism for the anxiolytic effects of cannabis use. Cannabis use was also associated with enhanced N2pc, an ERP component linked to selective attention, which indicates a greater spatial orientation of attention.

2.1.6 Attentional Disorders

Cannabis use at an early age can induce early development of ADHD and schizophrenia, exacerbate symptoms, or produce deficits in attention. One study observed the impact of chronic cannabis use on informational processing in abstinent individuals with a long-term history of cannabis use, both with and without schizophrenia (Rentzsch et al. 2016). The study confirms attentional deficits in schizophrenic patients with chronic cannabis use. However, cannabis use may lead to different cognitive sequelae in patients with schizophrenia than healthy controls, which reflects preexisting alterations in the endocannabinoid system in schizophrenia (Rentzsch et al. 2016). Blood oxygen level-dependent (BOLD) activation in the attention network of individuals who use cannabis is altered, which in some regions is reversible with prolonged abstinence (Chang et al. 2006). The pattern of findings relates well to the concept that long-term cannabis use exerts neurotoxic effects, which may be partly reversible with prolonged abstinence.

Hollis and colleagues observed whether adolescents with a high genetic risk for schizophrenia were more prone to cannabis' adverse psychological effects than healthy adolescents with low genetic risk or adolescents with ADHD (Hollis et al. 2008). Adolescents at high genetic risk for schizophrenia may have a hypersensitive dopaminergic system that increases vulnerability to cannabis-stimulated dopamine release in the mesolimbic system (Fergusson et al. 2006). Dopamine release could attenuate ADHD's behavioral symptoms by normalizing a hypo-functioning dopamine system (Swanson et al. 2007). Hollis and colleagues concluded that young adolescents with a high genetic risk of schizophrenia are particularly vulnerable to mental health problems associated with cannabis use, and these mental health problems may differ within the population.

Wallace and colleagues observed the effects of cannabis use and subclinical ADHD symptomatology on attention-based tasks in adolescents and young adults (Wallace et al. 2019). Subclinical ADHD symptoms may explain attentional deficits in those who use cannabis, with previous research discovering a comorbidity between ADHD and cannabis use (Gujska et al. 2023; Wimberley et al. 2020).

Individuals who use cannabis experience significant attentional deficits compared to their non-using counterparts (Hanson et al. 2010; Lisdahl and Price 2012). Overall, attention deficits are more related to the substance than ADHD symptomatology in those who use cannabis (Wallace et al. 2019).

2.2 *Preclinical Studies of Cannabis/THC and Attention*

Animal models with prenatal THC exposure (PTE) have altered behavior and deficits in attention, the CNS, neurotransmission systems, and psychomotor skills, all of which infer the presence of the equivalent of "mental disorders" in animals (Roncero et al. 2020). Using a CB1 agonist, WIN 55,212-2, in low or moderate doses also causes alterations in specific systems of the brain, some of which may be permanent (Antonelli et al. 2006; Saez et al. 2014). Even low doses of PTE result in atypical locomotor activity, cognitive impairments, altered emotional behavior (such as aggression or anxiety), and enhanced sensitivity to drugs of misuse in adult rodent offspring (Campolongo et al. 2010). Recent studies observing prenatal exposure in rodents demonstrate how rats exposed to cannabis during pregnancy or lactation have an increased likelihood of developing motor hyperactivity and impulsivity, both of which are characteristics of ADHD (Mereu et al. 2003; Navarro et al. 1995).

Preclinical studies are in line with clinical studies of PTE in that they both show deleterious effects on executive functions in the offspring during childhood and adolescence (Campolongo et al. 2010). Executive functions involve capacities such as cognitive flexibility, sustained and focused attention, and working memory (Campolongo et al. 2010). Deficits in executive functioning caused by PCE are long-lasting (Campolongo et al. 2010; Roncero et al. 2020). The areas in the brain involved with executive functioning do not fully mature until early adulthood (Sowell et al. 1999). PCE alters brain development and reduces grey matter in the PFC (Tirado-Muñoz et al. 2020). Some of these effects from prenatal exposure are also viewed in the mothers who are using cannabis. Both short-term (deficit in attention, executive functioning, coordination, etc.) and long-term effects of cannabis impair the mother's ability to care for her offspring and consequently, the well-being of the offspring (Tirado-Muñoz et al. 2020).

Paternal cannabis use is also important to consider. Holloway and colleagues found that chronic paternal THC exposure in rats causes changes in sperm methylation, locomotor activity, and attention (Holloway et al. 2020). THC was subcutaneously administered to male rats at concentrations of 0, 2, or 4 mg/kg THC per day for 28 days prior to mating with drug naïve females. At the 2 mg/kg dose of THC, locomotor hyperactivity was observed in the offspring during adolescence. At this dose, there was also a significant alteration in the sperm, which resulted in long-lasting attentional and behavioral impairments in the offspring relative to controls (see Table 3).

Previous preclinical research has examined the effects of THC and other CB1R agonists on attention in mice and rats (Table 5). One study looked at CB1Rs

Table 5 Preclinical THC and attention

Subjects	Age	ROA	THC dosage	Length of exposure	Attention test	Results and conclusions	Reference(s)
Male Long-Evans Rats ($n = 41$)	60 days	i.p. Injection	WIN 55,212-2 (1.0 mg/kg or 2.5 mg/kg) and SR141716A (0.1, 0.5, or 1.0 mg/kg)	Twice a week 15 min prior to testing	Lateralized reaction time task (LRT)	• SR141716A (1.0 mg/kg) did not have any measurable effects, but it was able to prevent impairment caused by WIN (2.5 mg/kg) when used in combination • This is analogous to the role and mechanism of CB1Rs with THC and CBD	Holloway et al. (2020), Laprairie et al. (2015), Niesink and van Laar (2013)
Male Lister hooded rats ($n = 25$) and Male C57B1/6 J mice ($n = 28$ in exp.1, $n = 26$ in exp.2, $n = 1$- in exp.3)	N/A	i.p. Injection	0.3, 1.0, or 3.0 mg/kg THC	30 min prior to testing	5-choice serial reaction time task (5-CSRTT), 5-choice continuous performance test (5C-CPT), Discrete trial task	• Both mice and rats use a temporal mediating strategy while performing the 5-CSRTT • Not all CB1R agonists slow temporal perception in the 5-CSRTT	Cope et al. (2016)
Male Wistar rats ($n = 8$/dose) for plasma evaluation, ($n = 36$) for 5-CSRTT	60 days	i.p. Injection	0.3 or 3.0 mg/kg THC	5-day drug administration (twice a day), followed by 9-day abstinence prior to behavioral testing	5-choice serial reaction time task (5-CSRTT)	• Long-term intermittent exposure to specific doses of THC produced persistent impairments in impulse control and attentional dysfunction • THC-exposed rats had significantly ↑ levels of motor impulsivity compared to drug-naïve controls when inter-trial intervals (ITI) had ↓ durations • These effects became more pronounced during cognitively challenging 5-CSRTT sessions with longer ITI	Irimia et al. (2015)

(continued)

Table 5 (continued)

Subjects	Age	ROA	THC dosage	Length of exposure	Attention test	Results and conclusions	Reference(s)
						• Chronic intermittent exposure to specific doses of THC in humans, may result in disruptions to academic and professional performance	
Male Sprague Dawley rats ($n = 7$/group)	N/A	Polyethylene catheter inserted into i.p. cavity	10 mg/kg THC, 0.5 mg/kg amphetamine	THC administered twice a day (10–14 h apart) for 14 days. A 2-day abstinent period was given prior to test training	Lateralized reaction time task (LRT)	• Rats were trained to detect four different target stimuli; then, performance with variable stimulus durations was tested by randomly presenting four different stimulus durations • THC-exposed rats had impairments in choice accuracy that continued for 14 days after the final dose of THC was administered • Sub-chronic THC exposure induces dysfunction of the PFC and impairs visuospatial divided attention • When THC-exposed rats were given a dose of amphetamine, they were able to perform just as well as the vehicle-treated rats • Motivational and motor functions were preserved with THC-exposed rats, but attentional and/or executive functions were negatively impacted	Verrico et al. (2004)

specifically to see how CB1R activation or inhibition mediated the impairment of visuospatial attention in rats during a lateralized reaction time task (LRT) (Arguello and Jentsch 2004). Rats were trained to perform the LRT, which measures the ability to detect visual stimuli. After the training period, researchers intraperitoneally injected either a CB1R agonist, WIN 55,212-2 (0–2.5 mg/kg), a CB1R antagonist, SR141716A (0–1.0 mg/kg), or a combination of the two. The CB1R agonist, WIN, disrupted cognition, impaired choice accuracy, and increased reaction times when administered at a dose of 2.5 mg/kg. SR141716A, the CB1R antagonist, did not have any measurable effects, but it prevented impairment caused by WIN (2.5 mg/kg) when they were used in combination.

Another study looked at attention/perception in rats using a 5-choice serial reaction time task (5-CSRTT) after exposure to a CB1R agonist (Cope et al. 2016). THC was administered via an i.p. injection, 30 min prior to testing. The ability of THC to slow temporal perception was observed using a discrete-trial temporal discrimination task. Both mice and rats were found to use a temporal mediating strategy while performing the 5-CSRTT. THC slowed premature response time and temporal perception. However, WIN 55,212-2, another CB1R agonist, does not alter premature responding in the 5-CSRTT (Han and Robinson 2001; Crystal et al. 2003).

Irimia and colleagues observed long-term THC exposure's effects on impulse control and attentional capacity using ten 14-day cycles of THC dosing and a 5-CSRTT (Irimia et al. 2015). Each cycle was comprised of five days of i.p. injections of either 0.3 or 3 mg/kg THC followed by the 5-CSRTT during the nine days of abstinence. Researchers continued the 5-CSRTT during five weeks of abstinence. The THC-exposed rats had significantly higher levels of motor impulsivity and behavioral disinhibition compared to drug-naïve controls, in a dose-dependent manner. In rats given THC, increased motor impulsivity and behavioral disinhibition persisted during the five weeks of abstinence, indicating that some deficits associated with cannabis use continue during abstinence.

Verrico and colleagues observed visuospatial attention in rats exposed to repeated, intermittent doses of THC via an i.p. catheter (Verrico et al. 2004). Previous studies found that long-term cannabis use in humans causes neurocognitive deficits in attention, cognitive flexibility, and an inability to shift or sustain attention (Meier et al. 2022). Three days after the catheter was surgically placed to facilitate drug exposure, the rats were given 10 mg/kg of THC twice daily for 14 days. A LRT was conducted seven days after drug exposure to assess visuospatial attention. Thirteen days after the final dose of THC, rats were given either an amphetamine or saline injection prior to testing. Results showed that THC-exposed rats had impairments in choice accuracy that continued for 14 days after the final dose of THC was administered. Amphetamine allowed THC-exposed rats to perform just as well as the vehicle-treated rats.

3 Nicotine Effects on Attention

Nicotine, a stimulant with anxiolytic properties, improves attention during cognitive tasks compared to nonsmoker controls (Valentine and Sofuoglu 2018; Anderson and Brunzell 2015). Attention is reduced, however, in humans and rats prenatally exposed to nicotine (Margolis et al. 2021; Schneider et al. 2011). Human offspring prenatally exposed to nicotine may develop attentional disorders like ADHD (Thakur et al. 2013; Pauly and Slotkin 2008). Despite this, nicotine may alleviate unwanted attentional problems in those with attentional disorders (Poltavski and Petros 2006). It is unknown precisely how attentional circuitry responds to nicotine: there are three attention networks within the brain (Table 1), and nicotine might enhance one or multiple networks while reducing the others. In upcoming sub-sections, we will review attention types, brain regions involved in attentional processing, and nicotine's effects on attention, inhibitory control, prefrontal attention circuitry, visuospatial attention, and ADHD.

3.1 Prenatal Effects of Nicotine on Attention

Previous clinical and preclinical research has assessed the prenatal effects of nicotine on attention in humans and rodents (Table 6).

3.1.1 Clinical Studies on Nicotine and Attention

Tiesler and Heinrich summarized findings on PNE's relationship with behavioral problems in offspring (Tiesler and Heinrich 2014). PNE was related to a higher likelihood of ADHD symptomatology and adverse health effects in newborns, infants, children, adolescents, and adults (Linnet et al. 2003; Tiesler et al. 2011). PNE leads to altered brain structure and function due to its effects on neurotransmitter systems during brain development (Bublitz and Stroud 2012; Dwyer et al. 2009). PNE also increases one's risk for pregnancy complications and preterm delivery, lower birth weight, reduced lung function, asthma, wheezing, airway hyperresponsiveness in children, obstructive lung disease, and accelerated lung aging in later years (CDC 2004). Offspring may also be more likely to develop conduct disorder (CD) (Gatzke-Kopp and Beauchaine 2007), depression and anxiety (Menezes et al. 2013), hypertension, obesity, cardiovascular disease, and type-2 diabetes (Bakker and Jaddoe 2011). The association between PNE and childhood behavior/cognition problems is due to reduced PFC function, which is also observed in patients with ADHD (Arnsten 2009; Arnsten and Rubia 2012). The PFC plays an important role in the levels of dopamine and norepinephrine, serotonin, and acetylcholine (Arnsten and Rubia 2012).

Table 6 Prenatal nicotine and attention

Subject	Age	ROA	Nicotine dosage	Length of exposure	Attention test	Results and conclusions	Reference
See review	See review	See review	See review	See review	See review	• Those prenatally exposed to nicotine are more likely to have externalizing problems (ADHD) than internalizing problems • 8 out of 11 studies mentioned in the review supported the presence of an increased risk for ADHD in the offspring	Tiesler and Heinrich (2014)
See review	See review	See review	See review	See review	See review	• Prenatal tobacco exposure is associated with ↓ academic achievement and cognitive abilities	Clifford et al. (2012)
Preschool children ($n = 209$)	3–4 years old	Exposure to maternal and paternal smoking during pregnancy	NA	Throughout pregnancy	ADHD-RS-IV	• Maternal, but not paternal, smoking during pregnancy was associated with inattention, hyperactivity/impulsivity, and ADHD symptoms in children	Nomura et al. (2010)
Mothers ($n = 133$), children ($n = 171$ total, $n = 136$ males)	7- to 15-year-old children	Exposure to either maternal smoking or maternal secondhand smoke exposure during pregnancy	N/A	Throughout pregnancy	Child Behavior Checklist (CBCL), Child Symptom Inventory (CSI)	• There is a link between maternal secondhand smoke exposure and child psychopathology in the offspring • Children exposed prenatally to cigarette smoke (direct or secondhand) demonstrated more severe symptoms of ADHD and CD	Gatzke-Kopp and Beauchaine (2007)

(continued)

Table 6 (continued)

Subject	Age	ROA	Nicotine dosage	Length of exposure	Attention test	Results and conclusions	Reference
Mothers ($n = 4886$), children ($n = 8889$)	4- to 10-year-old children ($n = 8889$)	Exposure to cigarette smoke during pregnancy	N/A	Throughout pregnancy	Child Behavior Checklist (CBCL)	• Children prenatally exposed to tobacco smoke had ↑ risk of developing ADHD symptoms compared to nonfamilial children who were not prenatally exposed • No effect of SDP on CD and ODD when comparing siblings who differed in their prenatal exposure to nicotine	D'Onofrio et al. (2008)
C57BL/6 mice	Mothers breeding age: 8–12 weeks, offspring began testing at PND90	Nicotine drinking water	100 μg nicotine/ml drinking water	Mothers began nicotine exposure 3 weeks before breeding and continued until PND21	Object-Based Attention (OBA), Cliff Avoidance Reflex (CAR; impulsive behavior)	• Perinatal nicotine exposure beginning 3 weeks before breeding caused significant impairment in attention (OBA) at PND 90 in male offspring only • No significant difference in impulsive behavior • Deficits in attention, as well as working memory, were sex-specific	Zhang et al. (2018)
Male ($n = 25$) and female ($n = 67$) Lister hooded rats, females divided into (a) NIC exposure ($n = 19$), (b) foster mother	3-month-old adult rat offspring ($n = 10$ NIC, $n = 12$ control) used in 5-CSRTT	Nicotine drinking water	0.06 mg nicotine/ml drinking water	NIC-exposed mothers began exposure 3 weeks prior to mating and continued throughout pregnancy	5-choice serial reaction time task (5-CSRTT)	• 5-CSRTT results indicated that adult rats had ↑ anticipatory responses, ↑omission errors, ↑ variable response times, and ↓ accuracy during the 1 s stimulus duration • Direct impact of prenatal	Schneider et al. (2011)

($n = 37$), or (c) blood nicotine levels assessed in foster mothers ($n = 11$)						nicotine exposure on cognitive and inhibitory control later in life • This study challenged the idea that the association between prenatal nicotine exposure and ADHD is mediated entirely by genetic effects	

Two hundred preschool children (3–4 years old) prenatally exposed to maternal smoking (not paternal smoking) had an increased risk of ADHD symptoms (Nomura et al. 2010). Paternal smoking generated an increased risk for maternal SDP. Therefore, it would be effective to offer interventions for both mothers and fathers during the mother's pregnancy to prevent ADHD-related behavior, symptoms, and diagnosis.

Gatzke-Kopp and Beauchaine investigated direct and passive PNE and the development of externalizing psychopathology (Gatzke-Kopp and Beauchaine 2007). They observed a higher risk for ADHD and CD symptoms in children with PNE and in children whose mother was exposed to environmental tobacco smoke during gestation.

D'Onofrio and colleagues explored genetic and environmental confounds influencing the relationship between SDP and offspring's risk of developing CD, oppositional defiant disorder (ODD), and ADHD between the ages of 4 and 10 (D'Onofrio et al. 2008). Researchers compared ADHD traits of siblings with and without PNE to account for familial and genetic effects. Children with PNE had an increased risk of developing ADHD symptoms compared to nonfamilial children without PNE. On the other hand, in siblings who differed in their exposure to maternal SDP, there was no association between SDP and offspring CD and ODD, while the relationship between SDP and ADHD was small and insignificant. This study emphasized that genetic and environmental influences may be at play when determining PNE's behavioral effects.

Nicotine, as mentioned previously, has some beneficial effects on attention and focus and alters attention in individuals with ADHD. This subsection will review, in more depth, nicotine's effect on selective attention, sustained attention, visuospatial attention, attentional bias, and attentional disorders. Please see Table 7 for a more comprehensive summary of the literature.

3.1.2 Selective Attention

Wignall and de Wit studied nicotine's effect on attention and inhibitory control in healthy nonsmokers (Wignall and de Wit 2011). They administered nicotine to individuals using a 7 mg transdermal nicotine patch, and attention was observed using various tests. Nicotine significantly improved performance when participants performed the Stroop task, but it impaired performance on the orienting aspect of the ANT, which is related to selective attention.

3.1.3 Sustained Attention

One study observed whether transdermal nicotine enhances alertness and attentiveness in nonsmokers without attentional deficits (Levin et al. 1998). Participants were given either a placebo or a 7 mg transdermal nicotine patch, and attention was assessed. Overall, nicotine significantly improved attention.

Table 7 Clinical nicotine and attention

Subject	Age	ROA	Nicotine dosage	Length of exposure	Attention test	Results and conclusions	Reference
40 Healthy non-smokers: ($n = 20$ females, $n = 20$ males)	18–40	Transdermal nicotine	7 mg	Given patch 15 min after arrival and wore it the entire day of testing (8 am-5 pm). Testing began 3 h after the patch was applied	The Stroop test (behavioral inhibition), Attention Network Task (tests all ranges of attention such as orienting, alerting, and executive function), signal detection task (reward responsiveness), and the balloon analogue risk task (risk-taking behavior)	• Nicotine ↑ performance only on the Stroop test, and ↓ performance on the Attention Network Test in the aspect of orienting attention. • Overall, nicotine ↓ selective attention during the ANT • The results indicated that nicotine affects subtypes of cognitive performance differently	Wignall and de Wit (2011)
Healthy smokers ($n = 8$ males, $n = 7$ females), non-smokers ($n = 7$ males, $n = 7$ females)	18–35	Transdermal nicotine	21 mg	Subjects smoked their last cigarette 15 min prior to the study, transdermal patch applied 2.5 h before testing	Rapid Visual Information Processing (RVIP) task	• Mildly abstinent smokers showed ↓ task-induced brain activation in the parietal cortex and caudate than nonsmokers. • Transdermal nicotine ↑ smokers' visual sustained attention during the RVIP task by bilaterally ↑ task-induced activation in the parietal cortex, thalamus, and caudate and inducing a generalized ↑ in occipital cortex activity • Nicotine improves attention by activating brain areas involved with visual attention	Lawrence et al. (2002)

(continued)

Table 7 (continued)

Subject	Age	ROA	Nicotine dosage	Length of exposure	Attention test	Results and conclusions	Reference
Male nonsmokers ($n = 61$), subjects split into 2 groups (low and high attentiveness)	18–34	Transdermal nicotine	7 mg	Patch applied before the study, 6 h after application, testing began	Conners' Continuous Performance Task (CPT), Stroop Task, Wisconsin Card Sorting Task (WCST)	• Participants were divided into one of two groups based on attention: adults with attentional deficits and adults with superior attention. These groups were then randomly divided into placebo and nicotine groups • The low-attention group performed better on Conners' CPT with transdermal nicotine • The Stroop task, on the other hand, showed no significant effects of nicotine on attention for either group • Nicotine significantly impaired performance on the WCST in the high-attention group • Overall, transdermal nicotine administration ↑ attention more significantly in the low-attention group of nonsmokers	Poltavski and Petros (2006)

Habitual smokers ($n = 27$ females, $n = 29$ males)	18–47	Transdermal nicotine	14 mg	Testing began 2 h after patch application and remained on for the duration of testing	Rapid Visual Information Processing (RVIP) task, performed twice (with and without nicotine)	• The RVIP task included four different variations, each of which had a different distractor. Each variation was performed on two different days: one day, participants wore a transdermal nicotine patch, and another day, participants wore a placebo patch • Nicotine enhanced performance, but these results varied with distractor type and genotype • Overall, nicotine modulates selective attention that depends partly on the emotional versus numeric nature of task distractors, DRD2 genotype, and the brain hemisphere that initially processes distractors	Gilbert et al. (2005)
Smokers ($n = 6$ males, $n = 11$ females), non-smokers ($n = 6$ males, $n = 11$ females)	19–49	Transdermal nicotine (given only to smokers)	21 mg	Patch applied 2–2.5 h prior to MRI scan, testing began ~3 h after patch application, smokers smoked a cigarette within 1 h before entering research facility	Spatial Attentional Resource Allocation Task (SARAT) (performed twice, smokers once with and without nicotine)	• Participants in the smoking group were split up and each received either a transdermal nicotine patch or a placebo patch. The following day, each smoker subgroup switched to the opposite patch • The SARAT showed that nicotine-induced task-related deactivation in brain regions	Hahn et al. (2007)

(continued)

Table 7 (continued)

Subject	Age	ROA	Nicotine dosage	Length of exposure	Attention test	Results and conclusions	Reference
						overlapped with the default network of resting brain function. These deactivations were associated with improvements in the visuospatial attention task • When attention toward a task diminishes and task-independent thought processes intrude, nicotine may hinder such internal processes and maintain the alerting properties of the task stimuli • Overall, nicotine enhanced activity in several brain areas associated with visual attention and arousal during the SARAT	
Nonsmokers without ADHD ($n = 10$ males, $n = 1$ female)	18–29	Transdermal nicotine	7 mg	Patch applied at 8:45 am, patch removed after final test at 1:15 pm, testing occurred twice, 10 min and 3 h after patch application	The profile of mood states (POMS), Conners' CPT, computerized interval timing task	• Nicotine ↑ self-perceived vigor as measured by the POMS, and the CPT indicated that nicotine significantly ↓ the number of omission errors, ↓ hit reaction time variability, and ↑ composite attentiveness scores • Nicotine significantly improved CPT performance in the transdermal nicotine patch group	Levin et al. (1998)

						• Overall, nicotine ↑ attention in nonsmoking subjects without any preexisting attentional impairment and may improve attentiveness beyond the normal state in nonsmokers. Therefore, nicotine administered via a transdermal patch can improve attentiveness in adult nonsmokers without preexisting attentional impairments	
Nonsmokers ($n = 35$ males, $n = 32$ females)	Mean age 19–19.5	Transdermal nicotine	7 mg	Patch applied 4 h before each session, each session separated by a week	Auditory attention task (startle eye blink reflex)	• Participants were given a 7 mg transdermal patch on one test day and a placebo patch on the next test day • Nicotine was found to affect aspects of automatic and controlled attention. Nicotine's enhancement of certain aspects of attention was observed in nonsmokers, indicating that nicotine's effect on attention is more than simply an improvement of withdrawal symptoms • Nicotine exerted its greatest cognitive-enhancing effects in those who had relatively weak baseline attentional processing	Baschnagel (2006)

(continued)

Table 7 (continued)

Subject	Age	ROA	Nicotine dosage	Length of exposure	Attention test	Results and conclusions	Reference
Healthy college undergraduates ($n = 9$ males, $n = 11$ females)	Mean age 19.5	Oral nicotine gum	0, 2, 4 mg	Gum was chewed for 15 min, subjects were to chew every 3 s, testing began immediately after	ANT task (alerting, orienting, and executive function), N-back test (working memory)	• Overall, nicotine produced dose-dependent ↑ in heart rate and dysphoria • No cognitive effects of nicotine were significant enough to report ↑ or ↓ in attention	Kleykamp et al. (2005)
Patients with ADHD ($n = 40$), smokers with nicotine dependence (ND) ($n = 40$), and healthy controls ($n = 40$)	18–65	N/A	N/A	N/A	Adult ADD/ADHD DSM-IV-based diagnostic screening and rating scale (ASRS), Wender Utah Rating Scale (WURS)	• The results showed that those with ADHD scored statistically ↑ in inattention • The ND and control groups had significantly different scores for hyperactivity, impulsivity, and related symptoms, but not for inattention • ND was not correlated with inattention/hyperactivity and related symptoms in both the ND and ADHD groups • Nicotine may act as a positive reinforcer for those with higher inattention symptoms due to nicotine's positive effects on attention and arousal	Bilgi et al. (2017)
Smokers with schizophrenia ($n = 20$), healthy comparison smokers ($n = 24$)	18–50	Transdermal nicotine	21 mg (for those who smoke <15 cigarettes/day), 35 mg (≥15 cigarettes/day)	Patch applied for 4.5 h, RVIP task began 4 h after patch application	RVIP task	• Nicotine was administered via a transdermal nicotine patch and a placebo patch which were given 5–14 days apart • Nicotine ↑ RVIP accuracy and processing speed in	Elliot Hong et al. (2011)

						schizophrenia patients but did not normalize their sustained attention deficit or their abnormal attention network • Individuals with schizophrenia also had ↓ activation during the RVIP task in the frontal-parietal-cingulate-thalamic attention network compared to healthy controls	
Smokers with schizophrenia ($n = 13$), control smokers ($n = 13$)	35–50	Transdermal nicotine	28 mg (BMI <26 kg/m^2), 35 mg (BMI >26 kg/m^2)	Patch applied for 6 h before fMRI scanning and left on during scanning	Auditory n-back task	• The participants were scanned twice with the fMRI: once after receiving a transdermal nicotine patch and once after receiving a placebo patch • While participants had the patch on, they performed an auditory n-back task with two levels of working memory and two levels of selective attention load • Schizophrenic patients performed more slowly and less accurately compared to controls across all task conditions except for the most difficult task • Nicotine ↑ performance on more difficult cognitive tasks by enhancing the activation and functional connectivity of cortical regions that	Jacobsen et al. (2004a)

(continued)

Table 7 (continued)

Subject	Age	ROA	Nicotine dosage	Length of exposure	Attention test	Results and conclusions	Reference
						mediate task performance. Nicotine enhanced activation in the ACC and bilateral thalamus and modulated thalamocortical functional connectivity in schizophrenic patients more so than controls	
ADHD patients ($n = 511$), unaffected siblings of patients with ADHD ($n = 286$), healthy controls (n = 220)	12–24	N/A	N/A	N/A	Parental Account of Childhood Symptoms (PACS), Diagnostic Interview Schedule for Children	• Researchers assessed (i) the prevalence and age of onset of polysubstance use disorder (PSUD) and ND in late adolescence, (ii) the effects of an additional childhood diagnosis of ADHD with ODD or CCD on developing PSUD and ND, (iii) and the familial risk of PSUD and ND in unaffected siblings of individuals with ADHD • Adolescents with a childhood ADHD diagnosis were 1.8 times more likely to develop PSUD later in life and 8.6 times more likely to develop ND compared to the healthy controls • Unaffected siblings were 2.9 times less likely to develop ND compared to their ADHD	Groenman et al. (2013)

						siblings, but they did not differ in the risk of developing PSUD. Unaffected siblings were also not at an ↑ risk of developing PSUD or ND compared to controls • There was an ↑ risk of developing PSUD in the group of participants with ADHD and CCD and for the ADHD group compared to controls. ADHD and ODD participants were not at ↑ risk for PSUD compared to the control subjects • All groups were at ↑ risk for developing ND compared to healthy controls • In summary, a childhood diagnosis of ADHD is a risk factor for PSUD and ND in adolescence, and ADHD with CDD but not ODD further ↑ the risk of PSUD and ND	

Another study examined the effects of transdermal nicotine, distractor type, and dopamine type-2 receptor (DRD2) genotype on RVIP task performance in habitual smokers (Gilbert et al. 2005). DRD2 genotype was chosen for the investigation because dopamine is the neurotransmitter most frequently hypothesized to play an important role in the attention (Robinson and Berridge 1993; Corrigall et al. 1992). RVIP task results showed that nicotine enhanced performance in habitual smokers; however, this effect varied with receptor genotype and distractor type. Nicotine significantly improved attention in individuals with a DRD2 A1 genotype.

Kleykamp and colleagues observed nicotine gum's effects on attention during the ANT in 20 never-smokers (Kleykamp et al. 2005). Participants were given either 0 mg, 2 mg, or 4 mg of nicotine. Nicotine did not influence any aspect of attention. Previous studies concluded that orienting and alerting aspects of attention are facilitated by nicotine exposure, whereas the executive function of attention is either enhanced or not affected by nicotine at all (Mumenthaler et al. 1998; Witte et al. 1997; Provost and Woodward 1991; Foulds et al. 1997; Ettinger et al. 2017). The lack of significance may reflect nicotine's true effect on attention or may be due to the study's small sample size.

3.1.4 Visuospatial Attention

Lawrence and colleagues observed nicotine's effect on visual attention in cigarette smokers using a 21 mg transdermal nicotine patch (Lawrence et al. 2002). Transdermal nicotine improved smokers' visual sustained attention during the RVIP by increasing the activation of posterior cortical and subcortical regions that are related to visual attention and arousal.

Another study found that nicotine enhances visuospatial attention by deactivating areas of the resting brain default network (Hahn et al. 2007). The spatial attentional resource allocation task (SARAT) was used to observe transdermal nicotine's effect on long-term smokers' attention and cognition in comparison to non-nicotine-exposed nonsmokers. The SARAT showed that nicotine-induced task-related deactivation in brain regions overlapped with the default network of resting brain function. These deactivations were associated with performance improvements.

3.1.5 Attentional Bias

Baschnagel and colleagues assessed a 7 mg transdermal nicotine patch's effect on attentional modification of the acoustic startle response in 67 nonsmokers (Baschnagel 2006). Overall, nicotine improved automatic attentional filtering and increased focused attentional processing in participants who had preexisting low levels of attention. This suggests that nicotine's effect on attention is more than simply an improvement of withdrawal symptoms.

3.1.6 Attentional Disorders

Poltavski and Petros examined a 7 mg transdermal nicotine patch's effect on attention in adult nonsmokers with and without attentional deficits (Poltavski and Petros 2006). Three separate methods were used to observe attention in this study (see Table 7). Nicotine treatment improved the low-attention group's sustained attention compared to the high-attention group.

Baschnagel and colleagues also studied transdermal nicotine's effect on nonsmoking participants with attentional deficits (Baschnagel 2006). A tone discrimination task assessed sustained attention, and it was concluded that nicotine exerted its greatest cognitive-enhancing effects in those who had weak baseline attentional processing.

Bilgi and colleagues assessed the clinical characteristics and smoking profiles of individuals with ADHD, smokers without any documented psychiatric disorders, and healthy controls (Bilgi et al. 2017). Compared to the general population, those with ADHD are more likely to begin smoking at a younger age, have a higher smoking prevalence, and have lower quitting rates (Molina and Pelham Jr. 2003). Genetic predispositions to psychosocial vulnerabilities and variants in genes involved in dopaminergic and serotonergic neurotransmission and metabolism contribute to the relationship between ADHD and smoking (Matthies et al. 2013; Gerra et al. 2005; Luciano et al. 2004; Skowronek et al. 2006). Nicotine's ability to improve attention and reduce deficits in dopaminergic function in individuals with inattention and executive dysfunction implies that these are some of the factors that promote smoking in those with ADHD (Molina and Pelham Jr. 2003; Burke et al. 2001). As a result, smoking represents a form of self-medication against ADHD symptomatology. Bilgi and colleagues concluded that nicotine may act as a positive reinforcer for participants with higher inattention symptoms due to nicotine's positive effects on arousal and attention.

Another study investigated how transdermal nicotine enhances, but does not normalize, visual sustained attention in smokers with schizophrenia as compared to nonschizophrenic smokers (Elliot Hong et al. 2011). After transdermal nicotine patches were administered, fMRI and RVIP task results were recorded. Previous research determined that nicotine improves sustained attention in schizophrenic patients (Harris et al. 2004; Sacco et al. 2005), so this fMRI study examined the neural basis of this relationship. Nicotine improved RVIP accuracy and processing speed in schizophrenia patients but did not normalize their sustained attention deficit or their abnormal attention network.

Jacobsen and colleagues investigated whether nicotine improves cognition in those with schizophrenia by facilitating the activation of brain regions mediating task performance or facilitating functional connectivity (Jacobsen et al. 2004a). The results indicated that schizophrenic patients performed more slowly and less accurately compared to controls across task conditions except for the most difficult task condition, in which nicotine improved performance accuracy in patients and worsened performance accuracy in controls.

Another study investigated the relationship between a childhood diagnosis of ADHD with and without ODD and/or comorbid CD (the combination of ODD and CD is CCD), and the later development of psychoactive SUD (PSUD)/alcohol/drug use disorder and/or ND in children and adolescents (Groenman et al. 2013). Overall, a childhood diagnosis of ADHD is a risk factor for polysubstance use disorder (PSUD) and ND during adolescence. Unaffected siblings are less likely to develop ND compared to their siblings with ADHD. In summary, a childhood diagnosis of ADHD is a risk factor for PSUD and ND in adolescence, and ADHD with CDD but not ODD further increases the risk of PSUD and ND.

3.2 *Preclinical Studies on Nicotine and Attention*

Zhang and colleagues examined attention and working memory deficits in mice perinatally exposed to oral nicotine (Zhang et al. 2018). Beginning three weeks before breeding and continuing until three weeks postpartum, female C57Bl/6 mice were given nicotine in drinking water that contained saccharin (2%) and 100 μg/ml of nicotine. On postnatal day 90, researchers conducted attention and memory tests. Spontaneous locomotor activity was used to assess anxiety and ADHD symptomatology. The results indicated that male mice were more prone to attentional and working memory deficits than female mice.

Another study with similar methods examined PNE's (administered via 0.06 mg/ml nicotine drinking water) effect on ADHD-like symptoms in adult rats (Schneider et al. 2011). Using a 5-CSRTT, indicators of attention and impulsivity deficits were observed through reaction times, accuracy errors, omission errors, and anticipatory responses. Nicotine drinking water was given three weeks prior to breeding and continued daily throughout pregnancy. 5-CSRTT results suggested that PNE delays learning in tasks that require high attention. Since the same rat strain was used for the control and experimental groups, genetic variance was controlled; therefore, the results show that PNE was directly responsible for the ADHD-like cognitive changes observed during the 5-CSRTT.

Previous rodent studies have examined the effects of nicotine on attention (Table 8). One study examined how chronic nicotine administration improves attention and how nicotine withdrawal induces performance deficits in the 5-CSRTT (Semenova et al. 2007). In humans, nicotine enhances attention and cognition, while nicotine withdrawal produces attentional deficits; yet, it is debatable whether nicotine improves performance under baseline conditions or ameliorates attentional deficits. To investigate this effect, Wistar and Sprague Dawley (SD) rats received nicotine via an osmotic minipump. 5-CSRTT results indicated that the Wistar rats performed more accurately than the SD rats. Overall, acute nicotine administration produced small increases in accuracy and impulsivity and decreases in omission errors. Nicotine withdrawal, however, decreased the number of correct responses, increased omissions, and increased latencies to respond.

Hahn and colleagues observed selective nicotinic receptor antagonists' effects on attention and nicotine-induced attentional enhancement (Hahn et al. 2011). To determine which nAChR subtype(s) mediate(s) nicotine's attention-enhancing effects, researchers tested whether the performance-enhancing effects of nicotine in rodent models of attention could be reversed by the α4β2, α4β4, α3β2, and α2β2 nAChR antagonist dihydro-β-erythroidine (DHβE) or the α7 antagonist methyl lycaconitine (MLA). Rats performed a 5-CSRTT after being injected with nicotine (0.1 mg/kg) or vehicle in the presence of increasing doses of DhβE (0.0, 0.3, 1, and 3 mg/kg) or MLA (0.0, 0.4, 1.3, and 4 mg/kg). This study yielded three main findings (see Table 8). Overall, nAChR subtypes are differentially involved in specific aspects of attention, which implies that a fine-tuned adjustment may be achievable with subtype-selective nicotinic compounds.

Counotte and colleagues observed lasting synaptic changes underlying attentional deficits caused by nicotine exposure during adolescence (Counotte et al. 2011). Nicotine interferes with PFC development and leads to cognitive impairments later in life, so researchers focused on the PFC (Slotkin 2004; Jacobsen et al. 2005; Mathers et al. 2006). Researchers exposed adolescent rats to nicotine vapor and assessed visuospatial attention, protein expression, and synaptic physiology in the PFC during adulthood. Nicotine treatment during adolescence increased impulsive behavior and impaired measures of attention in adulthood after five weeks of abstinence. Adolescent nicotine exposure induced lasting attentional disturbances, reduced mGluR2 protein, and reduced mGluR2 function on presynaptic terminals of PFC glutamatergic synapses.

4 Conclusion

In summary, cannabis and nicotine remain two of the most commonly used substances in the United States, with many individuals using both substances. Cannabis (and more specifically, THC) and nicotine impact attention, albeit by different mechanisms. THC, the main psychoactive component of cannabis, binds to CB1Rs in areas of the brain responsible for attention, producing deficits over time in the attention networks within the human brain (Harris et al. 2022) and impaired measures of attention in rodents (Arguello and Jentsch 2004; Cope et al. 2016; Irimia et al. 2015; Verrico et al. 2004) PCE is linked to attention problems and the onset of ADHD symptomatology in the offspring (Roncero et al. 2020). Paternal cannabis use is important to consider alongside maternal use since THC alters methylation of rat sperm DNA (Holloway et al. 2020) (Fig. 1). In contrast to cannabis/THC, nicotine improves attention for nonsmokers with preexisting low attention (Poltavski and Petros 2006; Baschnagel 2006), nonsmokers without preexisting attentional impairments (Levin et al. 1998), and those who are current cigarette smokers (Lawrence et al. 2002; Gilbert et al. 2005). Due to nicotine's enhancement of attention, smokers with lower than normal attention, such as those with ADHD and schizophrenia, may use nicotine-containing products as a form of self-

Table 8 Preclinical nicotine and attention

Subject	Age	ROA	Nicotine dosage	Length of exposure	Attention test	Results and conclusions	Reference
Male Wistar ($n = 22$) and Sprague Dawley rats ($n = 9$)	N/A	Subcutaneous delivery by 7-day osmotic pump	7 mg/kg/day base or 9 mg/kg/day nicotine salt or injected acutely at 0.017–0.14 mg/kg	Acute exposure for the duration of the day using osmotic minipump	5-choice serial reaction time task (5-CSRTT)	• In Wistar rats, nicotine had no effect on accuracy, yet there was a trend of ↑ correct responses and ↓ omissions only at the highest dose of nicotine • In Sprague Dawley rats, low-dose nicotine tended to ↓ accuracy, while high-dose nicotine ↑ accuracy • Overall, acute nicotine-induced small improvements in attentional performance, ↑ accuracy and impulsivity, and ↓ omission errors • Chronic nicotine ↑ accuracy during days 4–6 of nicotine administration after nicotine-induced impulsivity dissipated • Chronic, but not acute, nicotine improved accuracy, while nicotine withdrawal ↓ performance on some aspects of the 5-CSRTT	Semenova et al. (2007)

Male Lister hooded rats ($n = 12$)	~1 year old	Subcutaneous injection (nicotine and DhβE), intraperitoneal injection (MLA)	Nicotine (0.0 and 0.1 mg/kg) and DHβE (0.0, 0.3, 1, and 3 mg/kg), or nicotine (0.0 and 0.1 mg/kg) and MLA (0.0, 0.4, 1.3, and 4 mg/kg)	DHβE (10 min prior to testing), MLA (15 min prior to testing), nicotine (10 min prior to testing)	5-choice serial reaction time task (5-CSRTT)	• DhβE did not antagonize the attention-enhancing effects of nicotine. • MLA reversed the ↓ in omission errors by nicotine • MLA alone ↑ response accuracy at low doses	Hahn et al. (2011)
Adolescent rats ($n = 11$), adult rats ($n = 11$)	PND 34–43 (adolescents), PND 60–69 (adults)	Subcutaneous injection	0.4 mg/kg three times per day for 10 days	Testing began at 1 day abstinent and 5 weeks abstinent	5-choice serial reaction time task (5-CSRTT)	• Nicotine treatment during adolescence ↑ impulsive behavior and impaired measures of attention in adulthood after 5 weeks of abstinence • Adolescent nicotine exposure induced lasting attentional disturbances and ↓ mGluR2 protein and function on presynaptic terminals of PFC glutamatergic synapses • Local infusion of a group II mGluR agonist restored mGluR2 activity and rescued attentional disturbances in adult rats that received nicotine as adolescents	Counotte et al. (2011)

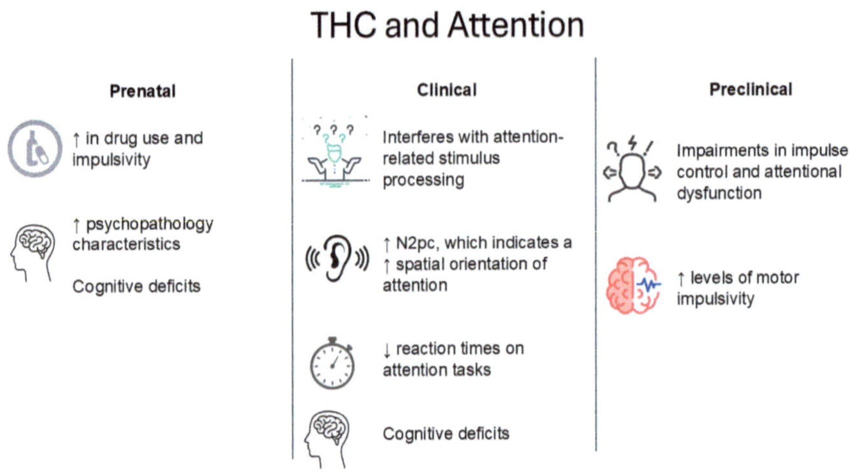

Fig. 1 Summary of the effects of THC on attention

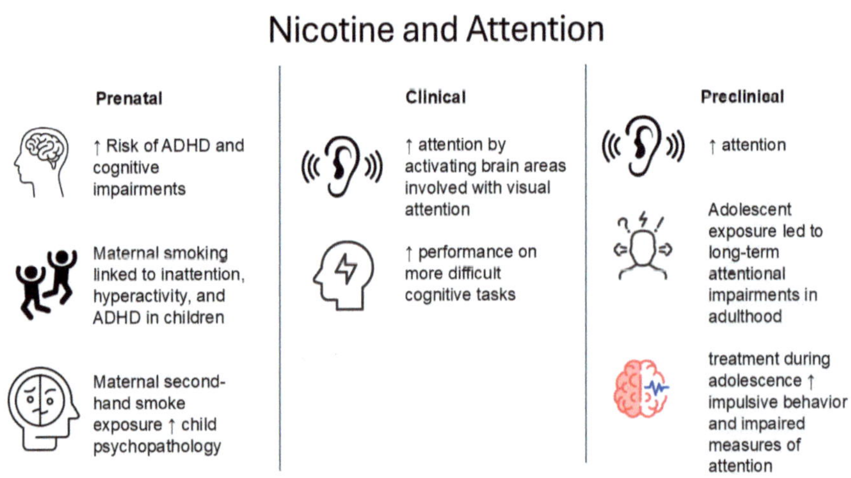

Fig. 2 Summary of the effects of nicotine on attention

medication (Bilgi et al. 2017; Elliot Hong et al. 2011). Although nicotine improves attention, PNE impairs attention in both humans (Clifford et al. 2012; Linnet et al. 2003; Tiesler et al. 2011; Gatzke-Kopp and Beauchaine 2007; Nomura et al. 2010; D'Onofrio et al. 2008; Krain and Castellanos 2006) and rodents (Schneider et al. 2011; Zhang et al. 2018) (Fig. 2).

Since ROA may be important when determining cannabis and nicotine's effects on attention, future research should investigate how different ROAs impact attention-related outcomes in both preclinical and clinical models. Additionally, there is much uncertainty about how prenatal nicotine and/or cannabis affect the alerting, orienting, and executive control attention networks in the brain, so this should be further explored. Lastly, the combined effect of cannabis/THC and tobacco/nicotine on attention should be further investigated, since most studies have focused on each substance alone. Currently, research suggests that nicotine masks cognitive impairment from cannabis since these substances have opposing effects on cognition (Hindocha et al. 2017); there, it is reasonable to hypothesize that nicotine may mask attentional impairments induced by cannabis. Despite this information, no studies to our knowledge specifically investigated the effect of cannabis and nicotine co-use on attention.

Acknowledgments This research was supported by NIH DA045640 and R01DA057228.

Ethics Declarations *Funding*: This work was supported by NIH DA045640 and R01DA057228.

Conflict of Interest: All authors certify that they have no affiliations with or involvement in any organization or entity with any financial interest or nonfinancial interest in the subject matter or materials discussed in this manuscript.

Human and Animal Rights and Informed Consent: No animal or human subjects were used by the authors in this study.

References

Abdullaev Y, Posner MI, Nunnally R, Dishion TJ (2010) Functional MRI evidence for inefficient attentional control in adolescent chronic cannabis abuse. Behav Brain Res 215(1):45–57. https://doi.org/10.1016/j.bbr.2010.06.023

Anderson SM, Brunzell DH (2015) Anxiolytic-like and anxiogenic-like effects of nicotine are regulated via diverse action at <scp>β</scp>2*nicotinic acetylcholine receptors. Br J Pharmacol 172(11):2864–2877. https://doi.org/10.1111/bph.13090

Antonelli T, Tomasini MC, Tattoli M, Cassano T, Finetti S, Mazzoni E et al (2006) Prenatal exposure to the cannabinoid receptor agonist WIN 55,212-2 and carbon monoxide reduces extracellular glutamate levels in primary rat cerebral cortex cell cultures. Neurochem Int 49(6): 568–576. Epub 20060614. https://doi.org/10.1016/j.neuint.2006.04.012

Arguello PA, Jentsch JD (2004) Cannabinoid CB1 receptor-mediated impairment of visuospatial attention in the rat. Psychopharmacology 177(1):141–150. https://doi.org/10.1007/s00213-004-1953-0

Arnsten AF (2009) Toward a new understanding of attention-deficit hyperactivity disorder pathophysiology: an important role for prefrontal cortex dysfunction. CNS Drugs 23(Suppl 1):33–41. https://doi.org/10.2165/00023210-200923000-00005

Arnsten AF, Rubia K (2012) Neurobiological circuits regulating attention, cognitive control, motivation, and emotion: disruptions in neurodevelopmental psychiatric disorders. J Am Acad Child Adolesc Psychiatry 51(4):356–367. Epub 20120303. https://doi.org/10.1016/j.jaac.2012.01.008

Arseneault L, Cannon M, Witton J, Murray RM (2004) Causal association between cannabis and psychosis: examination of the evidence. Br J Psychiatry 184(2):110–117. https://doi.org/10.1192/bjp.184.2.110

Babson KA, Sottile J, Morabito D (2017) Cannabis, cannabinoids, and sleep: a review of the literature. Curr Psychiatry Rep 19(4):23. https://doi.org/10.1007/s11920-017-0775-9

Bakker H, Jaddoe VWV (2011) Cardiovascular and metabolic influences of fetal smoke exposure. Eur J Epidemiol 26(10):763–770. https://doi.org/10.1007/s10654-011-9621-2

Baschnagel JS (2006) Effects of nicotine and incentives on attentional modification of startle in non-smokers/Joseph. State University of New York at Buffalo, Scott Baschnagel

Bernheim A, Halfon O, Boutrel B (2013) Controversies about the enhanced vulnerability of the adolescent brain to develop addiction. Front Pharmacol 4:118. Epub 20131128. https://doi.org/10.3389/fphar.2013.00118

Bilgi B, Aksoy UM, Şahmelikoğlu Onur Ö, Maner AF (2017) Attention-deficit/hyperactivity disorder and nicotine dependence in adults. Noro-Psikiyatri Arsivi 54(4):322–327. https://doi.org/10.5152/npa.2017.15882

Böcker KBE, Gerritsen J, Hunault CC, Kruidenier M, Mensinga TT, Kenemans JL (2010) Cannabis with high Δ 9-THC contents affects perception and visual selective attention acutely: an event-related potential study. Pharmacol Biochem Behav 96(1):67–74. https://doi.org/10.1016/j.pbb.2010.04.008

Breland A, Soule E, Lopez A, Ramôa C, El-Hellani A, Eissenberg T (2017) Electronic cigarettes: what are they and what do they do? Ann N Y Acad Sci 1394(1):5–30. https://doi.org/10.1111/nyas.12977

Bublitz MH, Stroud LR (2012) Maternal smoking during pregnancy and offspring brain structure and function: review and agenda for future research. Nicotine Tob Res 14(4):388–397. Epub 20111216. https://doi.org/10.1093/ntr/ntr191

Burke JD, Loeber R, Lahey BB (2001) Which aspects of ADHD are associated with tobacco use in early adolescence? J Child Psychol Psychiatry 42(4):493–502

Burns HD, Van Laere K, Sanabria-Bohórquez S, Hamill TG, Bormans G, Eng W-S et al (2007) [18F] MK-9470, a positron emission tomography (PET) tracer for in vivo human PET brain imaging of the cannabinoid-1 receptor. Proc Natl Acad Sci 104(23):9800–9805. https://doi.org/10.1073/pnas.0703472104

Campolongo P, Trezza V, Ratano P, Palmery M, Cuomo V (2010) Developmental consequences of perinatal cannabis exposure: behavioral and neuroendocrine effects in adult rodents. Psychopharmacology 214(1):5–15. https://doi.org/10.1007/s00213-010-1892-x

CDC (2004) Reports of the surgeon general. The health consequences of smoking: a report of the surgeon general. Centers for Disease Control and Prevention (US), Atlanta, GA

Cengelli S, O'Loughlin J, Lauzon B, Cornuz J (2012) A systematic review of longitudinal population-based studies on the predictors of smoking cessation in adolescent and young adult smokers. Tob Control 21(3):355–362. https://doi.org/10.1136/tc.2011.044149

Chang L, Yakupov R, Cloak C, Ernst T (2006) Marijuana use is associated with a reorganized visual-attention network and cerebellar hypoactivation. Brain (London, England: 1878) 129(5): 1096–1112. https://doi.org/10.1093/brain/awl064

Chayasirisobhon S (2020) Mechanisms of action and pharmacokinetics of cannabis. Perm J 25:1–3. https://doi.org/10.7812/tpp/19.200

Cioffredi L-A, Anderson H, Loso H, East J, Nguyen P, Garavan H et al (2022) Prenatal cannabis exposure predicts attention problems, without changes on fMRI in adolescents. Neurotoxicol Teratol 91:107089. https://doi.org/10.1016/j.ntt.2022.107089

Clifford A, Lang L, Chen R (2012) Effects of maternal cigarette smoking during pregnancy on cognitive parameters of children and young adults: a literature review. Neurotoxicol Teratol 34(6):560–570. https://doi.org/10.1016/j.ntt.2012.09.004

Cope ZA, Halberstadt AL, van Enkhuizen J, Flynn AD, Breier M, Swerdlow NR et al (2016) Premature responses in the five-choice serial reaction time task reflect rodents' temporal strategies: evidence from no-light and pharmacological challenges. Psychopharmacology 233(19–20):3513–3525. https://doi.org/10.1007/s00213-016-4389-4

Corbetta M, Shulman GL (2002) Control of goal-directed and stimulus-driven attention in the brain. Nat Rev Neurosci 3(3):201–215. https://doi.org/10.1038/nrn755

Corbetta M, Kincade JM, Ollinger JM, McAvoy MP, Shulman GL (2000) Voluntary orienting is dissociated from target detection in human posterior parietal cortex. Nat Neurosci 3(3):292–297. https://doi.org/10.1038/73009

Corbetta M, Patel G, Shulman GL (2008) The reorienting system of the human brain: from environment to theory of mind. Neuron 58(3):306–324. https://doi.org/10.1016/j.neuron.2008.04.017

Corrigall WA, Franklin KB, Coen KM, Clarke PB (1992) The mesolimbic dopaminergic system is implicated in the reinforcing effects of nicotine. Psychopharmacology 107:285–289. https://doi.org/10.1007/BF02245149

Counotte DS, Spijker S, van de Burgwal LH, Hogenboom F, Schoffelmeer ANM, de Vries TJ et al (2009) Long-lasting cognitive deficits resulting from adolescent nicotine exposure in rats. Neuropsychopharmacology (New York, NY) 34(2):299–306. https://doi.org/10.1038/npp.2008.96

Counotte DS, Goriounova NA, Li KW, Loos M, van der Schors RC, Schetters D et al (2011) Lasting synaptic changes underlie attention deficits caused by nicotine exposure during adolescence. Nat Neurosci 14(4):417–419. Epub 2011/02/22. https://doi.org/10.1038/nn.2770

Crystal JD, Maxwell KW, Hohmann AG (2003) Cannabinoid modulation of sensitivity to time. Behav Brain Res 144(1–2):57–66. https://doi.org/10.1016/S0166-4328(03)00062-7

Currie C, Hurrelmann K, Settertobulte W, Smith B, Todd J, World Health Organization (1999) Regional Office for E. Health and health behaviour among young people: international report/by C. Currie … [et al.]. WHO Regional Office for Europe, Copenhagen

DeBry SC, Tiffany ST (2008) Tobacco-induced neurotoxicity of adolescent cognitive development (TINACD): a proposed model for the development of impulsivity in nicotine dependence. Nicotine Tob Res 10(1):11–25. https://doi.org/10.1080/14622200701767811

D'Onofrio BM, Hulle CAV, Waldman ID, Rodgers JL, Harden KP, Rathouz PJ et al (2008) Smoking during pregnancy and offspring externalizing problems: an exploration of genetic and environmental confounds. Dev Psychopathol 20(1):139–164. https://doi.org/10.1017/S0954579408000072

Dosenbach NU, Fair DA, Cohen AL, Schlaggar BL, Petersen SE (2008) A dual-networks architecture of top-down control. Trends Cogn Sci 12(3):99–105. https://doi.org/10.1016/j.tics.2008.01.001

Drake P (2018) Cigarette smoking during pregnancy: United States, 2016. U.S. Department of Health and Human Services, Centers for Disease Control and Prevention, National Center for Health Statistics, Hyattsville, MD

Dwyer JB, McQuown SC, Leslie FM (2009) The dynamic effects of nicotine on the developing brain. Pharmacology & Therapeutics 122(2):125–139. https://doi.org/10.1016/j.pharmthera.2009.02.003

Eggan SM, Lewis DA (2007) Immunocytochemical distribution of the cannabinoid CB1 receptor in the primate neocortex: a regional and laminar analysis. Cereb Cortex (New York, NY 1991) 17(1):175–191. https://doi.org/10.1093/cercor/bhj136

El Marroun HMS, Tiemeier HPD, Steegers EAPPDMD, Jaddoe VWVPDMD, Hofman APDMD, Verhulst FCPDMD et al (2009) Intrauterine cannabis exposure affects fetal growth trajectories: the generation R study. J Am Acad Child Adolesc Psychiatry 48(12):1173–1181. https://doi.org/10.1097/CHI.0b013e3181bfa8ee

El Marroun H, Hudziak JJ, Tiemeier H, Creemers H, Steegers EA, Jaddoe VW et al (2011) Intrauterine cannabis exposure leads to more aggressive behavior and attention problems in 18-month-old girls. Drug Alcohol Depend 118(2–3):470–474. Epub 2011/04/08. https://doi.org/10.1016/j.drugalcdep.2011.03.004

Elliot Hong L, Schroeder M, Ross TJ, Buchholz B, Jo Salmeron B, Wonodi I et al (2011) Nicotine enhances but does not normalize visual sustained attention and the associated brain network in Schizophrenia. Schizophr Bull 37(2):416–425. https://doi.org/10.1093/schbul/sbp089

Ettinger U, Faiola E, Kasparbauer A-M, Petrovsky N, Chan RC, Liepelt R et al (2017) Effects of nicotine on response inhibition and interference control. Psychopharmacology 234:1093–1111. https://doi.org/10.1007/s00213-017-4542-8

Evans DE, Drobes DJ (2009) Nicotine self-medication of cognitive-attentional processing. Addict Biol 14(1):32–42. Epub 2008/10/16. https://doi.org/10.1111/j.1369-1600.2008.00130.x

Fergusson DM, Poulton R, Smith PF, Boden JM (2006) Cannabis and psychosis. BMJ 332(7534): 172–175. Epub 2006/01/21. https://doi.org/10.1136/bmj.332.7534.172
Foulds J, Stapleton JA, Bell N, Swettenham J, Jarvis MJ, Russell MA (1997) Mood and physiological effects of subcutaneous nicotine in smokers and never-smokers. Drug Alcohol Depend 44(2–3):105–115. https://doi.org/10.1016/s0376-8716(96)01327-0
Gatzke-Kopp LM, Beauchaine TP (2007) Direct and passive prenatal nicotine exposure and the development of externalizing psychopathology. Child Psychiatry Hum Dev 38(4):255–269. https://doi.org/10.1007/s10578-007-0059-4
Gerra G, Garofano L, Zaimovic A, Moi G, Branchi B, Bussandri M et al (2005) Association of the serotonin transporter promoter polymorphism with smoking behavior among adolescents. Am J Med Genet B Neuropsychiatr Genet 135B(1):73–78. https://doi.org/10.1002/ajmg.b.30173
Gilbert DG, Izetelny A, Radtke R, Hammersley J, Rabinovich NE, Jameson TR et al (2005) Dopamine receptor (DRD2) genotype-dependent effects of nicotine on attention and distraction during rapid visual information processing. Nicotine Tob Res 7(3):361–379. Epub 2005/08/09. https://doi.org/10.1080/14622200500125245
Glautier S (2004) Measures and models of nicotine dependence: positive reinforcement. Addiction 99(s1):30–50. https://doi.org/10.1111/j.1360-0443.2004.00736.x
Goldschmidt LPD, Richardson GAPD, Willford JPD, Day NLPD (2008) Prenatal marijuana exposure and intelligence test performance at age 6. J Am Acad Child Adolesc Psychiatry 47(3):254–263. https://doi.org/10.1097/CHI.0b013e318160b3f0
González S, Scorticati C, García-Arencibia M, de Miguel R, Ramos JA, Fernández-Ruiz J (2006) Effects of rimonabant, a selective cannabinoid CB1 receptor antagonist, in a rat model of Parkinson's disease. Brain Res 1073:209–219. https://doi.org/10.1016/j.brainres.2005.12.014
Goriounova NA, Mansvelder HD (2012) Short- and long-term consequences of nicotine exposure during adolescence for prefrontal cortex neuronal network function. Cold Spring Harb Perspect Med 2(12):a012120-a. https://doi.org/10.1101/cshperspect.a012120
Groenman AP, Oosterlaan J, Rommelse N, Franke B, Roeyers H, Oades RD et al (2013) Substance use disorders in adolescents with attention deficit hyperactivity disorder: a 4-year follow-up study. Addiction 108(8):1503–1511. https://doi.org/10.1111/add.12188
Guimarães K, Madureira DQM, Madureira AL (2017) The reward-attention circuit model: Nicotine's influence on attentional focus and consequences on attention deficit hyperactivity disorder. Neurocomputing (Amsterdam) 242:140–149. https://doi.org/10.1016/j.neucom.2017.02.072
Gujska JH, Silczuk A, Madejek R, Szulc A (2023) Exploring the link between attention-deficit hyperactivity disorder and cannabis use disorders: a review. Med Sci Monit 29:e939749. Epub 20230506. https://doi.org/10.12659/msm.939749
Hahn B, Ross TJ, Yang Y, Kim I, Huestis MA, Stein EA (2007) Nicotine enhances visuospatial attention by deactivating areas of the resting brain default network. J Neurosci 27(13): 3477–3489. Epub 2007/03/30. https://doi.org/10.1523/JNEUROSCI.5129-06.2007
Hahn B, Shoaib M, Stolerman IP (2011) Selective nicotinic receptor antagonists: effects on attention and nicotine-induced attentional enhancement. Psychopharmacology 217(1):75–82. Epub 2011/03/25. https://doi.org/10.1007/s00213-011-2258-8
Hall W, Degenhardt L (2008) Cannabis use and the risk of developing a psychotic disorder. World Psychiatry 7(2):68–71. https://doi.org/10.1002/j.2051-5545.2008.tb00158.x
Hammond D, Reid JL (2023) Trends in vaping and nicotine product use among youth in Canada, England and the USA between 2017 and 2022: evidence to inform policy. Tob Control 34(1). https://doi.org/10.1136/tc-2023-058241
Han CJ, Robinson JK (2001) Cannabinoid modulation of time estimation in the rat. Behav Neurosci 115(1):243–246. https://doi.org/10.1037/0735-7044.115.1.243
Hanson KL, Winward JL, Schweinsburg AD, Medina KL, Brown SA, Tapert SF (2010) Longitudinal study of cognition among adolescent marijuana users over three weeks of abstinence. Addict Behav 35(11):970–976. https://doi.org/10.1016/j.addbeh.2010.06.012

Harris JG, Kongs S, Allensworth D, Martin L, Tregellas J, Sullivan B et al (2004) Effects of nicotine on cognitive deficits in Schizophrenia. Neuropsychopharmacology 29(7):1378–1385. https://doi.org/10.1038/sj.npp.1300450

Harris JC, Wallace AL, Thomas AM, Wirtz HG, Kaiver CM, Lisdahl KM (2022) Disrupted resting state attentional network connectivity in adolescent and young adult cannabis users following two-weeks of monitored abstinence. Brain Sci 12(2):287. https://doi.org/10.3390/brainsci12020287

Hart C (2001) Effects of acute smoked Marijuana on complex cognitive performance. Neuropsychopharmacology 25(5):757–765. https://doi.org/10.1016/s0893-133x(01)00273-1

Harvey MA, Sellman JD, Porter RJ, Frampton CM (2007) The relationship between non-acute adolescent cannabis use and cognition. Drug Alcohol Rev 26(3):309–319. https://doi.org/10.1080/09595230701247772

Herkenham M, Lynn AB, Little MD, Johnson MR, Melvin LS, de Costa BR et al (1990) Cannabinoid receptor localization in brain. Proc Natl Acad Sci USA 87(5):1932–1936. https://doi.org/10.1073/pnas.87.5.1932

Hindocha C, Freeman TP, Xia JX, Shaban NDC, Curran HV (2017) Acute memory and psychotomimetic effects of cannabis and tobacco both 'joint' and individually: a placebo-controlled trial. Psychol Med 47(15):2708–2719. https://doi.org/10.1017/s0033291717001222

Hollis C, Groom MJ, Das D, Calton T, Bates AT, Andrews HK et al (2008) Different psychological effects of cannabis use in adolescents at genetic high risk for schizophrenia and with attention deficit/hyperactivity disorder (ADHD). Schizophr Res 105(1–3):216–223. Epub 2008/09/05. https://doi.org/10.1016/j.schres.2008.07.010

Holloway ZR, Hawkey AB, Pippin E, White H, Wells C, Kenou B et al (2020) Paternal factors in neurodevelopmental toxicology: THC exposure of male rats causes long-lasting neurobehavioral effects in their offspring. Neurotoxicology (Park Forest South) 78:57–63. https://doi.org/10.1016/j.neuro.2020.01.009

Irimia C, Polis IY, Stouffer D, Parsons LH (2015) Persistent effects of chronic Δ9-THC exposure on motor impulsivity in rats. Psychopharmacology 232(16):3033–3043. https://doi.org/10.1007/s00213-015-3942-x

Jacobsen LK, D'Souza DC, Mencl WE, Pugh KR, Skudlarski P, Krystal JH (2004a) Nicotine effects on brain function and functional connectivity in schizophrenia. Biol Psychiatry (1969) 55(8):850–858. https://doi.org/10.1016/j.biopsych.2003.12.023

Jacobsen LK, Mencl WE, Westerveld M, Pugh KR (2004b) Impact of cannabis use on brain function in adolescents. Ann N Y Acad Sci 1021(1):384–390. https://doi.org/10.1196/annals.1308.053

Jacobsen LK, Krystal JH, Mencl WE, Westerveld M, Frost SJ, Pugh KR (2005) Effects of smoking and smoking abstinence on cognition in adolescent tobacco smokers. Biol Psychiatry 57(1):56–66. https://doi.org/10.1016/j.biopsych.2004.10.022

Jager G, Kahn RS, Van Den Brink W, Van Ree JM, Ramsey NF (2006) Long-term effects of frequent cannabis use on working memory and attention: an fMRI study. Psychopharmacology 185(3):358–368. Epub 2006/03/08. https://doi.org/10.1007/s00213-005-0298-7

Kendler KS, Myers J, Damaj MI, Chen X (2013) Early smoking onset and risk for subsequent nicotine dependence: a monozygotic co-twin control study. Am J Psychiatry 170(4):408–413. https://doi.org/10.1176/appi.ajp.2012.12030321

Kim DJ, Schnakenberg Martin AM, Shin YW, Jo HJ, Cheng H, Newman SD et al (2019) Aberrant structural–functional coupling in adult cannabis users. Hum Brain Mapp 40(1):252–261. https://doi.org/10.1002/hbm.24369

Kleykamp BA, Jennings JM, Blank MD, Eissenberg T (2005) The effects of nicotine on attention and working memory in never-smokers. Psychol Addict Behav 19(4):433–438. Epub 2005/12/22. https://doi.org/10.1037/0893-164X.19.4.433

Ko JYP, Farr SLP, Tong VTMPH, Creanga AAMDP, Callaghan WMMDMPH (2015) Prevalence and patterns of marijuana use among pregnant and nonpregnant women of reproductive age. Am J Obstet Gynecol 213(2):201.e1–201.e10. https://doi.org/10.1016/j.ajog.2015.03.021

Krain AL, Castellanos FX (2006) Brain development and ADHD. Clin Psychol Rev 26(4): 433–444. Epub 20060209. https://doi.org/10.1016/j.cpr.2006.01.005

Laprairie RB, Bagher AM, Kelly MEM, Denovan-Wright EM (2015) Cannabidiol is a negative allosteric modulator of the cannabinoid-1 receptor. Br J Pharmacol 172(20):4790–4805. https://doi.org/10.1111/bph.13250

Lawrence NS, Ross TJ, Stein EA (2002) Cognitive mechanisms of nicotine on visual attention. Neuron (Cambridge, Mass) 36(3):539–548. https://doi.org/10.1016/S0896-6273(02)01004-8

Levin ED, Conners CK, Silva D, Hinton SC, Meck WH, March J et al (1998) Transdermal nicotine effects on attention. Psychopharmacology 140(2):135–141. https://doi.org/10.1007/s002130050750

Linnet KM, Dalsgaard S, Obel C, Wisborg K, Henriksen TB, Rodriguez A et al (2003) Maternal lifestyle factors in pregnancy risk of attention deficit hyperactivity disorder and associated behaviors: review of the current evidence. Am J Psychiatry 160(6):1028–1040. https://doi.org/10.1176/appi.ajp.160.6.1028

Lisdahl KM, Price JS (2012) Increased marijuana use and gender predict poorer cognitive functioning in adolescents and emerging adults. J Int Neuropsychol Soc 18(4):678–688. Epub 2012/05/23. https://doi.org/10.1017/s1355617712000276

Lisdahl KM, Wright NE, Medina-Kirchner C, Maple KE, Shollenbarger S (2014) Considering cannabis: the effects of regular cannabis use on neurocognition in adolescents and young adults. Curr Addict Rep 1:144–156. https://doi.org/10.1007/s40429-014-0019-6

Luciano M, Zhu G, Kirk KM, Whitfield JB, Butler R, Heath AC et al (2004) Effects of dopamine receptor D4 variation on alcohol and tobacco use and on novelty seeking: multivariate linkage and association analysis. Am J Med Genet 124B(1):113–123. https://doi.org/10.1002/ajmg.b.20077

Lundqvist T (2005) Cognitive consequences of cannabis use: comparison with abuse of stimulants and heroin with regard to attention, memory and executive functions. Pharmacol Biochem Behav 81(2):319–330. https://doi.org/10.1016/j.pbb.2005.02.017

Margolis AE, Pagliaccio D, Ramphal B, Banker S, Thomas L, Robinson M et al (2021) Prenatal environmental tobacco smoke exposure alters children's cognitive control circuitry: a preliminary study. Environ Int 155:106516. Epub 20210506. https://doi.org/10.1016/j.envint.2021.106516

Mark K, Saini J, Qato DM (2021) Serious psychological distress and cannabis use among pregnant women in the United States: findings from the National Survey of drug use and health, 2015–2018. Drug Alcohol Depend 229(Pt A):109116. https://doi.org/10.1016/j.drugalcdep.2021.109116

Mathers M, Toumbourou JW, Catalano RF, Williams J, Patton GC (2006) Consequences of youth tobacco use: a review of prospective behavioural studies. Addiction 101(7):948–958. https://doi.org/10.1111/j.1360-0443.2006.01438.x

Matheson J, Mann RE, Sproule B, Huestis MA, Wickens CM, Stoduto G et al (2020) Acute and residual mood and cognitive performance of young adults following smoked cannabis. Pharmacol Biochem Behav 194:172937. Epub 20200501. https://doi.org/10.1016/j.pbb.2020.172937

Matthies S, Holzner S, Feige B, Scheel C, Perlov E, Ebert D et al (2013) ADHD as a serious risk factor for early smoking and nicotine dependence in adulthood. J Atten Disord 17(3):176–186. Epub 20120127. https://doi.org/10.1177/1087054711428739

Meier MH, Caspi A, Knodt AR, Hall W, Ambler A, Harrington HL et al (2022) Long-term cannabis use and cognitive reserves and hippocampal volume in midlife. Am J Psychiatry 179(5): 362–374. https://doi.org/10.1176/appi.ajp.2021.21060664

Menezes AMB, Murray J, László M, Wehrmeister FC, Hallal PC, Gonçalves H et al (2013) Happiness and depression in adolescence after maternal smoking during pregnancy: birth cohort study. PLoS One 8(11):e80370. https://doi.org/10.1371/journal.pone.0080370

Mereu G, Fà M, Ferraro L, Cagiano R, Antonelli T, Tattoli M et al (2003) Prenatal exposure to a cannabinoid agonist produces memory deficits linked to dysfunction in hippocampal long-term potentiation and glutamate release. Proc Natl Acad Sci U S A 100(8):4915–4920. Epub 20030404. https://doi.org/10.1073/pnas.0537849100

Mezzacappa E (2004) Alerting, orienting, and executive attention: developmental properties and sociodemographic correlates in an epidemiological sample of young, urban children. Child Dev 75(5):1373–1386. https://doi.org/10.1111/j.1467-8624.2004.00746.x

Mikulskaya E, Martin F (2018) Visual attention to motion stimuli and its neural correlates in cannabis users. Eur J Neurosci 47(3):269–276. https://doi.org/10.1111/ejn.13810

Molina BS, Pelham WE Jr (2003) Childhood predictors of adolescent substance use in a longitudinal study of children with ADHD. J Abnorm Psychol 112(3):497–507. https://doi.org/10.1037/0021-843x.112.3.497

Morrison PD, Zois V, McKeown DA, Lee TD, Holt DW, Powell JF et al (2009) The acute effects of synthetic intravenous Δ9-tetrahydrocannabinol on psychosis, mood and cognitive functioning. Psychol Med 39(10):1607. https://doi.org/10.1017/s0033291709005522

Mumenthaler MS, Taylor JL, O'Hara R, Yesavage JA (1998) Influence of nicotine on simulator flight performance in non-smokers. Psychopharmacology 140(1):38–41. https://doi.org/10.1007/s002130050736

Murphy SK, Itchon-Ramos N, Visco Z, Huang Z, Grenier C, Schrott R et al (2018) Cannabinoid exposure and altered DNA methylation in rat and human sperm. Epigenetics 13(12):1208–1221. https://doi.org/10.1080/15592294.2018.1554521

Musso F, Bettermann F, Vucurevic G, Stoeter P, Konrad A, Winterer G (2007) Smoking impacts on prefrontal attentional network function in young adult brains. Psychopharmacology 191(1):159–169. https://doi.org/10.1007/s00213-006-0499-8

Nashed MG, Hardy DB, Laviolette SR (2021) Prenatal cannabinoid exposure: emerging evidence of physiological and neuropsychiatric abnormalities. Front Psych 11:624275. https://doi.org/10.3389/fpsyt.2020.624275

National Academies of Sciences Engineering and Medicine (2017) The health effects of cannabis and cannabinoids: the current state of evidence and recommendations for research. The National Academies Press, Washington, DC

Navarro M, Rubio P, de Fonseca FR (1995) Behavioural consequences of maternal exposure to natural cannabinoids in rats. Psychopharmacology 122(1):1–14. https://doi.org/10.1007/bf02246436

Newmeyer MN, Swortwood MJ, Taylor ME, Abulseoud OA, Woodward TH, Huestis MA (2017) Evaluation of divided attention psychophysical task performance and effects on pupil sizes following smoked, vaporized and oral cannabis administration. J Appl Toxicol 37(8):922–932. Epub 2017/02/01. https://doi.org/10.1002/jat.3440

Nicholls C, Bruno R, Matthews A (2015) Chronic cannabis use and ERP correlates of visual selective attention during the performance of a flanker go/nogo task. Biol Psychol 110:115–125. https://doi.org/10.1016/j.biopsycho.2015.07.013

Niesink RJM, van Laar M (2013) Does cannabidiol protect against adverse psychological effects of THC? Front Psychiatry 4:130. https://doi.org/10.3389/fpsyt.2013.00130

Nomura Y, Marks DJ, Halperin JM (2010) Prenatal exposure to maternal and paternal smoking on attention deficit hyperactivity disorders symptoms and diagnosis in offspring. J Nerv Ment Dis 198(9):672–678. https://doi.org/10.1097/NMD.0b013e3181ef3489

Odom GC, Cottler LB, Striley CW, Lopez-Quintero C (2020) Perceived risk of weekly cannabis use, past 30-day cannabis use, and frequency of cannabis use among pregnant women in the United States. Int J Women's Health 12:1075–1088. https://doi.org/10.2147/IJWH.S266540

O'Neill A, Bachi B, Bhattacharyya S (2020) Attentional bias towards cannabis cues in cannabis users: a systematic review and meta-analysis. Drug Alcohol Depend 206:107719. https://doi.org/10.1016/j.drugalcdep.2019.107719

Pattij T, Wiskerke J, Schoffelmeer ANM (2008) Cannabinoid modulation of executive functions. Eur J Pharmacol 585(2–3):458–463. https://doi.org/10.1016/j.ejphar.2008.02.099

Paul SE, Hatoum AS, Fine JD, Johnson EC, Hansen I, Karcher NR et al (2021) Associations between prenatal cannabis exposure and childhood outcomes: results from the ABCD study. JAMA Psychiatry 78(1):64–76. https://doi.org/10.1001/jamapsychiatry.2020.2902

Pauly JR, Slotkin TA (2008) Maternal tobacco smoking, nicotine replacement and neurobehavioural development. Acta Paediatr 97(10):1331–1337. Epub 20080612. https://doi.org/10.1111/j.1651-2227.2008.00852.x

Petersen SE, Posner MI (2012) The attention system of the human brain: 20 years after. Annu Rev Neurosci 35(1):73–89. https://doi.org/10.1146/annurev-neuro-062111-150525

Poltavski DV, Petros T (2006) Effects of transdermal nicotine on attention in adult non-smokers with and without attentional deficits. Physiol Behav 87(3):614–624. Epub 2006/02/10. https://doi.org/10.1016/j.physbeh.2005.12.011

Pomerleau O, Fagerstrom K-O, Marks J, Tate J, Pomerleau C (2003) Development and validation of a self-rating scale for positive- and negative-reinforcement smoking: the Michigan nicotine reinforcement questionnaire. Nicotine Tob Res 5(5):711–718. https://doi.org/10.1080/1462220031000158627

Power RA, Verweij KJH, Zuhair M, Montgomery GW, Henders AK, Heath AC et al (2014) Genetic predisposition to schizophrenia associated with increased use of cannabis. Mol Psychiatry 19(11):1201–1204. https://doi.org/10.1038/mp.2014.51

Provost SC, Woodward R (1991) Effects of nicotine gum on repeated administration of the Stroop test. Psychopharmacology 104(4):536–540. https://doi.org/10.1007/bf02245662

Rangel-Pacheco A, Lew BJ, Schantell MD, Frenzel MR, Eastman JA, Wiesman AI et al (2021) Altered fronto-occipital connectivity during visual selective attention in regular cannabis users. Psychopharmacology (Berl) 238(5):1351–1361. Epub 2020/11/27. https://doi.org/10.1007/s00213-020-05717-3

Rentzsch J, Stadtmann A, Montag C, Kunte H, Plockl D, Hellweg R et al (2016) Attentional dysfunction in abstinent long-term cannabis users with and without schizophrenia. Eur Arch Psychiatry Clin Neurosci 266(5):409–421. Epub 2015/07/18. https://doi.org/10.1007/s00406-015-0616-y

Robinson TE, Berridge KC (1993) The neural basis of drug craving: an incentive-sensitization theory of addiction. Brain Res Rev 18(3):247–291. https://doi.org/10.1016/0165-0173(93)90013-p

Roncero C, Valriberas-Herrero I, Mezzatesta-Gava M, Villegas JL, Aguilar L, Grau-Lopez L (2020) Cannabis use during pregnancy and its relationship with fetal developmental outcomes and psychiatric disorders. A systematic review. Reprod Health 17(1):25–29. https://doi.org/10.1186/s12978-020-0880-9

Sacco KA, Termine A, Seyal A, Dudas MM, Vessicchio JC, Krishnan-Sarin S et al (2005) Effects of cigarette smoking on spatial working memory and attentional deficits in schizophrenia: involvement of nicotinic receptor mechanisms. Arch Gen Psychiatry 62(6):649–659. https://doi.org/10.1001/archpsyc.62.6.649

Saez TM, Aronne MP, Caltana L, Brusco AH (2014) Prenatal exposure to the CB1 and CB2 cannabinoid receptor agonist WIN 55,212-2 alters migration of early-born glutamatergic neurons and GABAergic interneurons in the rat cerebral cortex. J Neurochem 129(4):637–648. Epub 20140106. https://doi.org/10.1111/jnc.12636

Schneider T, Ilott N, Brolese G, Bizarro L, Asherson PJE, Stolerman IP (2011) Prenatal exposure to nicotine impairs performance of the 5-choice serial reaction time task in adult rats. Neuropsychopharmacology 36(5):1114–1125. https://doi.org/10.1038/npp.2010.249

Semenova S, Stolerman IP, Markou A (2007) Chronic nicotine administration improves attention while nicotine withdrawal induces performance deficits in the 5-choice serial reaction time task in rats. Pharmacol Biochem Behav 87(3):360–368. Epub 2007/06/22. https://doi.org/10.1016/j.pbb.2007.05.009

Singer LT, Min MO, Minnes S, Short E, Lewis B, Lang A et al (2018) Prenatal and concurrent cocaine, alcohol, marijuana, and tobacco effects on adolescent cognition and attention. Drug Alcohol Depend 191:37–44. https://doi.org/10.1016/j.drugalcdep.2018.06.022

Skowronek MH, Laucht M, Hohm E, Becker K, Schmidt MH (2006) Interaction between the dopamine D4 receptor and the serotonin transporter promoter polymorphisms in alcohol and tobacco use among 15-year-olds. Neurogenetics 7(4):239–246. Epub 20060704. https://doi.org/10.1007/s10048-006-0050-4

Slotkin TA (2004) Cholinergic systems in brain development and disruption by neurotoxicants: nicotine, environmental tobacco smoke, organophosphates. Toxicol Appl Pharmacol 198(2): 132–151. https://doi.org/10.1016/j.taap.2003.06.001

Slotkin TA, Seidler FJ (2009) Nicotine exposure in adolescence alters the response of serotonin systems to nicotine administered subsequently in adulthood. Dev Neurosci 31(1–2):58–70. https://doi.org/10.1159/000207494

Smith LN, McDonald CG, Bergstrom HC, Brielmaier JM, Eppolito AK, Wheeler TL et al (2006) Long-term changes in fear conditioning and anxiety-like behavior following nicotine exposure in adult versus adolescent rats. Pharmacol Biochem Behav 85(1):91–97. https://doi.org/10.1016/j.pbb.2006.07.014

Solowij N, Grenyer BF (2002) Are the adverse consequences of cannabis use age-dependent? Blackwell Science Ltd, Oxford, pp 1083–1086

Sowell ER, Thompson PM, Holmes CJ, Jernigan TL, Toga AW (1999) In vivo evidence for post-adolescent brain maturation in frontal and striatal regions. Nat Neurosci 2(10):859–861. https://doi.org/10.1038/13154

Swanson JM, Kinsbourne M, Nigg J, Lanphear B, Stefanatos GA, Volkow N et al (2007) Etiologic subtypes of attention-deficit/hyperactivity disorder: brain imaging, molecular genetic and environmental factors and the dopamine hypothesis. Neuropsychol Rev 17(1):39–59. https://doi.org/10.1007/s11065-007-9019-9

Thakur GA, Sengupta SM, Grizenko N, Schmitz N, Pagé V, Joober R (2013) Maternal smoking during pregnancy and ADHD: a comprehensive clinical and neurocognitive characterization. Nicotine Tob Res 15(1):149–157. 10.1093/ntr/nts102

Tiesler CMT, Heinrich J (2014) Prenatal nicotine exposure and child behavioural problems. Eur Child Adolesc Psychiatry 23(10):913–929. https://doi.org/10.1007/s00787-014-0615-y

Tiesler CM, Chen CM, Sausenthaler S, Herbarth O, Lehmann I, Schaaf B et al (2011) Passive smoking and behavioural problems in children: results from the LISAplus prospective birth cohort study. Environ Res 111(8):1173–1179. Epub 20110720. https://doi.org/10.1016/j.envres.2011.06.011

Tirado-Muñoz J, Lopez-Rodriguez AB, Fonseca F, Farré M, Torrens M, Viveros M-P (2020) Effects of cannabis exposure in the prenatal and adolescent periods: preclinical and clinical studies in both sexes. Front Neuroendocrinol 57:100841. https://doi.org/10.1016/j.yfrne.2020.100841

Tiwari RK, Sharma V, Pandey RK, Shukla SS (2020) Nicotine addiction: neurobiology and mechanism. J Pharmacopuncture 23(1):1–7. https://doi.org/10.3831/kpi.2020.23.001

Torrence RD, Rojas DC, Troup LJ (2018) Residual effects of cannabis use on attentional bias towards fearful faces. Neuropsychologia 119:482–488. https://doi.org/10.1016/j.neuropsychologia.2018.09.008

Trezza V, Campolongo P, Manduca A, Morena M, Palmery M, Vanderschuren LJ et al (2012) Altering endocannabinoid neurotransmission at critical developmental ages: impact on rodent emotionality and cognitive performance. Front Behav Neurosci 6:2. Epub 20120124. https://doi.org/10.3389/fnbeh.2012.00002

Turner AR, Spurling BC, Agrawal S (2023) Marijuana toxicity. StatPearls Publishing LLC, Treasure Island (FL)

Valentine G, Sofuoglu M (2018) Cognitive effects of nicotine: recent Progress. Curr Neuropharmacol 16(4):403–414. https://doi.org/10.2174/1570159x15666171103152136

Verrico CD, Jentsch JD, Roth RH, Taylor JR (2004) Repeated, intermittent Δ9-tetrahydrocannabinol administration to rats impairs acquisition and performance of a test of visuospatial divided attention. Neuropsychopharmacology 29(3):522–529. https://doi.org/10.1038/sj.npp.1300316

Wallace AL, Wade NE, Hatcher KF, Lisdahl KM (2019) Effects of cannabis use and subclinical ADHD symptomology on attention based tasks in adolescents and young adults. Arch Clin Neuropsychol 34(5):700–705. Epub 2018/10/09. https://doi.org/10.1093/arclin/acy080

Wignall ND, de Wit H (2011) Effects of nicotine on attention and inhibitory control in healthy nonsmokers. Exp Clin Psychopharmacol 19(3):183–191. https://doi.org/10.1037/a0023292

Wimberley T, Agerbo E, Horsdal HT, Ottosen C, Brikell I, Als TD et al (2020) Genetic liability to ADHD and substance use disorders in individuals with ADHD. Addiction 115(7):1368–1377. https://doi.org/10.1111/add.14910

Witte EA, Davidson MC, Marrocco RT (1997) Effects of altering brain cholinergic activity on covert orienting of attention: comparison of monkey and human performance. Psychopharmacology 132(4):324–334. https://doi.org/10.1007/s002130050352

Xu J, Yin X, Ge H, Han Y, Pang Z, Tang Y et al (2015) Attentional performance is correlated with the local regional efficiency of intrinsic brain networks. Front Behav Neurosci 9:200. https://doi.org/10.3389/fnbeh.2015.00200

Yuan M, Cross SJ, Loughlin SE, Leslie FM (2015) Nicotine and the adolescent brain. J Physiol 593(16):3397–3412. https://doi.org/10.1113/JP270492

Yucel M, Lubman D, Solowij N, Brewer W (2007) Understanding drug addiction: a neuropsychological perspective. Aust N Z J Psychiatry 41(12):957–968. https://doi.org/10.1080/00048670701689444

Zhang L, Spencer TJ, Biederman J, Bhide PG (2018) Attention and working memory deficits in a perinatal nicotine exposure mouse model. PLoS One 13(5):e0198064-e. https://doi.org/10.1371/journal.pone.0198064

MIX
Papier aus verantwortungsvollen Quellen
Paper from responsible sources
FSC® C105338

If you have any concerns about our products,
you can contact us on
ProductSafety@springernature.com

In case Publisher is established outside the EU,
the EU authorized representative is:
Springer Nature Customer Service Center GmbH
Europaplatz 3, 69115 Heidelberg, Germany

Printed by Libri Plureos GmbH
in Hamburg, Germany